T0237097

Plant-Animal Interactions

Kleber Del-Claro • Helena Maura Torezan-Silingardi
Editors

Plant-Animal Interactions

Source of Biodiversity

 Springer

Editors
Kleber Del-Claro ⓘ
Inst de Biologia, Campus Umuarama
Universidade Federal de Uberlandia
Uberlandia, Brazil

Helena Maura Torezan-Silingardi ⓘ
Inst de Biologia, Campus Umuarama
Universidade Federal de Uberlandia
Uberlandia, Brazil

ISBN 978-3-030-66879-2   ISBN 978-3-030-66877-8   (eBook)
https://doi.org/10.1007/978-3-030-66877-8

© Springer Nature Switzerland AG 2021
This work is subject to copyright. All rights are reserved by the Publisher, whether the whole or part of the material is concerned, specifically the rights of translation, reprinting, reuse of illustrations, recitation, broadcasting, reproduction on microfilms or in any other physical way, and transmission or information storage and retrieval, electronic adaptation, computer software, or by similar or dissimilar methodology now known or hereafter developed.
The use of general descriptive names, registered names, trademarks, service marks, etc. in this publication does not imply, even in the absence of a specific statement, that such names are exempt from the relevant protective laws and regulations and therefore free for general use.
The publisher, the authors and the editors are safe to assume that the advice and information in this book are believed to be true and accurate at the date of publication. Neither the publisher nor the authors or the editors give a warranty, expressed or implied, with respect to the material contained herein or for any errors or omissions that may have been made. The publisher remains neutral with regard to jurisdictional claims in published maps and institutional affiliations.

This Springer imprint is published by the registered company Springer Nature Switzerland AG
The registered company address is: Gewerbestrasse 11, 6330 Cham, Switzerland

To Angela Helena Torezan Silingardi, our first mentor.

Preface

Biotic interactions are ubiquitous and have shaped the evolution of Earth's amazing biodiversity. Undoubtedly, plant-animal interactions have structured the majority of ecological networks and the biodiversity of interactions therein through evolutionary time. From antagonisms to mutualisms, plant-animal interactions are basic pieces of the evolutionary puzzle underpinning natural systems. Comprehending these relationships in all of their multidisciplinary aspects is fundamental to the future of life on a planet where negative human interference in natural systems is growing at an alarming pace.

Plant-Animal Interactions: Source of Biodiversity is a collaborative approach to this huge challenge, offering researchers and students new views, without leaving behind basic information. This book is an effort to pave the way for scientists interested in improving our knowledge of how plant-animal interactions shape biodiversity. Our book calls you to join us in studying and preserving plant-animal interactions, because they are sources of biodiversity. The book covers the most important theoretical aspects of this line of study, considering classical, basic, and naturalistic knowledge, but also presents advanced and applied approaches. Thus, in the opening chapter, we present a general view of plant-animal interactions. As editors, we considered it important to provide the foundations of plant-animal interactions from an evolutionary approach. A great deal of research in ecology and evolution has examined chemical mediation of plant-animal interactions. Thus, in ► Chap. 2, Lee A. Dyer and Chris S. Jeffrey discuss classical studies focused on plant compounds that reduce or deter insect damage (herbivory) and directly or indirectly affect secondary consumers. Here, the authors present a new view considering two focal theoretical frameworks that drive investigations of chemically mediated interactions, with a focus on phytochemical mixtures: coevolution and trophic interaction theory. This approach enables us to proceed to the field of herbivory with Robert J. Marquis and Renan F. Moura, who, in ► Chap. 3, discuss traits that enable plants to escape from their herbivores but have not been formerly considered part of plant resistance theory. They will brilliantly convince you that escape from herbivores can be used to effectively reduce herbivore pressure in agricultural systems, and that escape also contributes to biodiversity maintenance in preserved ecosystems. This chapter presents a full new perspective on the antagonistic relationships between plants and animals. However, to understand how plant defense against herbivory evolves, it is necessary to characterize the genetic underpinnings of resistance traits, quantify genetic variation in defense trait production, and characterize how natural selection is acting on these traits. We thank Liza M. Holeski for giving us ► Chap. 4, an amazing review of the genetic basis of plant-herbivore interactions and the evolutionary and ecological genetics of plant resistance against herbivory. Different aspects of defense against herbivory were considered in the previous chapters, and ► Chap. 5 continues this by presenting the role of biotic defenses in plant-animal interactions. Biotic defenses are relationships in which one organism (usually a plant or trophobiont herbivore) attracts predators of its own enemies. In ▶ Chap. 5, a team of young biologists—Renan F. Moura, Eva Colberg, Estevão Alves-Silva, Isamara Mendes-Silva, Roberth Fagundes, and Vanessa Stefani, joined by me (the old guy!)—deeply discuss all types of biotic defense systems and their mechanisms. Full of examples and exploring a very useful tool, experimental manipulation, this chapter illustrates how conditional the outcomes of biotic interactions may be, and how we still are in the infancy of these studies.

Starting with a holistic view of plant-animal interactions and their impact on biodiversity, the first five chapters of this book present the chemical and genetic aspects of plant-animal interactions and explore the most antagonistic relationships among these organisms, herbivory, and defenses against herbivores. However, recent reviews in plant-animal interactions suggest that mutualistic relationships (positive results to interacting organisms) are probably the strongest forces generating biodiversity. We will return to this issue later, after we present and competently exemplify the two main mutualistic relationships between plants and animals: pollination and seed dispersal. The following two chapters are very similar in structure, starting by covering the natural history and basic aspects of the main animal groups involved in these interactions, and then presenting new pathways for those interested in these lines of research. In ▶ Chap. 6, Helena Maura Torezan-Silingardi, Ilse Silberbauer-Gottesberger, and Gerhard Gottesberger draw on their backgrounds in pollination, and in ▶ Chap. 7, Richard T. Corlett covers seed dispersal and frugivory; these colleagues fulfilled the difficult mission of synthesizing in each of these chapters issues worthy of a whole book. In both chapters the authors go beyond characterizing and illustrating (with marvelous images) the most important mutualistic plant-animal interactions, also alerting us to the drastic problems caused by human impacts in natural systems. The reductions in populations and diversity of pollinators and seed dispersers are contributing to an enormous loss of ecological services, putting human food security at risk.

Plant belowground interactions with soil microbes alter plant fitness and physiology, affecting the performance of plant-associated aboveground organisms. Although this issue is clear to all biologists, especially field researchers, these aspects have only been superficially explored in previous books related to the evolutionary ecology of plant-animal interactions. So, we thank Frédérique Reverchon and Alfonso Méndez-Bravo in ▶ Chap. 8 for giving us a better understanding of the ecological interactions occurring within the phytobiome and their impacts on plant-animal interactions and associated biodiversity. This chapter opens up discussion into the main examples of facilitation in plant-animal interactions, that is, how these interactions can modify the environment by enlarging the niche for opportunistic organisms. In ▶ Chap. 9, an emerging group of very competent young ecologists, headed by Eduardo S. Calixto, and Danilo F. B. dos Santos, Diego V. Anjos, and Eva Colberg, discuss the concept of ecosystem engineering. This chapter addresses the concepts, applications, biodiversity implications, and future perspectives for the study of ecosystem engineers, especially regarding plant-arthropod interactions.

With these nine initial chapters, we are sure that the book provides all the basic, updated, and useful knowledge, including new approaches, for anyone interested in getting started in studying plant-animal interactions or settling previous fundamental questions. In the final part we have four chapters that place this book even further than the previous ones. In ▶ Chap. 10, Pedro Luna and Wesley Dáttilo start by explaining how interactive communities and populations generate organized networks and how these ecological networks vary over space and time. They close the chapter by calling attention to the importance of plant-animal networks in understanding the mechanisms and processes driving the geographic mosaic of coevolution, as proposed by John N. Thompson. This chapter complements the initial chapter in considering the geographic mosaic of coevolution theory as a key approach for understanding of origins and maintenance of biodiversity of interactions. Next, Judith L. Bronstein in ▶ Chap. 11 presents a new and very intriguing question in plant-animal interactions. She starts by considering that mutualisms are not only present, but are common and prominent interactions in every habitat on Earth. Thus, in a chapter full of wonderful examples from pollination, biotic defenses, and other mutualisms, she proposes an underlying rationale for why biological diversity tends to accumulate around mutualisms. And is mutualism a source of evolutionary innovation? ▶ Chap. 12, written by Rodrigo A. S. Pereira and Finn Kjellberg, explores this question by presenting examples of mutualisms that allowed insects and/or plants to expand their ecological niches. From a naturalistic up to a theoretical view, this book illustrates how plant-animal interactions are sources of biodiversity. However, in ▶ Chap. 13, Kleber Del-Claro and Rodolfo Dirzo close the book with a very disturbing topic. They discuss how in the Anthropocene, due to defaunation and deforestation, human interference in the structure of ecological networks may be forcing mass, global disruptions of ecological interactions, potentially leading to the end of the biodiversity of interactions.

All books have a singular history. *Plant-Animal Interactions: Source of Biodiversity* has a history mediated by a worldwide crisis, the SARS-CoV-2 or COVID-19 or simply the coronavirus pandemic. In normal times it is not easy to edit or to write a book or a book chapter. In a year of restrictions, suffering, loss of loved ones, a time when life was turned upside down, working was even harder. We thank each one of our authors for all of your dedication, resilience, and love of science. We know how difficult it was. Some of us have been closed in at home during all this time. Some of us lost loved ones and friends. One has a new baby (a piece of good news!). One retired and had to move to a new city during the pandemic crisis. One was forced to quarantine in a hotel room for 2 weeks. One housed the entire family of a colleague during the fires in California. We are sincerely thankful to you all.

We, in name of the whole group, thank our financial agencies, universities and employers. We sincerely thank our editor João Pildervasser and the marvelous Springer Nature team of collaborators.

Our very special acknowledgement goes to Ms. Eva Colberg for kindly revising the English of ▶ Chaps. 1, 5, 6, 9, and 10. There are no words to thank her collaboration.

We also thank our families for their support and patience. We thank each mutualistic organism living inside our bodies and cells for our lives, and plants and animals for their interactions that become this still wonderful world.

Kleber Del-Claro
Helena Maura Torezan-Silingardi
Uberlândia, MG, Brazil

Contents

Contributors

Estevão Alves-Silva Instituto Federal de Educação, Ciência e Tecnologia Goiano, campus Urutaí, Urutaí Goiás, Brazil

Diego V. Anjos Universidade de São Paulo, Ribeirão Preto, SP, Brazil
Universidade Federal de Uberlândia, Uberlândia, MG, Brazil

Judith L. Bronstein Department of Ecology and Evolutionary Biology, University of Arizona, Tucson, AZ, USA

Eduardo Soares Calixto Universidade de São Paulo, Ribeirão Preto, SP, Brazil

Eva Colberg University of Missouri-St. Louis, St. Louis, MO, USA
Harris World Ecology Center, St. Louis, MO, USA

Richard T. Corlett Xishuangbanna Tropical Botanical Garden, Chinese Academy of Sciences, Menglun, Yunnan, China

Wesley Dáttilo Red de Ecoetología, Instituto de Ecología A.C., Xalapa, Mexico

Kleber Del-Claro Laboratório de Ecologia Comportamental e de Interações (LECI), Instituto de Biologia, Universidade Federal de Uberlândia (UFU), Uberlândia, Minas Gerais, Brazil

Rodolfo Dirzo Department of Biology and Woods Institute for the Environment, Stanford University, Stanford, CA, USA

Lee A. Dyer Hitchcock Center for Chemical Ecology, University of Nevada, Reno, Reno, NV, USA

Roberth Fagundes Instituto de Ciências Exatas e da Natureza, Universidade da Integração Internacional da Lusofonia Afro-Brasileira, Redenção, Ceará, Brazil

Gerhard Gottsberger Universität Ulm, Ulm, Germany

Liza M. Holeski Department of Biological Sciences, Northern Arizona University, Flagstaff, AZ, USA

Christopher S. Jeffrey Hitchcock Center for Chemical Ecology, University of Nevada, Reno, Reno, NV, USA

Finn Kjellberg CEFE, CNRS, Université Montpellier, Université Paul Valéry, Montpellier, EPHE, IRD, Montpellier Cédex, France

Pedro Luna   Red de Ecoetología, Instituto de Ecología A.C., Xalapa, Mexico

Robert J. Marquis   Department of Biology and the Whitney R. Harris World Ecology Center, St. Louis, MO, USA

Isamara Mendes-Silva   Faculdade de Filosofia, Ciências e Letras de Ribeirão Preto, Universidade de São Paulo, Ribeirão Preto, São Paulo, Brazil

Alfonso Méndez-Bravo   CONACYT – Escuela Nacional de Estudios Superiores, Laboratorio Nacional de Análisis y Síntesis Ecológica, Universidad Nacional Autónoma de México, Morelia, Michoacán, Mexico

Renan F. Moura   Laboratório de Ecologia Comportamental e de Interações (LECI), Instituto de Biologia, Universidade Federal de Uberlândia (UFU), Uberlândia, Minas Gerais, Brazil

Rodrigo Augusto Santinelo Pereira   Depto de Biologia, Faculdade de Filosofia, Ciências e Letras de Ribeirão Preto, Universidade de São Paulo, São Paulo, Brazil

Frédérique Reverchon   Red de Estudios Moleculares Avanzados, Instituto de Ecología, A.C., Pátzcuaro, Michoacán, Mexico

Danilo Ferreira Borges dos Santos   Universidade de São Paulo, Ribeirão Preto, SP, Brazil

Ilse Silberbauer-Gottsberger   Universität Ulm, Ulm, Germany

Vanessa Stefani   Laboratório de História Natural e Reprodutiva de Artrópodes (LHINRA), Instituto de Biologia, Universidade Federal de Uberlândia (UFU), Uberlândia, Minas Gerais, Brazil

Helena Maura Torezan-Silingardi   Universidade Federal de Uberlândia (UFU) – Instituto de Biologia – Laboratório de Ecologia Comportamental e de Interações (LECI), Uberlândia, MG, Brazil

An Evolutionary Perspective on Plant-Animal Interactions

Kleber Del-Claro
and Helena Maura Torezan-Silingardi

Contents

© Springer Nature Switzerland AG 2021
K. Del-Claro, H. M. Torezan-Silingardi (eds.), *Plant-Animal Interactions*, https://doi.org/10.1007/978-3-030-66877-8_1

⊜ Learning Objectives

This chapter will help readers to understand the following:

1. The importance of biotic interactions in shaping the biodiversity of life;
2. A general characterization of the different types of plant-animal relationships;
3. Plant-animal interactions as the base of almost all ecological networks and the equal importance of basic and advanced studies in plant-animal interactions;
4. New pathways and the future of plant-animal interaction studies.

Life, as we can see looking out of the window, is a result of successive changes over time, a byproduct of evolution. Several biological processes are involved in the production and natural selection of these changes, of which perhaps the most important of all are biotic interactions.

1.1 Important Pieces of Ancient Natural History

Biotic interactions are present everywhere, in the air, earth, water and same inside the organism's body whether vertebrates, invertebrates, plants, fungi or any microorganism. From the pioneering studies of Lynn Margulis (e.g. Margulis and Fester 1991) we know that biotic interactions are one of the cornerstones that have shaped cells, organisms and populations, just as populations of distinct species have structured communities and ecosystems. Since learning that animal cells are a result of the interaction of an ancient anaerobic cell in symbiosis with an alien organism, the mitochondrion (and the mitochondrion and the chloroplast in vegetal cells), we have been studying deeply the importance and diversity of interactions shaping life. The mitochondria arose once during evolution, and its origin entailed an endosymbiosis accompanied by gene transfers from the endosymbiont to the host. The host that acquired the mitochondrion was an anaerobic nucleus-bearing cell, a full-fledged eukaryote that was able to engulf the mitochondrion actively via a phagocytosis which goes back 1.45 billion years in the fossil record given the coincidence of mitochondria with the eukaryotic state (Martin and Mentel 2010). A more recent but similar example of how interactions are ancient relationships that involve several organisms' features like self-recognition, communication and unexpected cooperation was presented by Broch et al. (2011), in a surprising evolutionary history. The microorganism *Dictyostelium* is a member of the Amoebozoa, a taxon that is basal to the Fungi-Metazoa branch. Broch et al. (2011) showed that *D. discoideum* is a social amoeba that has a primitive farming symbiosis with bacteria. The amoeba is a predator of the bacteria. However, instead of consuming all bacteria in their patch, the amoebae stop feeding early and incorporate bacteria into their fruiting bodies. Doing so, the amoebae carry bacteria during spore dispersal and can seed a new food crop, a major advantage when edible bacteria are lacking at the new site. But, when arriving at sites already containing appropriate bacteria, the costs of early feeding cessation are not compensated for. This example shows us how biotic interactions are complex even in simple organisms, and also how variable over time and evolutionary circumstances they can be. So, what are the conditions that indicate when bacteria play a more active role as prey, parasite or optional mutualist? How do circumstantial changes impact trait evolution in each interacting species?

Understanding how interactions evolve to shape the life and the amazing biodiversity that surround us is one of main goals of evolutionary biology. Biotic interactions are dynamic and their outcomes vary in space and time across a wide spectrum from positive to neutral to negative. All types of interactions (predation, parasitism, cooperation, mutualism, commensalism, etc.) have been present on Earth for more than 300 million years. In the evolutionary process the "actors" (species) can be replaced over time (i.e. through extinction and speciation), but independent of the actors the "theatrical play" (the interactions) continues (Del-Claro et al. 2016). These relationships have shaped biodiversity through creation, extinction, and coevolution of interactions mediated by a balance of loss and gains (Thompson 2013).

1.2 The Origin of Plant-Animal Interactions

In the early Paleozoic during the Devonian period, by 425 million years (myr) a tiny green coastal belt could be seen on Earth, the first land-dwelling vascular plants. Amazingly, less than 100 myr after the first ferns were used by the first herbivorous insects in the wetland forests of the Carboniferous period, almost all types of feeding strategies that an animal could use to eat a plant had developed (◘ Fig. 1.1; e.g. Slansky and Rodriguez 1987; Labandeira 2002). This was particularly true for insect-plant interactions.

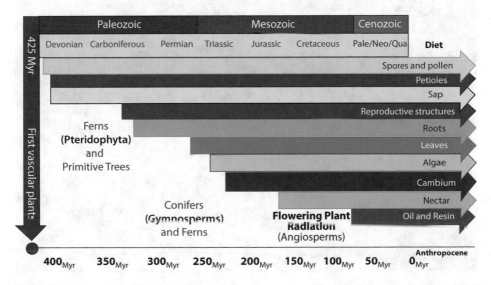

◘ **Fig. 1.1** Emergence of land-dwelling vascular plants and guilds of herbivores, according to the type of vegetal tissue consumed. Consider that the radiation of angiosperms may have occurred 70 myr prior to what is shown in the figure, still in the Jurassic

In the last decade, new methods of estimating global species richness have been developed to estimate that Earth contains 20 million species or more, including cryptic diversity (Stork 2018). Flowering plants (angiosperms) have been the dominant plant group since at least 150 myr (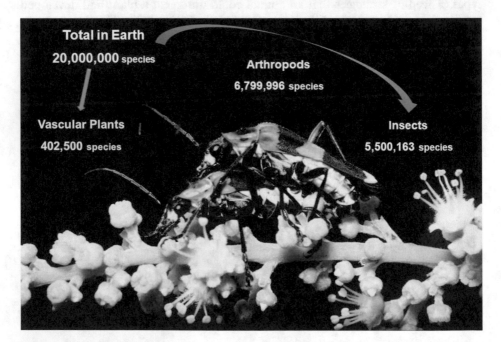 Fig. 1.1). Recent studies suggest that the origin of angiosperms may be still more ancient, in the Asian Jurassic (Fu et al. 2018). However, this view is controversial, and a thoughtful integration of fossil and molecular evidence could help resolve these conflicts (Coiro et al. 2019; Ramírez-Barahona et al. 2020).

Insects are the animals that most relate to flowering plants. These animals are small, abundant, reproduce quickly, and have amazing dispersive capabilities (i.e. fly), with a possible origin in the early Devonian (430 myr). Insects are extremely diverse and highly variable in morphology and feeding preferences, as reflected in the huge diversity of insect life histories. The first known relationships between insects and plants were antagonistic, with insects feeding on spores and pollen (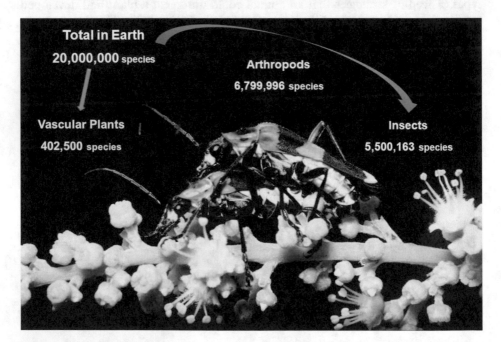 Figs. 1.1 and 1.2). Until the emergence of angiosperms, most plants reproduced and dispersed via wind, or sometimes by water, a very expensive process with high energetic costs in term of spores and pollen production. Considering that wind dispersal is totally random, wind-dispersed plants need to produce large quantities of reproductive structures to be successful (e.g. Novaes et al. 2020). This reproductive strategy of making immense quantities of spores and pollen full of amino acids (among other nutrients) did not go unnoticed by insects (see also Torezan-Siligardi et al., ▶ Chap. 6). Suggestive evidence of nectar feeding and pollination by insects

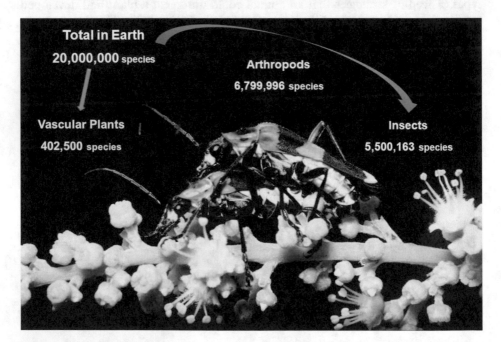

Total in Earth

20,000,000 species

Arthropods

6,799,996 species

Vascular Plants

402,500 species

Insects

5,500,163 species

◘ Fig. 1.2 Conservative approaches to global species richness (e.g. Stork 2018) estimate that the Earth contains 20 million species. Here, a couple of *Chauliognathus fallax* (Cantharidae) beetles copulating and feeding on pollen of Euphorbiaceae

dates to 300 myr, with great diversification in the feeding habits of herbivorous insects around 200–150 myr in the Mesozoic (◘ Fig. 1.1; Labandeira 1998, 2002). Indeed, the diversification and radiation of insects coincides with that of flowering plants in the Cretaceous (◘ Fig.1.1) (Crepet and Friis 1989; Bawa 1995; Grimaldi 1999; Hu et al. 2008). However, if angiosperms arose before the Jurassic, this timing has profound implications for our understanding how insect-plant interactions evolved, mainly pollination (van der Kooi and Ollerton 2020).

The flowering plants' conquering of land was undoubtedly related to the outcomes of their interactions with animals. According to a recent global estimate, 87.5% of angiosperms are pollinated by animals, primarily insects (Ollerton 2017), and even some gymnosperms are insect-pollinated (e.g. Gnetales; Kato et al. 1995) (◘ Figs. 1.2 and 1.3). The range of insect pollinators spans across orders and families (Ollerton 2017; Torezan-Siligardi et al., ► Chap. 6); even ants can be pollinators (Del-Claro et al. 2019, Del-Claro and Torezan-Silingardi 2020). Animals are also important seed dispersers (Simmons et al. 2018; Corlett, ► Chap. 7). The histories of these interactions are full of adaptations and counter-adaptations which in general unfolded gradually and slowly (but sometimes in jumps) throughout evolutionary time (Thompson 2012, 2013). But, how exactly have animals, especially insects, helped plants conquer the land environment?

Flying, the conquest of the aerial environment, as well as their small size and rapid reproductive cycle, were probably decisive factors for the success in the interactions of insects with plants. The ability to fly favored insects in the location of

◘ **Fig. 1.3** From antagonism (herbivory; left) to mutualism (pollination; right): adaptations and counter-adaptations in plant-animal interactions through evolutionary time. The Orthoptera *Scaphura nigra* (Tettigoniidae, left) eats pollen and other reproductive structures of a Fabaceae flower. On the right is sequence of a *Centris* bee landing and pollinating *Byrsonima intermedia* (Malpighiaceae)

food plants, and facilitated joint dispersion. The small size and the large reproductive capacity of insects, combined with a short reproductive and gestational cycle, often with multiple reproductive events in the same year, were also important. These factors probably facilitated the rapid evolution of different strategies in the associations between insects and other organisms, especially plants. Thus, favored genomes would be maintained and would remain in the game of the evolutionary process, while unfavorable ones, with little or no reproduction, would disappear from the game. The ability to fly benefitted not only insects, but also other winged animals, particularly birds and bats. There are many forms of interactions between these animal taxa and plants, such as bat and bird pollinators and/or seed dispersers (Corlett and Primack 2011, see also Corlett, ▶ Chap. 7).

1.3 Antagonism: Plants as a New Resource

The bases for the radiation of animals were provided in large part by the resources, mainly food, offered by plants, especially angiosperms. The vegetal cell is composed of water (80%), proteins (12%), nucleic acids (2%), carbohydrates (1%), inorganic salts (1%), lipids (0.5%) and other substances (3.5%) (Wayne 2010). Since vascular plants arose in the early Devonian (400 myr, ◘ Fig. 1.1), this new and valuable source of nutrients could attract a wide range of animals to dispute or share this new niche. In this way, plants began to suffer enormous pressure from the trophic level immediately above them, the primary consumers. As a reaction to this evolutionary pressure exerted by herbivores, plants responded by developing anti-herbivore defenses, which can be classified in different ways and categories (Price et al. 1980; Crawley 1983; Coley and Barone 1996; Marquis and Braker 1994; Marquis et al. 2016). The main types of defenses are: physical (e.g. spines, thorns, or trichomes in leaves, stems, and some fruits; leaf hardness); chemical (e.g. accumulation of alkaloids, tannins), developmental (e.g. flowering when there are fewer herbivores and/or more natural enemies) and biotic (e.g. association with protective species). Plants may exhibit defenses that are either constitutive (always expressed) or induced (expressed after damage or a risk of damage, or both; Karban and Baldwin 1997). Indeed, the plant defense system is dynamic and can combine different defenses simultaneously or at different times during development, where each defense is expressed and presents different peaks of effectiveness according to the plant stage (Calixto et al. 2015, 2020, see also Marquis and Moura, ▶ Chap. 3).

Virtually all parts of plants can serve as food for some type of animal. Vertebrates and invertebrates make use of leaves, roots, stems, flowers, fruits and seeds, which are all rich in a wide range of inorganic and organic compounds. It is clear that from the beginning, plants were a tempting food resource for animals and thus the first plant-animal relationships were antagonistic (e.g. herbivory; ◘ Fig. 1.3). Herbivory can be equated to predation when animals feed on the whole plant or destroy reproductive structures such as fruits and/or seeds. But herbivory can also be closer to parasitism, as in the case of those animals that remove only small pieces of leaves, petals, meristems or roots. The texts on herbivores are abundant and diverse, for over 150 years and we can consider as herbivores animals that feed on: leaves (folivores), roots, stems (e.g. borers) and buds, tendrils, sepals and petals, stamens and stigmas, flower buds and pollen (e.g. florivores), developing or ripe fruits (frugivores), seeds (granivores), nectar (nectarivores), oils and resins, galls; in short, collectively animals can eat any tissue, exudation or secretion of plants. Herbivores can also feed externally (exophytic) or internally (endophytic) in most plant parts (◻ Fig. 1.4).

Through evolutionary time plants have been selected to minimize the costs of producing defenses and maximize herbivore resistance, a premise of the Optimal Defense Theory (McKey 1974, 1979; Rhoades 1979). Thus, plants must allocate defenses to their tissues and structures according to their value and the probability of attack (e.g. Calixto et al. 2020). To surpass the plant arsenal of defenses, animals in turn evolved counter-adaptations. Animals have adapted to plant developmental phases (phenology), also changing their annual cycle; subdued chemical defenses, often using them for their own benefit; and developed behavioral strategies to evade or co-opt the protection of mutualistic protectors of the plants (Marquis and Braker 1994; Del-Claro et al. 2016).

Thus, in the last 400 million years, Earth has witnessed a true "arms race" between plants and their herbivores, with gene packages that confer advantages now and then. These gene pools are permanently challenged by the evolutionary forces of natural selection (Marquis 1992; Holeski et al. 2012, see also Marquis and Moura, ► Chap. 3 and Holeski, ► Chap. 4). Thus, antagonistic interactions can be thought of as the starting point for the evolution of harmonic relationships that led plants and animals to their great joint diversification, largely due to highly successful coevolutionary processses (Abrahamson 1989; Thompson 2014). However, recent evidence in the tropical Amazonian forest has demonstrated that although plants may evolve under selection by herbivores, those herbivores may not always show coevolutionary adaptations, suggesting a model more consistent with resource tracking than with the arms race model of coevolution (Endara et al. 2017).

1.4 Advantages of Mutual Benefits and Distribution of Plant-Animal Interactions

Mutually beneficial relations between animals and plants seem to be derived from antagonistic interactions, mainly herbivory (Abrahamson 1989). Whether in business or in nature, turning an enemy into a partner can pay off if the benefits outweigh the costs (see Bronstein, ► Chap. 11). In this sense, some authors (Price 2002a, b; Terborgh and Estes 2010) argue that mutualistic interactions, especially relationships between plants and their pollinators and/or seed dispersers, were possibly the major forces responsible for the conquest of land environments by plants. This hypothesis has great scientific potential, considering that: (a) pollination by flying animals allowed plants to have greater success and greater efficiency (saving energy resources present in pollen) in cross-breeding, even for plants that occur in

1

◨ Fig. 1.4 Herbivores may feed on virtually any plant part, eating pollen, petals and sepals, destroying entire flowers and same inflorescences, as this exophytic beetle **a**; endophytic beetles like this female **b** can insert eggs inside floral buds and flowers, producing larvae that will consume the reproductive structures of the plant. They can be extremely beautiful and conspicuous like hummingbirds getting floral nectar (**c**, photo of Eduardo S. Calixto), or cryptic like leaf-miners **d**. Some can eat fruits without destroying the seeds, serving as seed dispersers **e**, while others eat the entire nut and destroy the embryo completely **f**

patches or are very scattered; (b) the dispersion of seeds over long distances, mainly by birds and bats, resulted in the occupation and colonization of increasingly distant places, in addition to making it difficult for non-mutualistic herbivores (usually ruminant vertebrates and/or leaf-eating, granivorous and wood-boring insects) to locate these new individuals; (c) long-distance dispersion also resulted in the specialization of some plants to particular microhabitats (often leading to the evolution of new endemic species), also decreasing the pressure from original herbivores and parasites; (d) these interspecific interactions produced changes in population structures leading to rapid evolutionary changes, strong adaptive radiation and consequent speciation. These associative processes between plants and animals may have originated in or triggered the appearance of coevolutionary processes in geographic mosaics in terrestrial environments (Bronstein 2009; Thompson 2012, 2014).

Plants, as autotrophic organisms (producers), are the food base for almost all heterotrophic life on the planet, structuring trophic chains both in water and on land, but many of their evolutionary trajectories as well as their impacts on the evolution of other life forms have been shaped by their negative and positive interactions with animals. Ecological interactions have played a fundamental role in the processes of speciation and extinction in plant populations in the history of the Earth, and the fascinating diversity of life that we observe is dependent on the spectacular and complex network of interactions between plants and animals (see also Luna and Dáttilo, ▶ Chap. 10).

But why hasn't all this life, resulting from interactions between organisms, spread evenly, or at least not so heterogeneously, across the planet? In several current textbooks on ecology (e.g. Ricklefs 2001; Townsend et al. 2006) and evolution (Futuyma 2009) we find several classic arguments explaining why there is greater biodiversity in the tropical regions of the Earth than in the others. The most explored arguments are based on the fact that the tropics receive solar rays with lower variation throughout the year. Average temperatures oscillate around 25 °C in the continental and oceanic waters in these regions, causing great evaporation and regular and intense rains during most of the year. Under these conditions, microorganisms that decompose organic matter proliferate and act efficiently, increasing the amount and speed of nutrient cycling. The hypothesis is that these conditions are conducive to continuous growth and higher reproductive rates, which imply a greater number of mutations and recombinations, key items for speciation and consequent generation of diversity (Futuyma 2009 and references therein). Due to climatic conditions with well-marked seasons, especially with long periods of cold and drought, growth and reproduction in the temperate and polar regions experience periods of interruption every year, which in the long run puts them behind in the race for production of diversity relative to the tropical regions. In lower latitudes (tropical areas), the presence of more species in the same environment will produce more and greater changes in the physical and chemical environment, and will produce greater chances of contact, and new interactions. More interactions will lead the entire trophic mesh of the community to a greater degree of complexity, through increased connectivity (dependency links between organisms), compartmentalization (sub-groups within trophic chains) and omnivory

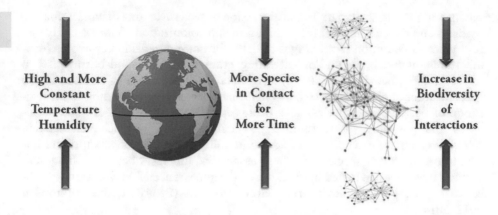

◘ Fig. 1.5 From the poles to the tropics the environment presents a more predictable and constant climate, with lower variation in temperature and humidity, that maintains regular hot and rainy seasons. These factors enable a quicker and stronger cycling of nutrients, plant growth and reproduction, which sustains more animals and therefore increases the biodiversity of life and interspecific interactions

(more organisms participating simultaneously at different trophic levels and/or sub-groups), all factors that contribute to an increase in diversity in the whole network of ecological interactions (e.g. Dáttilo and Rico-Gray 2018, ◘ Fig. 1.5).

Regardless of the geographic area on the planet, the discussion of which ecological-evolutionary processes shaped the joint diversification of plants and animals from the Paleozoic onwards remains. Three main hypotheses (Strong et al. 1984) are frequently discussed (Labandeira 2002, but see also Thompson 2014; Ollerton 2017). These hypotheses explain how ecological units, for example, functional groups related to food (i.e. guilds established by type of diet and feeding strategies), expanded in macroevolutionary time (Price 2002b).

The first hypothesis was established mainly by paleobiologists and called the "Ecological Saturation Hypothesis". Life experienced countless tagmoses (e.g. arthropods) in the great explosion of life forms of the Cambrian (500 million years), but few remained until today (Labandeira 2002). Ecological interactions must also have had a period of exponential growth in types of interactions, which generated numerous ecological roles and nodes in interactive networks, which also went through natural selection processes. After this stage of exponential growth and selection, the number of ecological positions or "roles" in the evolutionary scenario must have remained more or less constant. As such, species continued to enter (being created or immigrating) and leave (becoming extinct or emigrating) the ecological arenas of local biological communities, but the "roles" existing in the theater of ecological relations remained virtually the same (see Thompson 2014).

The second hypothesis, the "Resource Expansion Hypothesis", is supported mainly by biologists. It states that the availability and exploitation of ecological resources, such as food and niches, have gradually increased over time, allowing more and more plant-animal interactions to develop more recently. So, should we expect to find new or still unknown models of plant-animal interactions? What about "facilitation"? Reciprocal effects in plant-animal interactions are rarely investigated; facilitation might evolve like mutualism, commensalism or antagonism, depending on the effects on the facilitator species; but it is still only marginally studied (Bronstein 2009). For example, long term studies on the ecology of ecosystem engineers, a classic facilitator, are rare (e.g. Velasque and Del-Claro 2016; Calixto et al. ▶ Chap. 9). And "commensalism", which according to Mathis and Bronstein (2020) involves "interactions between two species in which one species benefits and the other experiences no net effect, are frequently mentioned in the ecological literature but are surprisingly little studied...commensalism should be of great interest in the study of species interactions due to its location at the center of the continuum between positive and negative outcomes".

The third hypothesis tries to find common ground between the two previous ones and not to make them exclusive hypotheses. The "Intrinsic Diversification Hypothesis" argues that the proportion of occupied ecological roles presents a disjointed global pattern, that is, the type and amount of interactions will vary depending on the groups involved, time and place. In fact, this is not far from the Ecological Saturation Hypothesis, and seems to be more of a variation, another way of looking at the same scenario (Labandeira 2002).

Thompson (1994, 2005, 2012, 2014) has also suggested a general hypothesis to explain the great profusion of ecological interactions that seem to sustain all the biodiversity of the Earth. His hypothesis is strongly influenced by the "Ecological Saturation Hypothesis" and was named "The Geographic Mosaic Theory of Coevolution" (Thompson 2005). This theory postulates that the long-term dynamics of coevolution may occur over large geographic ranges rather than within local populations. John N. Thompson predicts that "Interactive Biodiversity," that is, the diversity of biological interactions that sustain the diversity of life, has its origins in coevolutionary processes whose results and impacts act on a global scale. So, "a species may adapt and become specialized to another species differently in separate regions. A species that is involved in an interspecific interaction in one geographic area may not even be present in another geographic area. This geographic mosaic in evolving interactions provides the raw material for the overall direction of coevolution, which proceeds as genes that are favored in local interactions spread out into other populations" (Thompson 2020). Thus, certain regions would have a greater potential than others in generating interactions that can diversify and spread, structuring the trophic webs that spread throughout the planet. We are beginning to see studies exploring this theory based on plant-animal interactions. Indeed, Thompson's ideas have largely driven theories and studies of ecological networks of interactions over the past 20 years.

1.5 Future Directions

A persistent challenge in the study of plant-animal interactions is to understand how evolutionary mechanisms shape the interactions among individuals and assemblages of related species. How does trait evolution affect individual fitness and the outcomes of any evolutionary change in the ecological networks that structure the natural communities in which these organisms exist? Beyond the use of

1

experimental manipulations in natural conditions, the use of network models has proved a very useful and now widespread tool for evolutionary ecologists to test hypotheses related to trait evolution. Recently, de Andreazzi et al. (2020) successfully suggested that an adaptive network framework that combines the evolution of species traits and the assembly of interactions may allow us to better understand trait evolution and network structure in antagonistic and mutualistic networks (see also Dáttilo and Rico-Gray 2018). Considering technological advances in genetics (e.g. Endara et al. 2017; Barker et al. 2018) and chemistry (e.g. Razo-Belman et al. 2018), now we can also test evolutionary and ecological hypotheses of plant-herbivore interactions at increasingly precise levels. Belowground interactions with soil microbes alter plant fitness and indirectly affect the outcomes of interactions for plant-associated aboveground organisms. Révérchon and Méndez-Bravo (▶ Chap. 8) point out that a better understanding of the ecological interactions occurring within the phytobiome is now possible due to the advent of new technologies such as metatranscriptomics and metabolomics.

However, the future of studies in plant-animal interactions does not only require complex evolutionary questions. Basic studies are also still needed. Some types of plant-animal relationships deserve our attention and deeper studies to clarify the true character of the interaction. For example, facilitation (Bronstein 2009) and commensalism (Mathis and Bronstein 2020) are common yet barely studied relationships. These two types of interactions are common in plant-animal systems and need studies testing the true benefits or absence of benefit to each species involved, and testing how these outcomes vary with space, time (Thompson 1994, 2020) or specific features of related species (Del-Claro and Marquis 2015).

Whether of a broader evolutionary focus, such as network analyses and genetics, or a more basic focus, such as studies of direct outcomes and natural history, further research on the ecological interactions between animals and plants are still paramount.

Conclusion

Plant animal-interactions have shaped the majority of aquatic and terrestrial food chains, ecological networks, and the biodiversity of interactions therein. Throughout evolution the species involved have changed - some evolved, while others went extinct and were replaced - but the types of ecological interactions remained and structured the natural communities. From antagonisms to mutualisms, plant-animal interactions are fundamental pieces of the evolutionary puzzle underpinning the maintenance and viability of natural systems.

Key Points

- Origin and importance of ecological interactions to the maintenance of biodiversity;
- The distinct types of plant-animal interactions and their role shaping natural communities;

- Variable in space and time, the outcomes of plant-animal interactions are conditional depending on changes in abiotic and biotic features of the habitat.
- Considering technological advances, we can now test evolutionary and ecological hypotheses of plant-herbivore interactions at increasingly precise levels.

? Questions
- If biotic interactions are dynamic and their outcomes vary in space and time, covering a wide spectrum from positive to negative or neutral, how have evolutionary mechanisms shaped the interactions among plants and animals?
- How did insects contribute to the spread and diversification of plants in the terrestrial environment?
- What are the hypotheses that explain the ecological-evolutionary processes that shaped the joint diversification of plants and animals?

Acknowledgements   We thank to Eva Colberg for the excellent English review and comments in the chapter; and Renan Moura for comments and suggestions. CNPq for financial support.

References

Abrahamson WG (1989) Plant-animal interactions: an overview. In: Abrahamson WG (ed) Plant-animal interactions. Mc Graw-Hill Publishing, New York, pp 1–22

Barker HL, Holeski LM, Lindroth RL (2018) Genotypic variation in plant traits shapes herbivorous insect and ant communities on a foundation tree species. PLoS One 13:e0200954

Bawa KS (1995) Pollination, seed dispersal and diversification of angiosperms. Trends Ecol Evol 10(8):311–312

Broch DA, Douglas TE, Queller DC, Strassmann JE (2011) Primitive agriculture in social amoeba. Nature 469:393–396

Bronstein JL (2009) The evolution of facilitation and mutualism. J Ecol 97:1160–1170

Calixto ES, Lange D, Del-Claro K (2015) Foliar anti-herbivore defenses in Qualea multiflora Mart. (Vochysiaceae): changing strategy according to leaf development. Flora 212:19–23

Calixto ES, Lange D, Bronstein J, Torezan-Silingardi HM, Del-Claro K (2020) Optimal defense theory in an ant–plant mutualism: extrafloral nectar as an induced defence is maximized in the most valuable plant structures. J Ecol:1–12. https://doi.org/10.1111/1365-2745.13457

Coiro M, Doyle JA, Hilton J (2019) How deep is the conflict between molecular and fossil evidence on the age of angiosperms? New Phytol 223:83–99

Coley PD, Barone JA (1996) Herbivory and plant defenses in tropical forests. Annl Rev Ecol Syst 27:305–335

Corlett RT, Primack RB (2011) Tropical rain forests: an ecological and biogeographical comparison. Wiley-Blackwell, Sussex

Crawley MJ (1983) Herbivory, the dynamics of animal-plant interactions. Blackwell Science Publications, Oxford

Crepet WL, Friis EM (1989) The evolution of insect pollination in angiosperms. In: Friss EM, Chaloner WG, Crane PR (eds) The origins of angiosperms and their biological consequences. Cambridge University Press, Cambridge, UK, pp 181–201, 358

Dáttilo W, Rico-Gray V (2018) Ecological networks in the tropics: an integrative overview of species interactions from some of the most species-rich habitats on earth. Springer Nature, Cham de Andreazzi CS, Astegiano J, Guimarães PR Jr (2020) Coevolution by different functional mechanisms modulates the structure and dynamics of antagonistic and mutualistic networks. Oikos 129:224–237

Del-Claro K, Marquis RJ (2015) Ant species identity has a greater effect than fire on the outcome of an ant protection system in Brazilian Cerrado. Biotropica 47:459–467

Del-Claro K, Torezan-Silingardi HM (2020) In search of unusual interactions. A commentary on: 'Pollen adaptation to ant pollination: a case study from the Proteaceae'. Ann Bot. https://doi.org/10.1093/aob/mcaa126

Del-Claro K, Rico-Gray V, Torezan-Silingardi HM, Alves-Silva E, Fagundes R, Lnge D, Dáttilo W, Vilela AA, Aguirre A, Rodriguez-Morales D (2016) Loss and gains in ant–plant interactions mediated by extrafloral nectar: fidelity, cheats, and lies. Insect Soc 63:207–221

Del-Claro K, Rodriguez-Morales D, Calixto ES, Martins AS, Torezan-Silingardi HM (2019) Ant pollination of *Paepalanthus lundii* (Eriocaulaceae) in Brazilian savanna. Ann Bot 123:1159–1165

Endara MJ, Coley PD, Ghabash G, Nicholls JA, Dexter KG, Donoso DA, Stone GN, Pennington RT, Kursar TA (2017) Coevolutionary arms race versus host defense chase in a tropical herbivore–plant system. PNAS 114:E7499–E7505

Fu Q, Diez JB, Pole M et al (2018) Anunexpected noncarpellate epigynous flower from the Jurassic of China. Elife 7:e38827

Futuyma D (2009) Biologia evolutiva. Funpec, São Paulo

Grimaldi D (1999) The co-radiation of pollinating insects and angiosperms in the Cretaceous. Ann Mo Bot Gard 86:373–406

Holeski LM, Hillstrom ML, Whitham TG, Lindroth RL (2012) Relative importance of genetic, ontogenetic, induction, and seasonal variation in producing a multivariate defense phenotype in a foundation tree species. Oecologia 170:695–707

Hu S, Dilcher DL, Jarzen DM, Taylor DW (2008) Early steps of angiosperm–pollinator coevolution. PNAS 105(1):240–245

Karban R, Baldwin IT (1997) Induced responses to herbivory. Chicago University Press, Chicago

Kato M, Inoue T, Nagamitsu T (1995) Pollination biology of *Gnetum* (Gnetaceae) in a lowland mixed dipterocarp forest in Sarawak. Am J Bot 82(7):862–868

Labandeira CC (1998) Plant-insect associations from the fossil record. Geotimes 43:18–24

Labandeira CC (2002) The history of associations between plants and animals. In: Herrera CM, Pellmyr O (eds) Plant animal interactions, an evolutionary approach. Blackwell Science Ltd, Oxford, pp 26–76

Margulis L, Fester R (1991) Symbiosis as a source of evolutionary innovation: speciation and morphogenesis. MIT Press, Cambridge

Marquis RJ (1992) The selective impact of herbivores. In: Fritz RS, Simms EL (eds) Plant resistance to herbivores and pathogens. Univ. of Chicago Press, Chicago, pp 301–325

Marquis RJ, Braker HE (1994) Plant-herbivore interactions: diversity, specificity, and impact. In: McDade LA, Bawa KS, Hespenheide HA, Hartshorn GS (eds) La Selva: ecology and natural history of a Neotropical rain forest. Chicago Press, Chicago, pp 261–281

Marquis RJ, Salazar D, Baer C, Reinhardt J, Priest G, Barnett K (2016) Ode to Ehrlich and Raven or how herbivorous insects might drive plant speciation. Ecology 97:2939–2951

Martin W, Mentel M (2010) The origin of mitochondria. Nat Educ 3(9):58

Mathis KA, Bronstein JL (2020) Our current understanding of commensalism. Annu Rev Ecol Evol Syst 51:1

McKey D (1974) Adaptive patterns in alkaloid physiology. Am Nat 108:305–320

McKey D (1979) The distribution of plant secondary compounds within plants. In: Rosenthal GA, Janzen DH (eds) Herbivores: their interactions with secondary plant metabolites. Academic press, New York, pp 55–133

Novaes LR, Calixto ES, Oliveira ML, Lima LA, Almeida O, Torezan-Silingardi HM (2020) Environmental variables drive phenological events of anemocoric plants and enhance diaspore dispersal potential: a new wind-based approach. Sci Total Environ 730:139039

Ollerton J (2017) Pollinator diversity: distribution, ecological function, and conservation. Annu Rev Ecol Evol Syst 48:353–376

Price PW (2002a) Species interactions and the evolution of biodiversity. In: Herrera CM, Pellmyr O (eds) Plant animal interactions, an evolutionary approach. Blackwell Science Ltd, Oxford, pp 3–25

Price PW (2002b) Macroevolutionary theory on macroecological patters. Cambridge University Press, Cambridge

Price PW, Bouton CE, Gross P et al (1980) Interactions among three trophic levels: influence of plant on interactions between insect herbivores and natural enemies. Annu Rev Ecol Syst 11:41–65

Ramírez-Barahona S, Sauquet H, Magallón S (2020) The delayed and geographically heterogeneous diversification of flowering plant families. Nat Ecol Evol. https://doi.org/10.1038/s41559-020-1241-3

Razo-Belman R, Molina-Torres J, Martínez O, Heil M (2018) Plant-ants use resistance-related plant odours to assess host quality before colony founding. J Ecol 2018(106):379–390

Rhoades DF (1979) Evolution of plant defense against herbivores. In: Rosenthal GA, Janzen DH (eds) Herbivores: their interactions with secondary plant metabolites. Academic press, New York, pp 1–55

Ricklefs RE (2001) The economy of nature, 5th edn. W. H. Freeman, New York

Simmons BI, Sutherland WJ, Dicks LV, Albrecht J, Farwig N, García D, Jordano P, González-Varo JP (2018) Moving from frugivory to seed dispersal: incorporating the functional outcomes of interactions in plant–frugivore networks. J Anim Ecol 87:995–1007

Slansky F, Rodriguez JG (1987) Nutritional ecology of insects, mites, spiders and related invertebrates. Wiley Interscience, New York

Stork NE (2018) How many species of insects and other terrestrial arthropods are there on earth? Annu Rev Entomol 63:31–45

Strong DR, Lawton JH, Southwood TRE (1984) Insects on plants: community patterns and mechanisms. Harvard University Press, Cambridge

Terborgh J, Estes JA (2010) Trophic cascades. In: Predators, prey, and the changing dynamics of nature. Island Press, Washington, DC

Thompson JN (1994) The coevolutionary process. University of Chicago Press, Chicago

Thompson JN (2005) The geographic mosaic of coevolution. University of Chicago Press, Chicago

Thompson JN (2012) O futuro dos estudos plantas-animais. In: Del-Claro K, Torezan-Silingardi HM (eds) Ecologia das interações animais-plantas: uma abordagem ecológico-evolutiva. Technical Books Editoria, Rio de Janeiro, pp 307–318

Thompson JN (2013) Relentless evolution. University of Chicago Press, Chicago

Thompson JN (2014) Interaction and coevolution. University of Chicago Press, Chicago

Thompson JN (2020) Geographic mosaic theory of coevolution. Encyclopedia Britannica. https://www.britannica.com/science/community-ecology/Gene-for-gene-coevolution

Townsend CR, Begon M, Harper JL (2006) Fundamentos em Ecologia. ArtMed Editora, Porto Alegre

van der Kooi CJ, Ollerton J (2020) The origins of flowering plants and pollinators. Science 368:1306–1308

Velasque M, Del-Claro K (2016) Host plant phenology may determine the abundance of an ecosystem engineering herbivore in a tropical savanna. Ecol Entomol 41:421–430

Wayne R (2010) Plant cell biology. Academic Press, Elsevier, New York

Chemically Mediated Multi-trophic Interactions

Lee A. Dyer and Christopher S. Jeffrey

Contents

© Springer Nature Switzerland AG 2021
K. Del-Claro, H. M. Torezan-Silingardi (eds.), *Plant-Animal Interactions*, https://doi.org/10.1007/978-3-030-66877-8_2

2

🔵 **Learning Objectives**
- Learn about plant secondary metabolites and how they are synthesized.
- Develop an understanding of the importance of complex mixtures of phytochemicals in mediating trophic interactions.
- Be able to synthesize the existing hypotheses and methods relevant to chemically mediated coevolution.
- Appreciate the rapidly evolving methodology for studying chemical ecology.

2.1 Introduction

The current decline of plant and insect diversity is well documented, but specifics of declines and impacts on ecosystem services are far from clear (e.g., Cardinale et al. 2012; Johnson et al. 2017). For example, the recent uptick in the number of insect "population" or abundance declines provide no clear patterns for insects overall, rather there are geographic and taxonomic idiosyncrasies, with some taxa or regions exhibiting increases, others showing dramatic declines, and others with no real trends (Wagner et al. 2021). For insect-plant interactions, it is also true that we are losing plant and insect diversity, but it is difficult, or perhaps irrelevant, to say anything about abundances of these very broad taxa – instead, studies focused on abundances or other parameters related to population dynamics, it is more fruitful to examine declines of individual species. Similarly, while it is very likely that we are losing many additional axes of biodiversity, most notably genetic and functional diversity, the details of these losses are not well resolved, and chemical changes for individual plant species have not been documented. Among the many Anthropocene losses are the myriad tips of biochemical pathways that produce unique molecules, or secondary metabolites. Loss of phytochemical diversity is a tragedy similar to that of species loss for other biological units that are understudied, such as the insects, because like insects (Stork 2018), only a fraction of existing plant small molecules and their biochemical functions have been discovered and described (Dyer et al. 2018; Lautié et al. 2020). In this chapter we explore the importance of phytochemicals in mediating biotic interactions, with a focus on insect herbivory, and discuss two important theoretical frameworks for understanding origins and maintenance of trophic diversity: coevolution and trophic interaction theory. Empirical studies related to these theories will help answer questions about the consequences of global change on chemically mediated trophic interactions.

Chemically mediated biotic interactions are diverse and form the basis of large, complex networks (Richards et al. 2015; Sedio et al. 2018; Salazar et al. 2018). Phytochemicals are also the basis for interactions with abiotic factors, for example, by acting as photo-protectants or facilitating nutrient acquisition (e.g., Demmig-Adams et al. 2013). Very distinct categories of interactions within biotic networks, such as pollination and herbivory, can be connected by the same biochemical pathways. For example, iridoid glycosides can be both harmful to herbivores and beneficial to pollinators in the same plant species (Jacobsen and Raguso 2018). Many compounds important for flower coloration, pollinator attraction, or nectar rewards share biochemical pathways with compounds that defend against herbivores, and signaling molecules, such as jasmonic acid and salicylic acid are important for floral development, reward traits, and coordination of production of chemical defenses against herbivores (Jacobsen and Raguso 2018). In addition to shaping interactions with pollinators and herbivores, phytochemistry can determine outcomes of direct interactions, such as plant competition, growth of epiphytes, and seed dispersal (e.g., Baldwin et al. 2020). Indirect interactions shaped by phytochemistry are equally diverse, including attraction of predators and parasitoids (e.g., Greany et al. 1977), phytochemically mediated mate finding by insects (e.g., Erbilgin 2019), and bottom up trophic cascades (Dyer et al. 2004a, b). In this chapter we focus on anti-herbivore compounds that are traditionally termed "secondary metabolites" and focus on small molecules that are effective insect deterrents or toxins. Despite the focus on insect herbivory, the topics we cover are relevant to all biotic interactions: broad classes of secondary metabolites, trophic interaction theory, coevolutionary theory, and new approaches to chemical ecology.

2.2 Secondary Metabolites

Plants are an obvious necessity to terrestrial ecosystems. In addition to fixing carbon and producing oxygen, the biochemical pathways of plants provide a world of untouched metabolic diversity (◘ Fig. 2.1). Starting with the photosynthetic power of plants to harvest the energy of sunlight to fix carbon and produce organic compounds, these compounds are then converted into the common starting materials, or primary metabolites (nucleic acids, amino acids, lipids, and carbohydrates) that are in turn transformed via well-known enzymatic pathways to establish the diverse and complex molecules that are the outcome of plant secondary metabolism. Plant secondary metabolites, which have an important role in ecological success, encompass a seemingly limitless array of structural diversity that spans the dimensions of molecular shape, composition, and size. It has long been recognized that there is a continuum between primary and secondary metabolites, and there are secondary roles for primary metabolites (Berenbaum 1995) and vice versa (Seigler and Price 1976), yet the dichotomy is a useful semantic for referring to compounds that may not be essential for cell or tissue structure and function, but are ecologically important and may be necessary for the plant to persist. It is not possible to briefly review all of the biosynthetic pathways that produce molecules important for biotic interactions, and in fact, the endpoints of these pathways are only well resolved for primary metabolites and for gross classes of secondary compounds (◘ Fig. 2.1). For some compounds, such as cardenolides, the enzymes at the end of important pathways and candidate genes have been determined (e.g., Pandey et al. 2016), but most biochemical pathways are unknown for small molecules that affect biotic interactions. Even the full details of biosynthesis in particular plants of very well studied molecules, such as the iridoid glycoside, aucubin, are not resolved despite considerable effort (e.g., Damtoft et al. 1993; Dinda 2019). Known metabolic pathways have been determined for broad classes of compounds, and some of the more commonly studied in the context of biotic interactions are

2

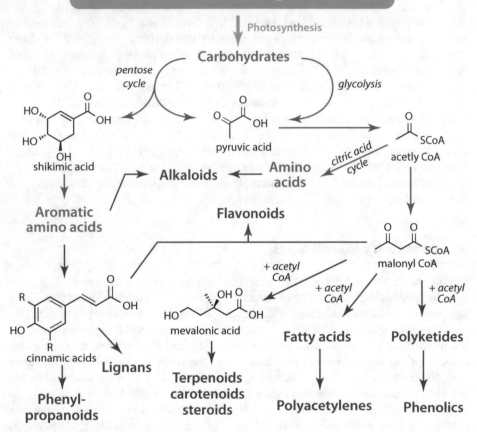

□ **Fig. 2.1**   General biosynthesis scheme for important secondary metabolites that mediate ecological interactions

the phenylpropanoids, flavonoids, terpenoids, polyketides, and alkaloids (□ Fig. 2.1). To provide a sense of phytochemical diversity, we provide a very brief review of these classes of compounds that have been important for development of chemical ecology. More complete reviews of plant secondary metabolites exist (Bennett and Wallsgrove 1994; Harborne 1999; Cseke et al. 2016), but a current multi-volume resource or online repository for ecologists would be quite useful.

Phenylpropanoids are ecologically important molecules that are derived from the amino acids phenylalanine and tyrosine and contain a three-carbon side chain attached to phenol (□ Fig. 2.2). They are the basis of polymers that have primary roles, such as the lignans in plant cell walls. The most notable phenylpropanoids and associated metabolites that affect trophic interactions include lignans, hydroxy-coumarins, and hydroxycinnamic acids. Phenylpropanoids are a good example of how phytochemistry can link diverse interactions, including herbivory and pollination, since many metabolites in this class are important for antiherbivore defense, pollen structure, floral pigments, and scent compounds (e.g., Deng and Lu 2017).

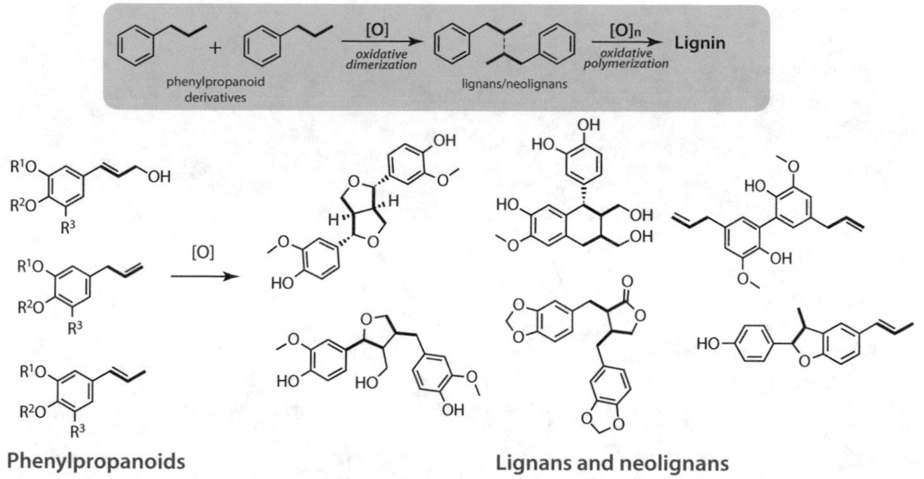

Phenylpropanoids　　　　　　　　　　　　**Lignans and neolignans**

□ **Fig. 2.2** General biosynthesis of phenylpropanoids with examples of specific well-studied phenylpropanoids, lignans, and neolignans that can mediate biotic interactions

Flavonoids

flavonols　　　　anthocyanin　　　　catechins　　　　flavones

flavanones　　　　chalcone　　　　dihydrochalcone　　　　isoflavone

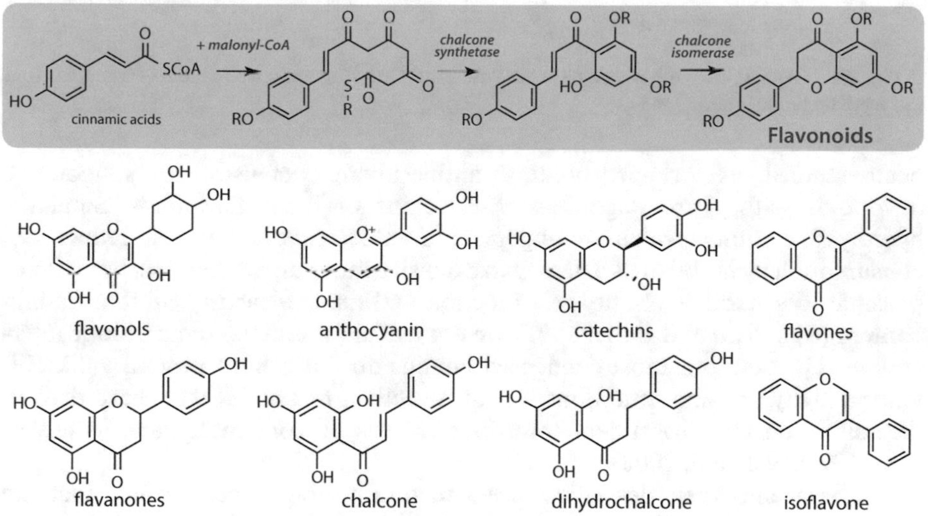

□ **Fig. 2.3** General biosynthesis of flavonoids with examples of specific well-studied molecules that can mediate biotic interactions

Like other classes of compounds reviewed here, the early steps in phenylpropanoid synthesis are clearly resolved – for example, the biochemistry is well known for Phenylalanine ammonia-lyase, the enzyme that transforms L-phenylalanine and tyrosine into trans-cinnamic acid and p-coumaric acid (Jun et al. 2018).

Flavanoids are derived from flavone and consist of two benzene rings joined by a propane unit; the diversity of flavonoids is generated by additions of hydroxyl groups and heterocyclic rings (□ Fig. 2.3). Some of the more notable flavonoids that mediate biotic interactions include the chalcones and anthyocyanins, which can be important for pollination. Condensed tannins, which have

◻ Fig. 2.4 General biosynthesis of terpenes with examples of specific well-studied terpenoids that can mediate biotic interactions

been studied as protein binding antiherbivore defenses, are oligomeric flavonoids – they are oligomers of catechins or other flavanoids. Similarly, hydrolysable tannins which are oligomers of gallic acid and can also reduce leaf consumption by herbivores. There is an extensive literature on tannins, but it is not possible to generalize about their function and gross measures of total tannin content (e.g., Price and Butler 1977) are not useful for understanding biotic interactions. For example, most condensed tannins do not exhibit general antiherbivore activity, raising questions about selective pressures that have led to diversification and investment towards synthesis of condensed tannins (Ayres et al. 1997; Heil et al. 2002).

Terpenes and their derivatives are structurally diverse compounds that are derived from isoprene and are classified based on their number of isopentane (5-carbon) units – most noatably the monoterpenes, sesquiterpenes, diterpenes, and triterpenes (10, 15, 20, and 30 carbons respectively) (◻ Fig. 2.4). *Terpenoids* include compounds with additional functional groups added to these, such as the iridoid glycoside, aucubin, which is a monoterpene with additional functionalities. The structural complexity, diversity, and ecological relevance of terpenes have yielded a number of model study molecules, such as catalpol (monoterpene), taxol (diterpene), and a variety of saponins (triterpene glycosides). There are examples of terpenoids mediating most important biotic interactions, including herbivory, competition, pollination, seed dispersal, and attraction of secondary consumers. Plant steroids are triterpenes that are not well studied in the context of mediating ecological interactions.

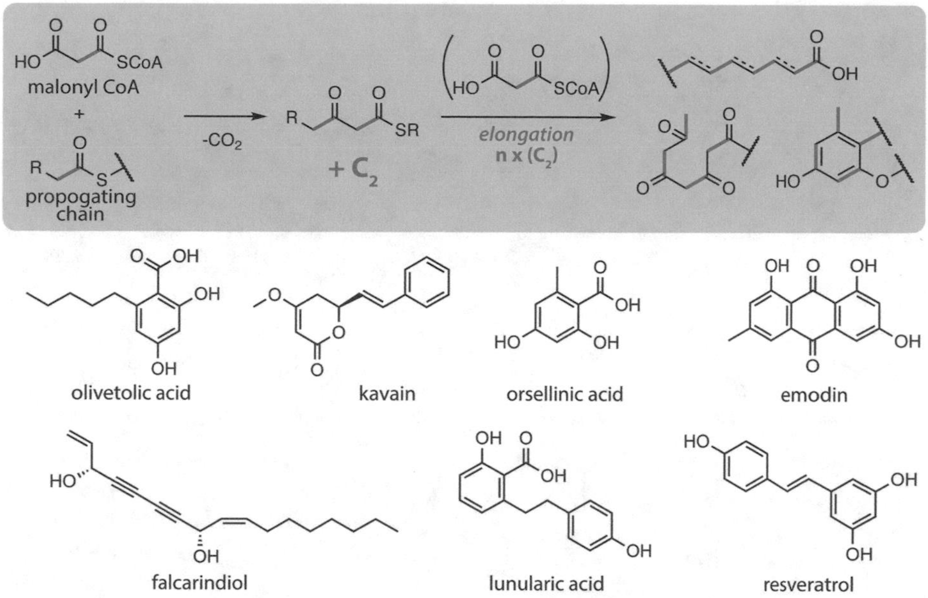

Fig. 2.5 General biosynthesis of polyketides with examples of plant derived polyketides that can mediate biotic interactions

A great deal of research on *polyketides* from microbes, plants, and fungi has focused on human health, based on their toxicity and clearly demonstrated antimicrobial, anticancer and immunosuppressive properties. The polyketides, derived from acetyl-CoA, are a large class of metabolites characterized by carbonyls and alcohols separated by methylene carbons (■ Fig. 2.5). Polyketide synthase enzymes responsible for catalytic mechanisms in polyketide synthesis are well studied (e.g., Morita et al. 2010) and operate somewhat like fatty acid synthase to generate a broad diversity of polyketides. Although much polyketide research is outside the realm of chemical ecology and focuses on production by non-plants, there is great potential for understanding the ecological roles of plant produced polyketides in complex mixtures (Gershenzon et al. 2012), and studies of model systems, like *Arabidopsis thaliana* (Brassicaceae) have discovered fascinating and important polyketide derivatives, such as the antibiotic macrobrevin, which affects multiple interactions on the phyllosphere of *A. thaliana* (Helfrich et al. 2018).

Alkaloids comprise a broad collection of natural products with varied biosynthetic pathways; these compounds contain nitrogen, are basic due to their amine functionality, and the nitrogen is usually part of a cyclic system (■ Fig. 2.6). They are most commonly derived from amino acids, aromatics, or terpenes. One method of categorizing alkaloids is based on their ring systems, for example: pyridine, pyrrolizidine, tropane, isoquinoline, quinolizidine, nicotine, or morphine. Alkaloids include many of the well-known examples of the pharmacological value of natural products, including morphine and cocaine, as well as giving physiological properties to recreational drugs, such as caffeine, nicotine, and cocaine. Examples of studies demonstrating the effects of alkaloids on diverse species interactions are very

2

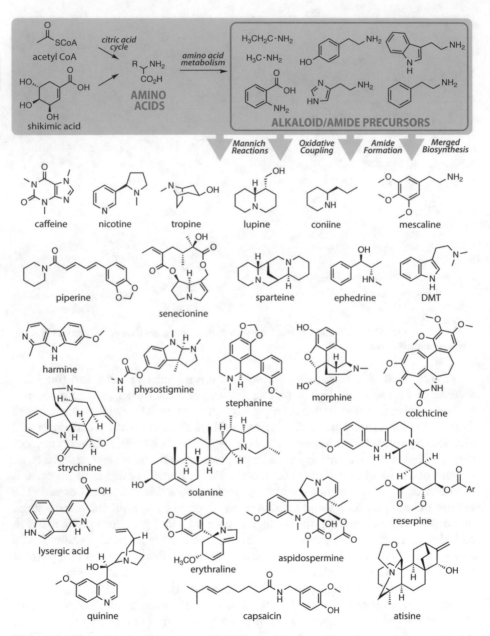

● **Fig. 2.6** General biosynthesis of alkaloids with examples of specific well-studied and unique molecules that can mediate biotic interactions

common, with some notable examples including quinine, nicotine, retronecine (and pyrrolizidine alkaloids in general), lupinine, and other compounds summarized in ● Fig. 2.6 (Wink 1998; Hartmann 1999; Aniszewski 2015).

Amines, which are derivatives of ammonia, can be important as antiherbivore, antifungal, and antimicrobial molecules – many of them are also categorized as alkaloids. For example, the aromatic amine, mescaline, derived from phenethyl-amine is biologically very active and ecologically important (e.g., De Panis et al. 2016), although most research has focused on its physiological effects on humans (Wink 1998). Some *amides*, which contain amine and carbonyl functional groups, and *imides*, which are comprised of two acyl groups bound to nitrogen, are also sometimes classified as alkaloids. The genus *Piper* (Piperaceae) includes biologically important amides and imides that affect multi-trophic interactions (Dyer et al. 2004a, b). Other nitrogen containing compounds that are relatively well studied and often quite toxic to herbivores are the biosynthetically related glucosinolates and cyanogenic glycosides, which are very well studied molecules in the context of trophic interactions (e.g., reviews by Agrawal et al. 2012; Gleadow and Møller 2014).

2.3 Phytochemical Diversity

Individual molecules never mediate ecological interactions in isolation, and what was once termed "redundancy" in chemical defense or in chemical attractants is now viewed as part of a complex chemical trait. Mixtures function as toxins, physiological disruptors, deterrents, attractants, facilitators, and other ecological effectors via additive effects, synergies, complimentary effects, or antagonistic effects. For example, one of the most effective mechanisms by which phytochemical mixtures can deter herbivores or other parasites is via synergistic effects, where the effects of combined compounds are greater than the effects predicted by their activity in isolation (Richards et al. 2016). In some cases, individual toxins will have no effect at all, such as antibiotics that are rendered ineffectual by bacterial toxin pumps – but synergies result when additional molecules in a mixture inhibit those pumps (Stermitz et al. 2000). Numerous studies have found that individual plant compounds have no effects or only weak influences on higher trophic levels, but synergy studies have demonstrated that those compounds exhibit potent biological activities when put in the right mixture (reviewed by Richards et al. 2016; Kessler and Kalske 2018).

So how do we measure mixture complexity? It is common to utilize ordination to summarize spectral data, which can collapse a great deal of data on molecules, fragments, and chemical features into a few factors that can be used as response variables in statistical models or that can provide insight into attributes of the mixture (reviewed by Dyer et al. 2018). An alternative is to use traditional entropy measures of diversity. The term, "phytochemical diversity," encompasses an emerging group of metrics that are useful for quantifying diversity of molecules and features in a plant extract (◘ Fig. 2.7). One dichotomy to consider is that mixtures are comprised of different structural complexities (based on unique features of molecules, such as size, degree of oxidation, degree of unsaturation, number of hydrogens, fraction of chiral centers, etc.) and distinct compositional diversities (i.e. a compound is the unit of diversity) (Philbin et al. 2021). A full measure of phytochemical diversity would both partition (i.e. calculate separately) as well as combine (i.e. calculate overall mixture diversity) these attributes of phytochemical diversity (Philbin et al. 2021). Understanding these aspects of phytochemical diversity

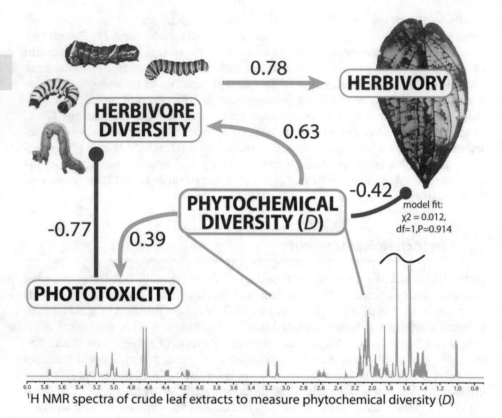

▫ Fig. 2.7   An example of how phytochemical diversity can affect plant toxicity and trophic interactions. Richards et al. (2015) used crude ^{1}H NMR spectra to quantify phytochemical diversity (Simpson's peak equivalents, D) for 22 Costa Rican *Piper* (Pipearaceae) shrub species (spectra from only one species is depicted here) and used as a predictor of function (phototoxicity), herbivory, and consumer diversity (Simpson's species equivalents, D). Crude ^{1}H NMR spectra reflect differences in phytochemical diversity by capturing structural, compositional, and metabolic complexity. Arrows are from a path analysis utilizing longterm caterpillar rearing data (Richards et al. 2015) – red arrows denote direct negative effects and green arrows are positive effects with associated path coefficients next to the arrows. Phytochemical diversity had substantive positive effects on toxicity and consumer diversity and a large negative effect on herbivory

is central to hypothesis tests in chemical ecology and these metrics should be examined in the context of a number of important theories in chemical ecology, including trophic interactions and coevolution, which are summarized below.

Because the tools of organic chemistry continue to develop greater resolution and accessibility, it is entirely possible to focus on the effects of particular mixtures of plant secondary metabolites on ecological interactions, and quantifying phytochemical diversity is certainly feasible, provided that the research approach includes fruitful collaborations with chemists. Herbivores, and higher trophic levels (Dyer et al. 2018). This approach can expand the focus of ecologists and evolutionary biologists beyond a particular compound, or crude measures of entire classes of compounds. For example, Richards et al. (2015) utilized a ^{1}NMR measure of phytochemical diversity to accurately predict overall diversity of herbivores, levels of parasitism on those herbivores, and variation in the distribution of diet breadth among herbivores. This focus on phytochemical diversity will facilitate understanding the evolutionary relationships between phytochemical diversity and biodiversity (see below), will provide rich insight into the origin and function of secondary metabolites, and will help estimate the extent of phytochemical losses that are part of the Anthropocene.

2.4   Trophic Interaction Theory and Chemical Defense

Plant secondary metabolites mediate ecological interactions from the smallest spatial scales, such as localized shifts in chemistry within a plant due to a pulse in nutrients, to larger scales, such as transfer of secondary metabolites across ecosystems via leaching into waterways. The effects of phytochemicals on arthropods are also evident at diverse temporal scales, from the quick release of volatiles to attract insects, to long term coevolutionary processes that are mediated by plant chemistry. Trophic interaction theory includes analytical models (An et al. 1993), conceptual models (e.g., plant defense theory, Stamp 2003), simulations (e.g., Puente et al. 2008), and empirical tests of hypotheses about how plant secondary metabolites or specific nutritional mixes of primary metabolites affect herbivores and enemies of those herbivores (reviewed by Burkepile and Parker 2017; Giron et al. 2018). The main hypotheses relevant to chemically mediated trophic interactions are focused on specialization, sequestration, bottom-up trophic cascades, and interaction diversity, and we provide brief working definitions here. *Specialization* can be simply defined as the number of taxa consumed by an herbivore or secondary consumer, but definitions can also correct for phylogeny or geographic scale. *Sequestration* refers to secondary metabolites associated with insect tissue and can be casual (phytochemical mixtures are found in the guts of herbivorous insects) or involve mechanisms for moving plant compounds across the insect gut into the hemolymph or fat bodies. Chemically mediated *bottom-up trophic cascades* consist of indirect trophic effects of changes in plant chemistry (Dyer et al. 2004b) – for example increased defensive chemistry could lower generalist herbivore abundances, which could negatively affect populations of specialist parasitoids. Trophic *interaction diversity* can be defined as the richness and relative abundance of direct and indirect links between plants, herbivores, and natural enemies. There are a wide range of mechanisms by which cascades and interaction diversity, and in turn, overall community diversity, can be affected by phytochemistry, but sequestration of secondary metabolites provides a clear pathway to affecting other trophic levels and entire arthropod communities.

The traditional view of plant-herbivore interactions is that insect herbivores avoid, excrete, process, or sequester secondary metabolites, or those metabolites may be toxic or otherwise affect the herbivore's fitness. Any of these herbivore responses potentially affect their physiology, population dynamics, competitive abilities, and interactions with natural enemies. Although it is difficult to find much generality in herbivore responses to plant chemistry, there is evidence that herbivore responses are partitioned by diet breadth, with specialists being more likely to process or sequester and generalists more frequently avoiding or excreting more toxic compounds (Dyer 1995). With respect to trophic interactions, sequestered metabolites can negatively affect predators and parasitoids in the same way that those metabolites negatively affect herbivores (Ode 2013), or sequestration can have positive effects on parasitoids via disruption of encapsulation and melaniza-tion or other anti-parasite responses (Smilanich et al. 2009a, b). According to the safe haven hypothesis, specialist herbivores are more toxic to predators than gener-alists, but as a result of that predator avoidance, specialists are a safe oviposition choice for parasitoids and host higher numbers and diversities of parasites (e.g., Lampert et al. 2010; Dyer 2018). If these effects of chemistry on herbivores and their enemies cascade up to affect a large proportion of consumer communities, then well defended plant communities should be characterized by greater compart-mentalization, more reticulate networks, and higher interaction diversity. These bottom-up effects on arthropod diversity are more likely to occur if there are greater abundances of specialist versus generalist herbivores and natural enemies, as is the case for tropical forests. Consistent with this idea, both theoretical and empirical data suggest that specialist herbivores in the tropics are locally more abundant, while densities of generalists are higher at larger scales (Dyer et al. 2010; Dyer et al. unpublished data).

Two other important hypotheses related to the chemical ecology of trophic interactions is that herbivory can induce production of plant secondary com-pounds, and that plants can produce volatile organic compounds (VOCs) to attract natural enemies. Induced defenses have been reviewed extensively in multiple sub-disciplines (e.g., for agriculture Olivoto et al. 2017) and are the subject of an inor-dinate number of empirical studies (Dyer et al. 2018). It is hard to estimate the importance of induced defenses, because the focus on many empirical studies has been on critical p-values rather than estimating biologically meaningful effect sizes. It is not surprising that disruptions to plant tissue, or any other stress to a plant, will cause shifts in entire metabolic pathways or changes in concentrations of indi-vidual compounds. It is not clear how much of a change is necessary for fitness effects or other biologically meaningful consequences for plant or herbivore. VOCs provide chemical cues that natural enemies use for locating herbivores (reviewed by numerous authors, including Aljbory and Chen 2018) and clear effects on upper trophic levels have been well documented, but primarily in agricultural and labora-tory settings (Kersch-Becker et al. 2017; Kalske et al. 2019). Remarkably, natural enemies can determine the number and types of herbovores present, which could have substantial effects on tritrophic interactions, but it is not at all clear how important VOCs might be in more complex systems, particularly complex net-works such as the understory of a tropical forest. Like other aspects of chemically-mediated plant-herbivore interactions, there is a lot to be learned about how well-studied laboratory or greenhouse phenomena translate to affecting entire communities or interacting networks of plants and arthropods.

Can variation in phytochemistry structure entire arthropod communities? It is certainly true that changes in nutrient availability, which affect both primary and secondary metabolism, are accompanied by striking differences in entire biotic communities (reviewed by Dyer and Letourneau 2013). For secondary metabolites alone, some studies on with model systems have demonstrated that entire plant-associated communities are in fact influenced by phytochemicals. For example, higher concentrations of tannins in distinct genotypes of cottonwood trees (*Populus* spp., Salicaceae) can support greater diversities of arthropods, but the mechanisms are not well resolved (Martinsen et al. 1998; Wimp et al. 2007; Barbour et al. 2015). Similarly, shifts in the saponin and sapogenin content of alfalfa (*Medicago sativa*) plants were associated with large differences in arthropod diversity (Pearson et al. 2008), and some mechanisms underlying those changes have been documented in laboratory experiments – most notably, increases in saponins contributed to caterpillar phenological changes that were detrimental to parasitoids (Dyer et al. 2013). A goal of chemical ecology is to elucidate these connections, from processes such as sequestration or production of VOCs to community level parameters, like interaction diversity.

2.5 Coevolution: Variation of Chemically Mediated Interactions Over Space and Time

Coevolution, when interacting taxa exert, and evolve in response to, reciprocal natural selection (Carmona et al. 2015), remains a centerpiece of chemically mediated interaction theory since the concept was formally introduced by Ehrlich and Raven (1964) using butterflies and their host plants as a model system for understanding "community evolution" and as one general mechanism for the diversification of insects and angiosperms. Ehrlich and Raven's concept of a coevolutionary arms race between plants and herbivores, producing increases in diversity of plant secondary compounds and adaptive radiations of interacting taxa, has long been an important theoretical framework for studies in plant-animal interactions (Thompson 2005; Althoff et al. 2014). Coevolutionary processes are predicted to yield high levels of consumer or mutualist specialization due to trade-offs in physiological responses to unique phytochemistry among different plant taxa. There is limited support for antagonistic pleiotropy in animal adaptations to phytochemistry, which is expected to underly such trade-offs (Fry 1990; Dyer 2011), but with new genomics approaches, it is possible to test carefully for genetic trade-offs as constraints on the evolution of specialization (Gompert and Messina 2016). Detecting these trade-offs is still a relevant goal of research on plant-animal interactions and should be investigated with modern genomic and chemical approaches.

Studies of geographic variation in the putative mechanisms responsible for chemically mediated coevolution have steadily increased in recent decades. These studies on the geographic mosaic of coevolution (Thompson 2005) have contributed to our understanding of processes that contribute to geographic variation in phytochemistry and animal adaptations to that chemistry (e.g., Berenbaum and Zangerl 1998; Hague et al. 2020; Dyer 2011). As a result, a great deal of literature on chemistry and biotic interactions has focused on adaptations that are both causes and consequences of coevolution, and these adaptations help shape interaction diversity and other community parameters (Raguso et al. 2015). Furthermore, with respect to the causal relationships between biotic interactions and biodiversity, broadly defined coevolutionary interactions between interacting parasites and hosts and interacting mutualists may have generated a significant portion of existing biodiversity. Coevolutionary adaptations that have shaped chemically mediated biotic interactions include production of toxic or deterrent phytochemical mixtures, feeding specialization of animals on plants with specific mixtures of compounds, detoxification or sequestration of secondary metabolites, phytochemical attractants to predators and parasitoids, and antennal or ovipositor sensilla sensitive to specific compounds and mixtures (e.g., Zacharuk 1980; Agrawal et al. 2012). Chemical ecology research on geographic variation in coevolutionary plant-herbivore interactions has great potential to increase our understanding of the processes generating and maintaining phytochemical variation across the landscape.

A remaining challenge for studies of chemically mediated plant-animal interactions is to establish the link between microevolutionary processes driving coevolution and macroevolutionary patterns of biological diversification (Althoff et al. 2014). Modern phylogenetic comparative methods can allow for hypothesis tests about how chemical traits have evolved within plant lineages, allowing for insights into the evolution of plant insect interactions. In particular, new modeling approaches can provide estimates of phylogenetic signal (e.g., Pagel's λ^2 or Blomberg's K^3) versus trait lability across plant species. Clear phylogenetic signal may be pronounced for some of the broad compound classes reviewed above, such as subclasses of alkaloids, while lack of phylogenetic signal might be expected for features that contribute to structural complexity and are detected via spectroscopy. Many suites of features may evolve convergently if they are effective in deterring herbivores or attracting pollinators. It is likely that chemical traits that vary substantially among related species might be influential in current ecological interactions. In contrast, strong signal and positive correlations among different compounds could be due to either metabolic constraints or possible adaptations via synergistic (e.g., Richards et al. 2016) or other mixture defensive attributes. Another macroevolutionary pattern could be characterized by negative correlations in chemical traits across related species, which could be indicative of a biosynthetic or evolutionary tradeoff between traits – for example, one might expect negative correlations between kavalactones and p-alkenylphenols in *Piper* species. These compounds are synthesized via a shared biosynthetic pathway with a branch point, where one chain extension pathway from cinnamic acid or coumaric acid can yield the long-chain lipophilic substituent characteristic of the p-alkenylphenols, and an alternative chain extension pathway conserves oxidation states through the chain extension process to produce lactones through cyclization reactions. Both positive and negative correlations have been uncovered among chemical traits (e.g. Johnson et al. 2014; Kariñho-Betancourt et al. 2015) after accounting for evolutionary distances among species, but examining the evolution of spectroscopic features and mixtures of compounds is more likely to provide insight into long-term evolutionary changes in plant chemistry and their relationships to animal adaptations. Finally, estimates of heritability and degree of plasticity for individual traits and for measures of mixture complexity, such as structural diversity, are necessary for better understanding macroevolutionary patterns. Highly plastic chemical defenses could obscure any phylogenetic comparative methods, and while a fair number of studies have estimated heritability for individual compounds (e.g., Baghalian et al. 2010; Bresadola et al. 2019), this has not been accomplished for any measures of mixture complexity or suites of features or compounds.

2.6 Novel Approaches to Chemical Ecology and Statistical Inference

Many tests of hypotheses related to coevolution and chemically mediated trophic interactions require analysis of very large datasets derived from field observations, experiments, and comparative analytical chemistry. There are a number of hurdles that prevent generating high quality data and good hypothesis tests, including the lack of availability of pure compounds, a dearth of characterizations of full phytochemical phenotypes of plant species or individual plants, a reliance on outdated or crude methods, (e.g., colorimetric measures of total phenolics), and the overuse of old statistical approaches (e.g., principle components analysis). There are easy solutions to these issues, and collaborating with chemists and statisticians can facilitate such solutions with benefits to multiple fields of investigation. For example, a synthesis approach to acquiring pure compounds for experiments or standards involves examining metabolite transformation in vitro, allowing for thorough exploration of biosynthetic hypotheses and the potential chemical space of natural products. One notable addition to the chemical ecology toolbox that requires a collaborative approach is the rapid advancement of non-targeted metabolomics approaches (e.g., Richards et al. 2018) that are now being applied to ecological systems, replacing single or multi-component targeted approaches, and this is accompanied by enhanced metrics of phytochemical diversity (Matsuda et al. 2011; Wetzel and Whitehead 2020; Philbin et al. 2021). Full reviews of the many advances in approaches to isolation, synthesis, and structure determination are beyond the scope of this chapter but are summarized in a recent review (Dyer et al. 2018 and supplements to that paper). It is clear from this review (and references therein) that the need for diverse metabolic libraries is increasing, especially as plant diversity and phytochemical diversity rapidly decline due to global change; furthermore, synthesis, isolation, structure determination, and resolution of biochemical pathways are clear priorities for research on ecological phytochemistry. Using these approaches, it is possible to determine the biological function and metabolic fate of individual compounds and natural mixtures, allowing for rigorous hypothesis tests in chemical ecology (◘ Fig. 2.8).

Similarly, modern mathematical, computational, and statistical models are indispensable tools for chemical ecology – it is exciting to move beyond old null hypothesis testing frameworks and to explore more sophisticated modeling approaches, including advances in statistical, predictive analytical, or simulation models (e.g., DeAngelis and Yurek 2017; Rangel et al. 2018). These approaches can shape our understanding of mechanistic processes relevant to chemically mediated

2

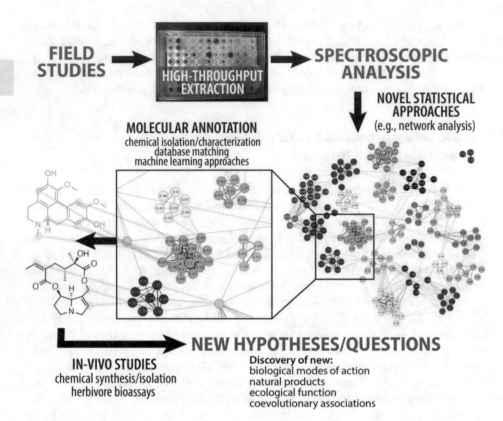

☐ Fig. 2.8 Novel approaches to studying chemically mediated biotic interactions merge natural history, field and mesocosm experiments, modern spectroscopic analysis, recent statistical advances, and detailed chemical studies to provide molecular resolution at ecologically relevant scales. Specifically, ecological methods yield harvested plant or insect tissues that are extracted and examined with spectroscopy, such as NMR and MS methods coupled with a myriad of sources. Raw spectroscopic data can be analyzed using a number of new approaches, including powerful multivariate approaches, machine learning, and network analyses, depicted here. Nodes from those networks can be used as variables in statistical analyses to examine relationships with associated ecological data. These methods, accompanied by accurate molecular annotations can lead to the discovery of molecules, insight into coevolutionary associations, and greater understanding of biological functions

interactions and coevolution and can stimulate new hypotheses. Once appropriate quantitative and conceptual models have guided research in chemical ecology, options for statistical analyses are myriad – the toolbox for analysis of large amounts of spectroscopic data generated from crude extracts or solvent fractions continues to grow rapidly (Dyer et al. 2018). The most important methods for statistical analysis that will help push chemical ecology forward include combinations of traditional multivariate generalized linear models, hierarchical Bayesian models, Bayesian analyses using informative priors, novel ordination methods, network analyses, machine learning approaches, and artificial intelligence (e.g., Dobson and Barnett 2018; Bohan et al. 2017; McElreath 2020). Several recent reviews focus on these approaches for ecology and chemical ecology (Humphries et al. 2018; Dyer et al. 2018 – supplemental materials, Christin et al. 2019).

It is becoming more common to employ weighted networks (Horvath 2011) to analyze spectral data and construct a network – for example with nodes that are binned chemical shifts from ^{1}H-NMR and edges that are determined based on the correlations between the chemical shifts (Richards et al. 2018). Therefore, chemical shifts that co-vary across samples are more connected, and highly connected chemical shifts can be organized into modules, which can correspond to distinct subpopulations of a plant species, individual plants from experimental treatments, or different species of pollen collected from pollinating insects. Calculating eigen-vectors allows one to examine correlations between modules and the individual samples collected from the field or from experiments. This method can be validated with artificial mixtures of known compounds and extracts from plant species with known chemistry (Richards et al. 2018). One advantage of these methods is that they can produce detailed insight into chemical features (e.g., modules from the networks) that are associated with plant subpopulations, experimental treatment level combinations, or other samples. Similarly, MS/MS-based molecular network-ing can be employed using cosine similarities between spectra based on differences in relative intensities of all fragment ions and the difference in masses of the parent ions between two spectra. In these networks, each node is a parent ion mass and the edges represent the cosine score between the two corresponding spectra. This approach can be used to characterize molecular similarities among samples from the field or from experiments and also provide useful information about functional differences or the presence of structurally novel metabolites. This approach has been used to link bioactivity, genomic, metabolomics, and transcriptomic datasets, allowing for inferences about links between genetic variation, chemical novelty, and ecological relevance (e.g., Kurita et al. 2015).

It is clear that the machine learning, artificial intelligence, and deep learning methods mentioned above have great potential to transform approaches to under-standing complex phytochemical mixtures and their effects on interactions. For example, convolutional neural networks (CNN) provide a powerful method for handling large data outputs from nontargeted metabolomics or combined -omics datasets (Brodrick et al. 2019; Zhu et al. 2020). Generally, the CNN method involves developing algorithms that can automatically extract discriminant fea-tures from big data without human involvement. This is an improvement over stan-dard ordination (PCA, NMDS) or Support Vector Machine (SVM) analyses, that feature extraction and preprocessing steps that require user iterations and are time-consuming. The possibilities for linking these and similar approaches to advances in organic chemistry are exciting and will contribute to leaps in progress in studies of chemically mediated interactions.

2

Conclusion

Three related questions that have produced rich theory or could eventually contribute to understanding how phytochemicals mediate biotic interactions are: *what are the precise metabolic pathways by which plants produce high diversities of secondary metabolites, how does that diversity function to mediate biotic interactions, and how much phytochemical diversity is being lost in the Anthropocene?* Recent advances in theory and recent methodological improvements will contribute substantively to answering these questions. A strong theoretical framework for trophic interactions, coevolution, and other topics in the realm of chemical ecology is well established, but what we lack are empirical data, modern approaches to hypothesis testing, and an influx of creative statistical and mathematical models. Do we lack new ideas? Not so much. Really, none of the existing theories or hypotheses in chemical ecology are at a point where major portions could be rejected outright or labeled as "dead" (Rausher 1988; Hamilton et al. 2001). A focus on these well-established theories, using high quality empirical data, modeling, and quantitative synthesis will contribute to answering the pressing question of what we are losing as we destroy plant chemistry and biotic interactions across all ecosystems in the Anthropocene.

Key Points

- All plants have complex mixtures of secondary metabolites that can function as defenses, attractants, or physiological regulators, and chemical ecologists must determine how to characterize and study these complex mixtures and how they affect other organisms.
- Effects of plant chemistry can cascade up through their direct effects on interacting animals to affect entire biotic communities.
- The amount of phtyochemical diversity being lost in the Anthropocene is unknown and probably will never be known.

❓ Questions

- What factors drive the variation of phytochemicals in time and space and how does that variation affect mutualists, parasites, and competitors?
- What are the physiological and ecological costs to plants of phytochemical defenses?
- How do phytochemicals affect interactions between plants and animals at the community level?

References

Agrawal AA, Petschenka G, Bingham RA, Weber MG, Rasmann S (2012) Toxic cardenolides: chemical ecology and coevolution of specialized plant–herbivore interactions. New Phytol 194:28–45

Aljbory Z, Chen M-S (2018) Indirect plant defense against insect herbivores: a review. Insect Sci 25:2–23

Althoff DM, Segraves KA, Johnson MTJ (2014) Testing for coevolutionary diversification: linking pattern with process. Trends Ecol Evol 29:82–89

An M, Johnson IR, Lovett JV (1993) Mathematical modeling of allelopathy: biological response to allelochemicals and its interpretation. J Chem Ecol 19:2379–2388

Aniszewski T (2015) Alkaloids: chemistry, biology, ecology, and applications. Elsevier, Amsterdam

Ayres MP, Clausen TP, MacLean SF Jr, Redman AM, Reichardt PB (1997) Diversity of structure and antiherbivore activity in condensed tannins. Ecology 78:1696–1712

Baghalian K, Sheshtamand MS, Jamshidi AH (2010) Genetic variation and heritability of agro-morphological and phytochemical traits in Iranian saffron (Crocus sativus L.) populations. Ind Crop Prod 31:401–406

Baldwin JW, Dechmann DKN, Thies W, Whitehead SR (2020) Defensive fruit metabolites obstruct seed dispersal by altering bat behavior and physiology at multiple temporal scales. Ecology 101:e02937

Barbour MA, Rodriguez-Cabal MA, Wu ET, Julkunen-Tiitto R, Ritland CE, Miscampbell AE, Jules ES, Crutsinger GM (2015) Multiple plant traits shape the genetic basis of herbivore community assembly. Funct Ecol 29:995–1006

Bennett RN, Wallsgrove RM (1994) Secondary metabolites in plant defence mechanisms. New Phytol 127:617–633

Berenbaum MR (1995) Turnabout is fair play: secondary roles for primary compounds. J Chem Ecol 21:925–940

Berenbaum MR, Zangerl AR (1998) Chemical phenotype matching between a plant and its insect herbivore. Proc Natl Acad Sci 95:13743–13748

Bohan DA, Vacher C, Tamaddoni-Nezhad A, Raybould A, Dumbrell AJ, Woodward G (2017) Next-generation global biomonitoring: large-scale, automated reconstruction of ecological networks. Trends Ecol Evol 32:477–487

Bresadola L, Caseys C, Castiglione S, Buerkle CA, Wegmann D, Lexer C (2019) Admixture mapping in interspecific Populus hybrids identifies classes of genomic architectures for phytochemical, morphological and growth traits. New Phytol 223:2076–2089

Brodrick PG, Davies AB, Asner GP (2019) Uncovering ecological patterns with convolutional neural networks. Trends Ecol Evol 34:734–745

Burkepile DE, Parker JD (2017) Recent advances in plant-herbivore interactions. F1000Res 6:119

Cardinale BJ, Duffy JE, Gonzalez A, Hooper DU, Perrings C, Venail P, Narwani A, Mace GM, Tilman D, Wardle DA, and others.(2012) Biodiversity loss and its impact on humanity. Nature 486:59–67

Carmona D, Fitzpatrick CR, Johnson MTJ (2015) Fifty years of co-evolution and beyond: integrating co-evolution from molecules to species. Mol Ecol 24:5315–5329

Christin S, Hervet É, Lecomte N (2019) Applications for deep learning in ecology. Methods Ecol Evol 10:1632–1644

Cseke LJ, Kirakosyan A, Kaufman PB, Warber S, Duke JA, Brielmann HL (2016) Natural products from plants. CRC Press Boca Raton, FL.

Damtoft S, Jensen SR, Jessen CU, Knudsen TB (1993) Late stages in the biosynthesis of aucubin in Scrophularia. Phytochemistry 33:1089–1093

De Panis DN, Padró J, Furió-Tarí P, Tarazona S, Milla Carmona PS, Soto IM, Dopazo H, Conesa A, Hasson E (2016) Transcriptome modulation during host shift is driven by secondary metabolites in desert Drosophila. Mol Ecol 25:4534–4550

DeAngelis DL, Yurek S (2017) Spatially explicit modeling in ecology: a review. Ecosystems 20:284–300

Demmig-Adams B, Cohu CM, Amiard V, van Zadelhoff G, Veldink GA, Muller O, Adams WW III. (2013) Emerging trade-offs – impact of photoprotectants (PshS, xanthophylls, and vitamin E) on oxylipins as regulators of development and defense. New Phytol 197:720–729

Deng Y, Lu S (2017) Biosynthesis and regulation of phenylpropanoids in plants. Crit Rev Plant Sci 36:257–290

Dinda B (2019) Chemistry and biosynthesis of iridoids. In: Pharmacology and applications of naturally occurring iridoids. Springer International Publishing, Cham, pp 119–143

Dobson AJ, Barnett AG (2018) An introduction to generalized linear models. CRC Press, Boca Raton

Dyer LA (1995) Tasty generalists and nasty specialists? Antipredator mechanisms in tropical Lepidopteran larvae. Ecology 76:1483–1496

Dyer LA (2011) New synthesis-back to the future: new approaches and directions in chemical studies of coevolution. J Chem Ecol 37:669

Dyer LA (2018) Multidimensional diversity associated with plants: a view from a plant–insect interaction ecologist. Am J Bot 105:1439–1442

Dyer LA, Letourneau DK (2013) Can climate change trigger massive diversity cascades in terrestrial ecosystems? Diversity 5:479–504

Dyer LA, Dodson CD, Richards J (2004a) In: Dyer LA, Palmer AN (eds) Piper. A model genus for studies of evolution, chemical ecology, and trophic interactions. Kluwer Academic Publishers, Boston

Dyer LA, Letourneau DK, Dodson CD, Tobler MA, Stireman JO III, Hsu A (2004b) Ecological causes and consequences of variation in defensive chemistry of a neotropical shrub. Ecology 85:2795–2803

Dyer LA, Walla TR, Greeney HF, Stireman JO III, Hazen RF (2010) Diversity of interactions: a metric for studies of biodiversity. Biotropica 42:281–289

Dyer LA, Richards LA, Short SA, Dodson CD (2013) Effects of CO2 and temperature on tritrophic interactions. PLoS ONE 8:e62528

Dyer LA, Philbin CS, Ochsenrider KM, Richards LA, Massad TJ, Smilanich AM, Forister ML, Parchman TL, Galland LM, Hurtado PJ, and others.(2018) Modern approaches to study plant-insect interactions in chemical ecology. Nat Rev Chem 2:50–64

Ehrlich PR, Raven PH (1964) Butterflies and plants: a study in coevolution. Evolution 18:586–608

Erbilgin N (2019) Phytochemicals as mediators for host range expansion of a native invasive forest insect herbivore. New Phytologist 221:1268–1278

Fry JD (1990) Trade-offs in fitness on different hosts: evidence from a selection experiment with a phytophagous mite. Am Nat 136:569–580

Gershenzon J, Fontana A, Burow M, Wittstock UTE, Degenhardt J (2012) Mixtures of plant secondary metabolites: metabolic origins and ecological benefits. In: The ecology of plant secondary metabolites: from genes to global processes, p 56

Giron D, Dubreuil G, Bennett A, Dedeine F, Dicke M, Dyer LA, Erb M, Harris MO, Huguet E, Kaloshian I, Kawakita A, Lopez-Vaamonde C, Palmer TM, Petanidou T, Poulsen M, Sallé A, Simon J-C, Terblanche JS, Thiéry D, Whiteman NK, Woods HA, Pincebourde S (2018) Promises and challenges in insect–plant interactions. Entomol Exp Appl 166:319–343

Gleadow RM, Møller BL (2014) Cyanogenic glycosides: synthesis, physiology, and phenotypic plasticity. Annu Rev Plant Biol 65:155–185

Gompert Z, Messina FJ (2016) Genomic evidence that resource-based trade-offs limit host-range expansion in a seed beetle. Evolution 70:1249–1264

Greany PD, Tumlinson JH, Chambers DL, Boush GM (1977) Chemically mediated host finding byBiosteres (Opius) longicaudatus, a parasitoid of tephritid fruit fly larvae. J Chem Ecol 3:189–195

Hague MTJ, Stokes AN, Feldman CR, Brodie ED Jr, Brodie ED III. (2020) The geographic mosaic of arms race coevolution is closely matched to prey population structure. Evol Lett 4:317–332

Hamilton JG, Zangerl AR, DeLucia EH, Berenbaum MR (2001) The carbon – nutrient balance hypothesis: its rise and fall. Ecol Lett 4:86–95

Harborne JB (1999) Classes and functions of secondary products from plants, in Chemicals from Plants—Perspectives on Plant Secondary Products, ed by Walton JN and Brown DE, Imperial College Press, London, UK, pp 1–25

Hartmann T (1999) Chemical ecology of pyrrolizidine alkaloids. Planta 207:483–495

Heil M, Baumann B, Andary C, Linsenmair EK, McKey D (2002) Extraction and quantification of "condensed tannins" as a measure of plant anti-herbivore defence? Revisiting an old problem. Naturwissenschaften 89:519–524

Helfrich EJN, Vogel CM, Ueoka R, Schäfer M, Ryffel F, Müller DB, Probst S, Kreuzer M, Piel J, Vorholt JA (2018) Bipartite interactions, antibiotic production and biosynthetic potential of the Arabidopsis leaf microbiome. Nat Microbiol 3:909–919

Horvath S (2011) Weighted network analysis: applications in genomics and systems biology. Springer Science & Business Media, New York

Humphries GRW, Magness DR, Huettmann F (2018) Machine learning for ecology and sustainable natural resource management. Springer, Cham

Jacobsen DJ, Raguso RA (2018) Lingering effects of herbivory and plant defenses on pollinators. Curr Biol 28:R1164–R1169

Johnson MTJ, Ives AR, Ahern J, Salminen J-P (2014) Macroevolution of plant defenses against herbivores in the evening primroses. New Phytol 203:267–279

Johnson CN, Balmford A, Brook BW, Buettel JC, Galetti M, Guangchun L, Wilmshurst JM (2017) Biodiversity losses and conservation responses in the anthropocene. Science 356:270–275

Jun S-Y, Sattler SA, Cortez GS, Vermerris W, Sattler SE, Kang C (2018) Biochemical and structural analysis of substrate specificity of a phenylalanine ammonia-lyase. Plant Physiol 176:1452–1468

Kalske A, Shiojiri K, Uesugi A, Sakata Y, Morrell K, Kessler A (2019) Insect herbivory selects for volatile-mediated plant-plant communication. Curr Biol 29:3128–3133

Kariñho-Betancourt E, Agrawal AA, Halitschke R, Núñez-Farfán J (2015) Phylogenetic correlations among chemical and physical plant defenses change with ontogeny. New Phytol 206:796–806

Kersch-Becker MF, Kessler A, Thaler JS (2017) Plant defences limit herbivore population growth by changing predator-prey interactions. Proc R Soc B Biol Sci 284:20171120

Kessler A, Kalske A (2018) Plant secondary metabolite diversity and species interactions. Annu Rev Ecol Evol Syst 49:115–138

Kurita KL, Glassey E, Linington RG (2015) Integration of high-content screening and untargeted metabolomics for comprehensive functional annotation of natural product libraries. Proc Natl Acad Sci 112:11999–12004

Lampert EC, Dyer LA, Bowers MD (2010) Caterpillar chemical defense and parasitoid success: cotesia congregata parasitism of ceratomia catalpae. J Chem Ecol 36:992–998

Lautié E, Russo O, Ducrot P, Boutin JA (2020) Unraveling plant natural chemical diversity for drug discovery purposes. Front Pharmacol 11:397

Martinsen GD, Driebe EM, Whitham TG (1998) Indirect interactions mediated by changing plant chemistry: beaver browsing benefits beetles. Ecology 79:192–200

Matsuda F, Nakabayashi R, Sawada Y, Suzuki M, Hirai MY, Kanaya S, Saito K (2011) Mass spectra-based framework for automated structural elucidation of metabolome data to explore phytochemical diversity. Front Plant Sci 2:40

McElreath R (2020) Statistical rethinking: a Bayesian course with examples in R and Stan. CRC Press, Boca Raton

Morita H, Shimokawa Y, Tanio M, Kato R, Noguchi H, Sugio S, Kohno T, Abe I (2010) A structure-based mechanism for benzalacetone synthase from Rheum palmatum. Proc Natl Acad Sci 107:669–673

Ode PJ (2013) Plant defences and parasitoid chemical ecology. In: Chemical ecology of insect parasitoids. Wiley, Chichester, West Sussex, pp 11–28

Olivoto T, Nardino M, Carvalho IR, Follmann DN, Szareski VIJ, Ferrari M, de Pelegrin AJ, de Souza VQO (2017). Plant secondary metabolites and its dynamical systems of induction in response to environmental factors: A review. African Journal of Agricultural Research, 12(2), pp.71–84

Pandey A, Swarnkar V, Pandey T, Srivastava P, Kanojiya S, Mishra DK, Tripathi V (2016) Transcriptome and metabolite analysis reveal candidate genes of the cardiac glycoside biosynthetic pathway from Calotropis procera. Sci Rep 6:34464

Pearson CV, Massad TJ, Dyer LA (2008) Diversity cascades in alfalfa fields: from plant quality to agroecosystem diversity. Environ Entomol 37:947–955

Philbin CS, Dyer LA, Jeffrey CS, Glassmire AE, Richards LA. (2021) Structural and compositional dimensions of phytochemical diversity in the genus Piper reflect distinct ecological modes of action. J Ecol

Price ML, Butler LG (1977) Rapid visual estimation and spectrophotometric determination of tannin content of sorghum grain. J Agric Food Chem 25:1268–1273

Puente M, Magori K, Kennedy GG, Gould F (2008) Impact of herbivore-induced plant volatiles on parasitoid foraging success: a spatial simulation of the Cotesia rubecula, Pieris rapae, and Brassica oleracea system. J Chem Ecol 34:959–970

Raguso RA, Agrawal AA, Douglas AE, Jander G, Kessler A, Poveda K, Thaler JS (2015) The raison d'être of chemical ecology. Ecology 96:617–630

Rangel TF, Edwards NR, Holden PB, Diniz-Filho JAF, Gosling WD, Coelho MTP, Cassemiro FAS, Rahbek C, Colwell RK (2018) Modeling the ecology and evolution of biodiversity: biogeographical cradles, museums, and graves. Science 361:eaar5452

Rausher MD (1988) Is coevolution dead? Ecology 69:898–901

Richards LA, Dyer LA, Forister ML, Smilanich AM, Dodson CD, Leonard MD, Jeffrey CS (2015) Phytochemical diversity drives plant-insect community diversity. Proc Natl Acad Sci U S A 112:10973

Richards LA, Glassmire AE, Ochsenrider KM, Smilanich AM, Dodson CD, Jeffrey CS, Dyer LA (2016) Phytochemical diversity and synergistic effects on herbivores. Phytochem Rev 15:1153–1166

Richards LA, Oliveira C, Dyer LA, Rumbaugh A, Urbano-Muñoz F, Wallace IS, Dodson CD, Jeffrey CS (2018) Shedding light on chemically mediated tri-trophic interactions: A^{1}H-NMR network approach to identify compound structural features and associated biological activity. Front Plant Sci 112(35):10973–10978

Salazar D, Lokvam J, Mesones I, Pilco MV, Zuñiga JMA, de Valpine P, Fine PVA (2018) Origin and maintenance of chemical diversity in a species-rich tropical tree lineage. Nat Ecol Evol 2:983–990

Sedio BE, Parker JD, McMahon SM, Wright SJ (2018) Comparative foliar metabolomics of a tropical and a temperate forest community. Ecology 99:2647–2653

Seigler D, Price PW (1976) Secondary compounds in plants: primary functions. Am Nat 110:101–105

Smilanich AM, Dyer LA, Chambers JQ, Bowers MD (2009a) Immunological cost of chemical defence and the evolution of herbivore diet breadth. Ecol Lett 12:612–621

Smilanich AM, Dyer LA, Gentry GL (2009b) The insect immune response and other putative defenses as effective predictors of parasitism. Ecology 90:1434–1440

Stamp N (2003) Out of the quagmire of plant defense hypotheses. Q Rev Biol 78:23–55

Stermitz FR, Lorenz P, Tawara JN, Zenewicz LA, Lewis K (2000) Synergy in a medicinal plant: antimicrobial action of berberine potentiated by 5′-methoxyhydnocarpin, a multidrug pump inhibitor. Proc Natl Acad Sci 97:1433–1437

Stork NE (2018) How many species of insects and other terrestrial arthropods are there on earth? Annu Rev Entomol 63:31–45

Thompson JN (2005) The geographic mosaic of coevolution. University of Chicago Press

Wagner DL, Fox R, Salcido DM, Dyer LA (2021) A window to the world of global insect declines: moth biodiversity trends are complex and heterogeneous. Proc Natl Acad Sci 118(2)

Wetzel WC, Whitehead SR (2020) The many dimensions of phytochemical diversity: linking theory to practice. Ecol Lett 23:16–32

Wimp GM, Wooley S, Bangert RK, Young WP, Martinsen GD, Keim P, Rehill B, Lindroth RL, Whitham TG (2007) Plant genetics predicts intra-annual variation in phytochemistry and arthropod community structure. Mol Ecol 16:5057–5069

Wink M (1998) Chemical ecology of alkaloids. In: Alkaloids. Springer, Boston, MA pp 265–300

Zacharuk RY (1980) Ultrastructure and function of insect chemosensilla. Annu Rev Entomol 25:27–47

Zhu D, Zhang F, Wang S, Wang Y, Cheng X, Huang Z, Liu Y (2020) Understanding place characteristics in geographic contexts through graph convolutional neural networks. Ann Am Assoc Geogr 110:408–420

Escape as a Mechanism of Plant Resistance Against Herbivores

Robert J. Marquis and Renan F. Moura

Contents

© Springer Nature Switzerland AG 2021
K. Del-Claro, H. M. Torezan-Silingardi (eds.), *Plant-Animal Interactions*, https://doi.org/10.1007/978-3-030-66877-8_3

⊜ Learning Objectives
1. What are escape traits and how they differ from other resistance mechanisms.
2. Several types of escape traits exhibited by plants.
3. The evidence and relative importance of each escape trait.
4. The challenges and benefits of studying all resistance traits combined.

3

3.1 Introduction

Traits that contribute to the success of plants in the presence of herbivores and pathogens (hereafter herbivores) are called resistance traits (Boege and Marquis 2005; Agrawal et al. 2006). Traditionally, resistance has been considered to have two major components, defense and tolerance. Defense traits are those that reduce the amount of tissue loss to natural enemies once the herbivore has encountered the plant. Some plants, once found, will have a combination of defense traits that result in high levels of tissue loss, while other plants will have a combination of defense traits that might result in little or no tissue loss. These defense traits can either be direct (directly mediating the interaction between the plant and the herbivore, typically secondary compounds, nutrient content of tissue, architecture, and physical characteristics of the tissue) or indirect (mediating the interaction via impacts of the third trophic level on the herbivores) (Pearse et al. 2020).

Plant tolerance traits reduce the impact of herbivores on plant fitness subsequent to tissue loss, that is, after the herbivores have consumed plant tissue. Given equal amounts of tissue loss, plants that are able to grow more and produce more seeds following attack would be considered to be more tolerant of herbivore attack. Mobilization and allocation of stored resources, activation of meristems, and up-regulation of photosynthesis are common mechanisms by which plants can recuperate fitness following tissue loss (Strauss and Agrawal 1999; Stowe et al. 2000).

There are plant traits, however, that influence whether plants are found initially by their herbivores (escape in space), and when found, whether the plants are in a vulnerable state (escape in time). These traits can be classified as a third component of resistance, called escape. Plants escape in space if their herbivores cannot find them. Most herbivore species have a relatively small list of plant species from which they will feed. Even species with long lists of potential host plants may feed on only a few species at any one location or at any one time (Vidal et al. 2020). Thus both generalist and specialist herbivore species alike must find their individually preferred host plant species. Plants can also escape in time. Herbivores must not only find their host plants, but they must find them when they are palatable. Most plants go through developmental changes at the whole plant level as they grow, shifting allocation to defenses that influence palatability (Boege and Marquis 2005; Barton and Koricheva 2010; Boege et al. 2011). In addition, individual plant parts, leaves in particular, also change dramatically in toughness, nutrient composition, pubescence, and defense compound concentration as they age (e.g., Coley 1983a; Kursar and Coley 2003).

We define escape traits as those traits that decrease the likelihood that an herbivore will find an otherwise consumable plant in space or time. We would predict that any traits that decrease the ability of the herbivore to find its host plant in space and time would be selected for, and thus increase in frequency in the plant population. We argue that escape is a component involved in plant-herbivore interactions separate from defense and tolerance but like defense and tolerance, contributes to the success of the plant in the face of herbivore pressure. Escape traits are quantifiable and have a heritable basis, and their impact on plant fitness can be quantified. Many if not all the traits we discuss in this chapter are recognized as contributing to the success of plants in a world of herbivores, by mediating plant-herbivore interactions. There has been, however, no formal discussion of escape as integral to the impact of herbivores on plant fitness, and therefore integral to the evolution of plants. The first goal of this chapter is to bring together a disparate literature on plant traits that reduce the ability of herbivores to find their host plants or to find them when they are most vulnerable. The second goal is to provide an example of how to measure the three components of plant resistance to herbivores, and to quantify their independent contribution. The third goal is to suggest how escape might be used in agriculture, in conjunction with more traditional approaches focusing on defense and tolerance, to reduce pest pressure. The fourth goal is to describe how escape may contribute to biodiversity maintenance in communities. The overall intended result is a more thorough understanding how herbivores influence the ecology and evolution of plants, their associated plant communities, and their attendant food webs.

3.2 History of the Concept

Feeny (1976) was the first to write of plant apparency as an explanation for why short-lived herbaceous plant species might differ in their defense profiles from long-lived woody plant species. He suggested that small, ephemeral, annual herbs might escape discovery from their specialist herbivores. He contrasted these annuals with trees, which would be readily apparent to their herbivores because they were large, and long-lived, occupying one spot for many generations of their insect herbivores.

Importantly, these differences in apparency were proposed to have consequences for defense (Feeny 1976; Rhoades and Cates 1976; see also Stamp 2003; Agrawal et al. 2006; Strauss et al. 2015). Feeny (1976) proposed that apparent plants would be defended by qualitative defenses, and unapparent species by quantitative defenses. He suggested that highly apparent plants are more likely to suffer continuous herbivore pressure, and that selective pressure would give rise to generalized defenses that are most effective at increasingly higher concentrations. In contrast, unapparent plants presumably suffer less herbivore pressure because they are less likely to be found. If found, their defense compounds would be effective at low concentration, toxic to all but a few specialist herbivore species. In his example, Feeny (1976) chose annual species of Brassicaceae to represent plants of low apparency. They are defended by glucosinolates, which prevent feeding by most species of potential herbivores. As such, they are deterrents at low concentrations against all but a few very specialized species. In contrast, Feeny chose trees of the genus *Quercus* (Fagaceae) to represent plants of high apparency. Because they are likely to be found by their herbivores, due to their size and their longevity (i.e., they are easy to find and will be in one place for a long time), they have evolved phenolics that are most effective at relatively high concentration.

The apparency hypothesis has been criticized in the most recent extensive review of the plant defense theories (Stamp 2003, but see Agrawal et al. 2006), as not able to explain broad patterns of plant defense. Two more recent reviews of the relationship between apparency and defensive chemistry also found a general lack of correlation (Endara and Coley 2011; Smilanich et al. 2016), particularly calling into question the quantitative-qualitative dichotomy of defense (Smilanich et al. 2016). But what of the traits themselves that confer "non-apparency", that is, escape[1]? Implied in Feeny's discussion is that plant size, and possibly timing of activity, are both important. Since then, researchers have suggested that there are two components of escape: escape in space and time (Kogan 1986). Since the initial discussions in the 1970's, thousands of papers have been written on plant-herbivore interactions. These studies provide additional mechanisms, and tests of those mechanisms, that might lead to escape from herbivores. In addition, we have learned much about the mechanisms by which herbivores find their host plants (▶ Box 3.1). Here we describe what can be gathered from the literature about traits that might contribute to escape.

Box 3.1   How Do Herbivores Find Host Plants?

Herbivorous insects use several sensorial organs to find their host plants, including vision, smell, and hearing, and specific organs to sense vibration (Hilker and Meiners 2008). However, the olfactory system is the most important sense used by non-vertebrate herbivores to perceive the chemical compounds emitted by host plants (Kessler and Heil 2011). Chemicals emitted by plants can be volatile or not, and they are usually complex blends that vary depending on the producing species and taxonomical group (Knudsen et al. 2006). Furthermore, different species of herbivores sharing the same host plant often use distinct chemical cues to find their host. For example, while Papilio polyxenes *butterflies use a combination of glycosides and chlorogenic acids produced by one of its host plants,* Daucus carota, *as cues for oviposition (Brooks et al. 1996), the carrot fly,* Psila rosae, *uses phenylpropenes and polyacetylenes as oviposition cues (Degen et al. 1999). Some volatiles indicate that a plant is a potential host, while others signal that it is a non-host (Bruce and Pickett*

---

1   Although 'apparency' was first used to describe the concept, we choose to use the term 'escape' as the umbrella term. We prefer "natural selection for escape traits contributes to reduced herbivory", rather than the awkward wording of "natural selection for non-apparency traits contributes to reduced herbivory." As a result, a similar language construction results for all three components of resistance: the evolutionary response by plants to natural selection by herbivores leads to the evolution of traits that increase escape, increase defense, and increase tolerance.

2011). *Visual stimuli are also used by some herbivores* (Prokopy and Owens 1983), *especially vertebrates, although some insect herbivores such as flies have well-developed visual systems, able to distinguish colors and even UV* (Heisenberg and Buchner 1977). *In natural conditions, however, most herbivores likely use more than one sensorial organ to search for and select hosts* (e.g., Stutz et al. 2017). *For example,* Battus philenor *butterflies use leaf shape and chemical stimuli information to select host plants on which to oviposit* (Papaj 1986). *The* Delia radicum *fly integrates distinct senses: as it approaches the host plant it relies on visual cues, but after arrival, it uses chemical and mechanical cues to decide if the host is appropriate, and in a positive case, to find a suitable place to oviposit* (Nottingham 1988).

3.3 Escape Traits

Our goal here is to review the gamut of plant traits that contribute to the inability of the herbivore to find its host plant, or to find it when it is palatable. A large body of experimental data, starting with Root (1973) and onward, demonstrates that plants can escape in space (for example, see reviews by Trenbath 1993; Agrawal et al. 2006; Underwood et al. 2014), and in time (e.g., Valdés and Ehrlén 2017). In some cases, this escape is due to the vegetation context in which a plant is found, leading to associational resistance (Barbosa et al. 2009; Kim 2017). Here we focus on specific traits of the plants themselves that lead to escape. Most plant traits that contribute to escape can be classified into one of these six categories. An exception might be underground flowering, proposed as an adaptation to escape aboveground herbivores (Rausher and Feeny 1980). We focus on terrestrial plant-herbivore systems, although certainly the same processes occur in freshwater and marine systems (e.g., Hay 1984; Rengefors et al. 1998).

Predictability in space: adaptation to ephemeral habitats   Feeny (1976) proposed that plant species that are predictably found in space would suffer greater herbivory than less predictable species (see also Rhoades and Cates 1976, Chew and Courtney 1991). For example, species that are adapted to grow only in ephemeral habitats (e.g., animal disturbances, light gaps in forest canopies) are more likely to escape herbivore attack than species adapted to grow in more permanent habitat types. The lack of predictability arises from the habitat being relatively short-lived and unpredictable in space. The Platte thistle, *Cirsium canescens*, occurs in greatest density in temporary washes or blowouts in prairie (Louda et al. 1990). Consistent with the idea that these disturbances are unpredictable in space, vertebrate-caused mortality of seedlings of this species was lower in wind-caused disturbances compared to undisturbed prairie (Louda et al. 1990). In a Panamanian forest, Coley (1983b) hypothesized that if Feeny is correct, herbivory levels from one time to the next would be more variable in *Cecropia insignis* (Moraceae), a light gap specialist, than in *Trichilia cipo* (Meliaceae), a shade tolerant, closed forest species. Coley found equally variable levels of insect damage in the two species. At this time, the support for this mechanism of escape is not very strong.

Predictability in time: phenology   Here, escape occurs when herbivores find the plant but the entire plant or parts of it are not vulnerable (e.g., all leaves are mature) or are unavailable (e.g., flowers not available for florivores). Individual plant species (Wiklund and Friberg 2009) and individual plants (Fogelström et al. 2017) can escape attack by flowering before herbivores are active. The same is true for leaf flushing. Larvae of *Operophtera brumata* on *Quercus robur* must be active when young leaves are available. If they hatch from eggs too soon, then no leaves are available. If they hatch too late, the leaves are too tough, low in nitrogen and water, and high in pheno-lics (Feeny 1970). Experimental studies that manipulate temperature in which woody plants and their spring-feeding caterpillars develop show this to be a common phe-nomenon in temperate forests (Schwartzberg et al. 2014; Abarca and Lill 2015; Kharouba et al. 2015; Foglestrӧm et al. 2017; review by Bale et al. 2002).

In wet to seasonally wet neotropical forests, mature leaves of shade-tolerant tree species are well-defended against most insect herbivores (Coley 1983a). The vulnerable stage is when leaves are expanding (Coley 1983a; Coley and Barone 1996). Aide (1993) showed that for 10 tree species, individual plants that produced leaves at the peak of conspecific leaf production suffered significantly less damage than individuals producing leaves in the non-peak. These results together suggest that plants can escape their herbivores by leafing out when the rest of the popula-tion is doing the same, and that satiation of herbivores during the peak is the mech-anism by which most leaves escape (Calderón-Sanou et al. 2019). Season is also important, as leaves produced during the tropical dry season suffer less damage than those produced in the wet season (Aide 1988, 1992; Murali and Sukumar 1993). Augspurger (1981) demonstrated the importance of timing experimentally by manipulating flowering and fruiting time in *Hybanthus prunifolius* via watering to cause plants to flower and fruit outside of the main peak of the species. Watered (asynchronous) plants were more heavily attacked by seed predators than were control (synchronous) plants.

Masting is a multi-annual pattern of flowering and fruiting, hypothesized to be an adaptation to escape seed predators. Kelly and Sork (2002), in their meta-analysis, found evidence that both seed predator satiation and adaptation to cli-mate (specifically rainfall) have influenced the evolution of masting. A study conducted by Calderón-Sanou et al. (2019) found that leaf herbivory on *Quercus costaricensis* decreased at higher densities of seedlings and that seedling survival increased 5–8% after a masting event. Other adaptive scenarios for specific timing of reproductive events (e.g., taking advantage of light and water availability) may be more relevant than adaptations to herbivory (e.g., Chaves and Avalos 2006).

Stature   The larger the plant, the more likely it is to be encountered by its herbivores, a hypothesis originally proposed by Feeny (1976). In an early test of this hypothesis, Rausher (1981) demonstrated that as the surrounding native vegetation increased in height across a season, plants of *Aristolochia reticulata* were less likely to be found by adult females of the specialist *Battus philenor*, which search for host plants by flying above the vegetation. There was no seasonal change in the probability of escape from searching caterpillars, which search by traveling along the ground. Thus, the effect of decreasing size on escape is at least partially dependent on the height and density of the surrounding vegetation and the searching behavior of the herbivore, as influenced by herbivore ontogeny. Unfortunately, we do not have measures of the height of individual plants in this study. This information would allow us to know how a plant's height might have affected its susceptibility to discovery and subsequent oviposition by searching females.

There is a system that does provide strong support for the stature hypothesis. *Solidago altissima* is attacked by two gall makers of the main stem, the tephritid galler *Eurosta solidaginis* and the gall midge *Rhopalomyia solidaginis* (Wise and Abrahamson 2008; Wise 2009). Some genotypes of this herbaceous perennial have a bent terminal stem (the "candy-cane" trait) while growing up through the surrounding vegetation. During development, bent terminal stems are 36% shorter than straight stems, but reach the same height at maturity. Such candy-cane stems were 30% less likely to be galled by by *E. solidaginis* and 50% less likely to be attacked by *R. solidaginis* (Wise and Abrahamson 2008). Both this system (Wise et al. 2009) and the *Aristolochia* system (Rausher 1981) demonstrate the need for careful experimentation that distinguishes the impact of the individual plant traits from the effect of the surrounding vegetation (i.e., associational resistance and associational susceptibility) on plant escape from its herbivores.

Size *per se* may not be the ultimate cause for greater attack, given that any number of influential factors may be correlated with plant size. Larger plants of *Pastinaca sativa* are more heavily attacked by the parsnip webworm, *Depressaria radiella* (formerly *D. pastinacella*), but this vulnerability is tied more to the number and timing of production of inflorescences, the food source of the webworm (Thompson and Price 1977; Thompson 1978). Larger plants of *Cardamine pratensis* escape attack by *Anthocharis cardamines* because their large size allows them to flower early, before ovipositing females begin their activity (Fogelström et al. 2017). Thus, the effect of plant size may be tightly correlated with plant phenology, and the impact of phenology on escape, but not in predictable ways or always by the same mechanism.

Mimicry   Leaf mimicry in plants (Wiens 1978) occurs when leaves of one species of plant are similar in shape to that of an unrelated species. Herbivore pressure is often invoked as an adaptive explanation. Mimicry would allow the mimicking species to escape its herbivores by looking like a model species whose chemistry would be significantly different, making the plant seemingly unpalatable. For example, selection by possums has been suggested to explain similarity in leaf shapes between mistletoes (epiphytic hemiparasites in the Loranthaceae) and that of their host *Eucalyptus* tree (Barlow and Wiens 1977). Divergence of leaf shape in *Passiflora* resulting in mimicry of distantly related plant species has been suggested as an adaptation to escape oviposition by visually orienting *Heliconius* butterflies (Gilbert 1980). Gilbert (1980) took this a step further to suggest that the number of *Passiflora* species that could co-occur in a given location was limited by the number of non-*Passiflora* species found in a location that could serve as models. Perhaps the most remarkable example of leaf mimicry is that of the vine *Boquila trifoliolata*, which mimics the host plant on which it climbs, changing leaf shape with different host species (Gianoli and Carrasco-Urra 2014). Vines growing on other plants have reduced herbivory compared to free-standing plants, and plants growing on leafless stems.

3

Other cases of mimicry could also engender escape. Portions of plant parts that mimic eggs dissuade oviposition by searching females (Shapiro 1981a, b; Williams and Gilbert 1981; Lev-Yadun 2016). Females avoid ovipositing on a plant already containing eggs to avoid competition or cannibalism by caterpillars soon to hatch from the "eggs" (Williams and Gilbert 1981). The amount of white variegation on leaves of the tropical forest aroid *Caladium steudneriifolium,* is variable from plant to plant, and appears to mimic damage by leaf miners (Soltau et al. 2009). Experiments demonstrate that artificial white variegation reduces leaf mining from 7.9% of leaves in control, naturally non-variegated leaves to 0.4% in painted experimental leaves (Soltau et al. 2009).

Background matching   Blending in with the background vegetation and/or substrate may increase the likelihood that a plant could escape its herbivores (Strauss and Cacho 2013; Strauss et al. 2015; Niu et al. 2018; Cacho and McIntyre 2020, and references therein). Blending in (or crypsis) could come about chemically by reducing volatile emissions, or visually. The former does not seem to have been studied, but there have a few studies of visual background matching. Strauss et al. (2015) predicted damage should be more variable in plants adapted to blend in with their environment than those that are not. This is because cryptically colored species are less likely to be found, but if found, will suffer more damage because of selection to reduce defense. This hypothesis harkens back to the apparency hypothesis of Feeny (1976; see also Coley 1983b). Strauss and Cacho (2013) manipulated the substrate of naturally occurring *Streptanthus breweri* and found that plants in mismatching backgrounds had 8% more leaf damage than control plants. Nonetheless, in a more recent study, Strauss et al. (2015) measured 300–700 nm reflectance patterns of various plant species and their associated microhabitat in New Zealand and California (USA) rocky, open habitats. They then classified plants as either blending in or not based on the similarity in reflectance between plant and background. Species classified as background matching received 7X more damage than did non-matching species. This is consistent with the prediction of greater defense in the latter species. However, there was no evidence of greater variability in damage among individuals of the former species, suggesting that they are not more likely to escape detection.

Strong tests of the background-matching hypothesis are difficult to conduct. Exceptions include that by LoPresti and Karban (2016), who tested whether sand trapping of sticky stems and leaves of *Abronia latifolia* decreases herbivory. They found that the presence of sand reduced herbivory, but that the color of supplemented sand had no effect. The authors suggested sand does not decrease the likelihood of discovery but perhaps acts as a physical deterrent to herbivores once they encounter the plant. Seeds of numerous plant species are covered in mucilage that traps soil particles (LoPresti et al. 2019). LoPresti et al. (2019) found that this soil covering reduced predation by ants in 48/53 plant species tested, but again there was no evidence for a crypsis effect.

Dispersal   Dispersal of seeds away from the parent plant, sometimes in combination with colonization of particular sites with higher likelihood of survival (directed dispersal), is hypothesized as an adaptation to escape natural enemies (Howe and

Smallwood 1982; Wenny 2001). Numerous experimental and observational studies demonstrate that dispersal of seeds away from the parent plant increases the likelihood of escape from distance and density-responsive seed and seedling herbivores, including pathogens (the "Janzen-Connell effect", see reviews by Clark and Clark 1984; Hyatt et al. 2003; Terborgh 2015). Depending on the behavior of the disperser, seeds may be dispersed non-randomly without regard to distance from the parent plant. This is directed dispersal. These non-random locations may represent increased likelihood of escape from natural enemies. Wenny and Levey (1998) showed that seeds of *Ocotea endresiana* taken by three-wattled bellbirds (*Procnias tricarunculata*) were more likely to be dispersed to light gaps, and if so, the resulting seedlings were more likely to escape attack by pathogens in light gaps compared to shaded forest understory.

Directed dispersal, if resulting in lowered herbivory, can have implications for maintaining plant species richness. Salazar et al. (2013) studied the impact of directed seed dispersal by bats on the coexistence of *Piper* shrub species in a Costa Rican wet forest. *Piper* seeds are dispersed by bats who often defecate those seeds as they leave their roosts. The roosts are empty trunks of large trees. Salazar et al. (2013) found that stands of mixed species of *Piper* near to real and artificial bat roosts suffered less herbivory and were more species rich than stands near randomly chosen large trees without a bat roost. This result is consistent with the hypothesis that directed dispersal from bat roosts leads to the local accumulation of *Piper* species due to reduced herbivory.

3.4 Research Plan

Establishing that a particular plant trait contributes to the success of a plant in the presence of herbivores is a difficult undertaking (Erb 2018). To do so for the entire suite of traits that might contribute to plant resistance to herbivores would be a daunting task. Nevertheless, to achieve a more complete understanding of the plant trait under selection by herbivores, it would be important to know which plant traits are under selection. If there are no ecological, allocation, or genetic tradeoffs among traits (e.g., a trait that contributes to escape versus one that contributes to defense), ignoring escape traits simply means the picture is incomplete for understanding the full set of traits contributing to plant success in the presence of herbivores. If tradeoffs occur in one form of resistance versus another (e.g., Heschel and Riginos 2005; Franks 2011), ignorance or omission of one subset of traits would result in misinterpretation of the range of evolutionary responses available to the plant. Escape and defense, in particular, are likely to interact. Studying a group of 55 shade-tolerant species from 39 genera and 24 families, Kursar and Coley (2003) found that although the leaves of fast-growing plants are vulnerable to herbivores for only brief periods, they are less chemically defended and thus suffer higher damage from herbivores.

Two approaches can be taken to understand how escape traits contribute to plant success relative to defense and tolerance traits. The first is to compare multiple species with the presumption that the species will vary along these three axes (escape, defense, and tolerance). Given a relatively large sample of species, analyses should be able to determine the relative contribution of each set of traits for each species, and how these species vary in resistance strategies (see Moreira et al. (2020) and references therein, for a similar approach, but only for defense traits). The second approach is to specifically choose genotypes within a species that vary in the relevant traits, or to modify the phenotype through selection (e.g., Stowe and Marquis 2011) or genetic engineering (e.g., Bergelson et al. 1996; Erb 2018). The environment of these genotypes can then be manipulated in a replicated design to test the relative contribution of the traits to plant success. We describe two literature examples (Norghauer et al. 2014, Endara et al. 2017) that take the first approach, and then propose an experimental setup as an example of the second.

Multi-species comparisons   Endara et al. (2017) sought to determine which plant traits determine attack of 33 species of the tropical genus *Inga* (Fabaceae) by three clades of Lepidoptera herbivores. They found that both timing of leaf production and production of amines predicted attack by Erebidae caterpillars. Specifically, plant species that did not produce leaves at the beginning of the wet and dry seasons, and those that did not produce amines were less likely to be attacked by caterpillar species in the Erebidae. Thus, both escape and defense traits influence attack in this system.

   To date, there has been a single study that has considered all three components of resistance, via an interspecific comparison (Norghauer et al. 2014). The researchers contrasted attack on seedlings of three tree species of Fabaceae. They found that all three components (escape, measured as distance from parent trees; defense, measured as nutrient and total phenolic content of leaves; and tolerance, regrowth following experimental removal of leaves) contributed to the success of the three species. Furthermore, the relative importance of each component varied by tree species. This relative contribution was correlated with the shade-tolerance of the species. The least shade tolerant species relied on escape and tolerance, while the most shade tolerant species relied on defense. The third species was intermediate between the two.

Experimental comparison of genotypes/phenotypes within a species   The best systems to study the relative importance of escape, defense, and tolerance within a species are ones in which the candidate traits for each type are known. The general experimental plan would be to expose different plant genotypes that vary in relevant resistance traits to herbivores. Measuring seed production by each of these genotypes would then reveal the relative contribution of each trait to the fitness of the plant in the face of herbivore attack. We have chosen the wild parsnip (Apiaceae: *Pastinaca sativa*) and the parsnip webworm (Depressaridae: *Depressaria radiella*) system in North America, as an example (◻ Fig. 3.1). The plant is an herbaceous biennial, producing a single flowering stalk from a rosette of basal leaves. The insect is a specialist on the host plant, attacking developing and mature flowers, and maturing but not mature seeds. Flowers are clustered in umbels, of which the primary umbel is the first to flower and contains the most flowers. The primary umbel sits at the terminus of the main stalk, and contains only hermaphroditic flowers. Side branches of the

☐ **Fig. 3.1** The *Pastinaca-Depressaria* system: **a** roadside population in River Falls, WI, USA; **b** primary inflorescence with a *D. radiella* larva under silk webbing; **c** unattacked plant showing positions of the primary (1), secondary (2), and tertiary (3) inflorescences; **d** terminal inflorescence after being attacked, and the caterpillar has completed its development; **e** tertiary (all male flower) inflorescence from an unattacked plant. (All photos by RJ Marquis)

main stalk support secondary, then tertiary inflorescences, with decreasing numbers of hermaphroditic flowers and increasing numbers of male flowers from secondary to tertiary umbels. The plant occurs in the North America, where it is not native, in fields, meadows, and prairies, and along roadsides and railroad tracks, usually grow-ing in relatively dense stands. Fruits are winged and are dispersed by wind. See ▶ Box 3.2 for current understanding of *P. sativa* traits that contribute to escape, defense, and tolerance in this plant species.

3

> **Box 3.2   Relevant Traits that Contribute to Escape, Defense, and Tolerance in** *Pastinaca sativa* **in the Face of Herbivory by** *Depressaria radiella*
>
> *Escape traits. Both plant size and density are correlated with the likelihood of attack. Large, isolated plants are more heavily attacked than plants growing in dense clusters because the flowering phenology is extended over a longer time period for larger plants. However, plants of the same size found in higher densities are more likely to be attacked than isolated plants. These results* (Thompson and Price 1977; Thompson 1978) *suggest that flowering time, plant size, and seed dispersal, as they affect the density of conspecifics, all influence the likelihood that plants will be found by female moths of* D. radiella.
>
> *Defense traits. Flowers and developing seeds are defended by at least six kinds of furanocoumarins. The absolute concentrations of each and their relative concentra-tions in mixtures influence resistance against both* Heliothis zea, *a generalist* (Berenbaum and Neal 1985), *and* D. radiella, *a specialist* (Berenbaum et al. 1986; Nitao and Zangerl 1987). *Reproductive parts usually have the highest relative con-centration of chemical compounds, although these parts are the main target of herbi-vores* (Zangerl and Rutledge 1996). *Furthermore, the concentration of at least two compounds produced by* P. sativa *is negatively correlated with the presence of* Copidosoma sosares, *a parasitoid wasp of the parsnip webworm. This illustrates a possible ecological cost conveyed by chemical defenses* (Lampert et al. 2008).
>
> *Tolerance traits. Upon damage to the primary umbel by* D. radiella, *there is a gender switch from male to hermaphroditic flowers in secondary and tertiary umbels* (Hendrix 1979). *This change in sex expression results in no decline in seed number despite complete destruction of the seed production by the primary umbel* (Hendrix 1979).

The information in ▶ Box 3.2 suggests an experiment in which the three com-ponents of resistance are simultaneously manipulated to determine the relative impact of each on the success of a given genotype. Escape can be manipulated by plant density (growing alone vs. growing in patches) and flowering time (early vs. late, manipulated by controlling temperature during development). Escape would be measured as the number of eggs laid on a plant and the number of webworms attacking the plant. Defense (low versus high resistant genotypes) can be manipu-lated by planting genotypes of known furanocoumarin chemistry. Defense would be measured as the number of flowers eaten. Finally, tolerance can be manipulated by planting genotypes that exhibit weak vs. strong changes in sexual expression in response to damage. These genotypes would be identified prior to the main experi-ment (Stowe et al. 2000). Tolerance would be measured as the percentage of flowers changing gender in response to attack of the primary umbel. Fitness would be measured as the number of viable seeds produced. Structural equation modeling can then be used to determine the relative contribution of each defense component to fitness (seed production), and the relative contribution of each individual trait to each component of defense (e.g., the relative importance of each furanocoumarin compound to defense and fitness).

3.5 Implications for Agriculture

Both defense and tolerance have been incorporated into conventional breeding strategies to reduce the impact of pest herbivores on crop yield (Russell 2013). We see two mechanisms by which escape can and has been used as a strategy to reduce pest attack in agriculture and plantation forestry.

Human-Mediated Dispersal   Here, planting crop species across the globe and outside of their native range has the advantage of making food crops available to a wide span of the human population, increasing access to calories and improved nutrition. The plants themselves are not dispersing, but humans are aiding their dispersal. At least initially, movement of plants across the globe has had the serendipitous result that the crop could be planted pest free. The history of the exploitation of the rubber tree (Schultes 1977) is a classic example of a story that has been repeated many times over. Seeds or cuttings are taken from the native range and planted across the globe, often enjoying great success because they are growing in a pest free environment at least initially (Strong 1974). This approach can be exploited intentionally in more recent agricultural strategies. For example, Yang and Feng (2001) summarized the effect of longitude and latitude on the number of diseases found on soybean plants in North America. As one goes both north and west from the southeastern U.S., the number of diseases drops precipitously.

Phenology   Breeding for and planting early or late varieties of crops that escape the phenology of their herbivore pests is a long-developed strategy of pest management. For example, early varieties of blueberries completely escape the spotted wing fruitfly, *Drosophila suzuki* (Hampton et al. 2014). Similar examples exist for apples (Miñarro and Dapena 2014) and cranberries (McMahan et al. 2017).

> **Conclusions: Interactions and Contribution to Biodiversity**
> To the human observer, the world appears to be a jumble of plants of various shapes, sizes, and shades of green. Some species are common and comprise the majority of the biomass of a given location, some are covered by a canopy of other plants, others are rare within and under that canopy, and a few species protrude through the canopy. All produce a few to hundreds of compounds that are volatile and increase in concentration in the air as the plant is neared. These chemical cues, in addition to visual cues, are used by herbivores to locate their plant prey (see ▶ Box 3.1). One

stumbling block with any theory of escape from herbivores is being able to perceive the world as the herbivore perceives it. Measuring the amount of damage per plant individual is far from satisfying, as both escape and defense have already come into play. A better approach, one that would not confound escape and defense, would be to score plants as having escaped ('yes' or 'no') by the presence of one or more eggs, larvae, or feeding adult. Any resultant damage would be a measure of defense independent of escape. There might be ambiguous cases in which the herbivores finds the plant, but rejects it before ovipositing or feeding. We suggest that such cases could be scored as "chemical escape" or "morphological escape" (Williams and Gilbert 1981). In the agricultural pest literature, these two together comprise antixenosis, or a non-preference reaction to an otherwise healthy plant (e.g., Boiça et al. 2015). There are likely to be interactions between escape and defense (e.g., Euler and Baldwin 1996), just as there are interactions between defense and tolerance traits (Pilson 2000; Stowe et al. 2000) and interactions between escape and tolerance (e.g., Pilson and Decker 2002; Wise 2009).

In as much as escape from herbivore attack actually leads to increasing population growth rate of the escaping plant species, we would expect that the escape mechanisms proposed here allow coexistence of higher species numbers than in the absence of the herbivores (Marquis 2005). We reviewed a number of mechanisms that potentially contribute to escape and therefore species coexistence. For some, the evidence is stronger than others. On one hand, background matching enjoys weak support but perhaps because it has been only studied a relatively few times. On the other hand, the Janzen-Connell Hypothesis is a mechanism of escape through dispersal that has been exhaustively investigated since the 1970's. Many studies have shown the escape effect to occur for some species but not others. Research by Wenny and Levey (1998) and Salazar et al. (2013) strongly suggest that directed dispersal has a positive effect on species coexistence. Dispersal traits increase the probability of survival of many plant species, which is particularly important for weak competitors, plants that have few defensive mechanisms, and species that demand specific environmental conditions in which to grow (e.g., Wenny and Levey 1998).

Many plant traits can influence the occurrence of herbivore predators and parasitoids, affecting the community composition of ecological systems through bottom-up effects (Agrawal et al. 2006; Poelman et al. 2008). Direct and indirect defenses of plants not only increase the diversity of natural enemies of plant arthropods, but their effects can also be extended to the highest trophic levels (Gruner and Taylor 2006). Although escape traits certainly contribute to the maintenance of biodiversity, they have received less attention than the defense and tolerance traits of plants. Furthermore, as we mentioned earlier in this chapter, only a single study considered the effects of all three resistance traits combined (Norghauer et al. 2014). We argue that including escape traits in ecological models will provide researchers a more integrated view of several ecological processes, with potential applications for ecosystem management, including agriculture, and conservation practices.

> **Key Points**
> - Escape traits are those that decrease the likelihood that an herbivore will find a consumable plant in space or time.
> - Escape is a component involved in plant-herbivore interactions separate from defense and tolerance, but like defense and tolerance, contributes to the success of plants in the face of herbivore pressure.
> - Escape traits are quantifiable and have a heritable basis, and their impact on plant fitness can be quantified.

❓ Questions
- What is the relative contribution of escape traits to overall plant resistance?
- How does climate change affect escape traits?
- What are the possible tradeoffs involving all three resistance mechanisms?

Acknowledgements RJM thanks Steve Hendrix and May Berenbaum for introducing him to the *Pastinaca* system. RJM and RFM thank their respective institutions for providing facilities during the conceptualization and realization of this chapter.

References

Abarca M, Lill JT (2015) Warming affects hatching time and early season survival of eastern tent caterpillars. Oecologia 179:901–912

Agrawal AA, Lau JA, Hambäck PA (2006) Community heterogeneity and the evolution of interactions between plants and insect herbivores. Q Rev Biol 81:349–376

Aide TM (1988) Herbivory as a selective agent on the timing of leaf production in a tropical understory community. Nature 336:574–575

Aide TM (1992) Dry season leaf production: an escape from herbivory. Biotropica 24:532–537

Aide TM (1993) Patterns of leaf development and herbivory in a tropical understory community. Ecology 74:455–466

Augspurger CK (1981) Reproductive synchrony of a tropical shrub: experimental studies on effects of pollinators and seed predators on *Hybanthus prunifolius* (Violaceae). Ecology 62:775–788

Bale JS, Masters GJ, Hodkinson ID et al (2002) Herbivory in global climate change research: direct effects of rising temperature on insect herbivores. Glob Chang Biol 8:1–16

Barbosa P, Hines J, Kaplan I et al (2009) Associational resistance and associational susceptibility: having right or wrong neighbors. Annu Rev Ecol Evol Syst 40:1–20

Barlow BA, Wiens D (1977) Host-parasite resemblance in Australian mistletoes: the case for cryptic mimicry. Evolution 31:69–84

Barton KE, Koricheva J (2010) The ontogeny of plant defense and herbivory: characterizing general patterns using meta-analysis. Am Nat 175:481–493

Berenbaum M, Neal JJ (1985) Synergism between myristicin and xanthotoxin, a naturally cooccurring plant toxicant. J Chem Ecol 11:1349–1358

Berenbaum MR, Zangerl AR, Nitao JK (1986) Constraints on chemical coevolution: wild parsnips and the parsnip webworm. Evolution 40:1215–1228

Bergelson J, Purrington CB, Palm CJ, Lopez-Gutierrez JC (1996) Costs of resistance: a test using transgenic *Arabidopsis thaliana*. Proc Roy Soc Lond SB Biol Sci 263:1659–1663

Boege K, Marquis RJ (2005) Facing herbivory as you grow up: the ontogeny of resistance in plants. Trends Ecol Evol 20:441–448

Boege K, Barton KE, Dirzo R (2011) Influence of tree ontogeny on plant-herbivore interactions. In: Meinzer FA, Lachenbruch B, Dawson TE (eds) Size-and age-related changes in tree structure and function. Springer, Dordrecht, pp 193–214

Boiça AL, Costa EN, de Souza BHS et al (2015) Antixenosis and tolerance to *Diabrotica speciosa* (Coleoptera: Chrysomelidae) in common bean cultivars. Fla Entomol 98:464–472

Brooks JS, Williams EH, Feeny P (1996) Quantification of contact oviposition stimulants for black swallowtail butterfly, *Papilio polyxenes*, on the leaf surfaces of wild carrot, *Daucus carota*. J Chem Ecol 22:2341–2357

Bruce TJ, Pickett JA (2011) Perception of plant volatile blends by herbivorous insects–finding the right mix. Phytochemistry 72:1605–1611

Cacho NI, McIntyre PJ (2020) The role of enemies in bare and edaphically challenging environments. In: Nunez-Farfan J, Valverde PL (eds) Evolutionary ecology of plant-herbivore interaction. Springer, Cham, pp 249–267

Calderón-Sanou I, Ríos LD, Cascante-Marín A et al (2019) The effect of conspecific density, herbivory, and bamboo on seedling dynamics of a dominant oak in a Neotropical highland forest. Biotropica 51:817–825

Chaves OM, Avalos G (2006) Is the inverse leafing phenology of the dry forest understory shrub *Jacquinia nervosa* (Theophrastaceae) a strategy to escape herbivory? Rev Biol Trop 54:951–963

Chew FS, Courtney SP (1991) Plant apparency and evolutionary escape from insect herbivory. Am Nat 138:729–750

Clark DA, Clark DB (1984) Spacing dynamics of a tropical rain forest tree: evaluation of the Janzen-Connell model. Am Nat 124:769–788

Coley PD (1983a) Herbivory and defensive characteristics of tree species in a lowland tropical forest. Ecol Monogr 53:209–234

Coley PD (1983b) Intraspecific variation in herbivory on two tropical tree species. Ecology 64:426–433

Coley PD, Barone JA (1996) Herbivory and plant defenses in tropical forests. Annu Rev Ecol Syst 27:305–335

Degen T, Buser HR, Städler E (1999) Patterns of oviposition stimulants for carrot fly in leaves of various host plants. J Chem Ecol 25:67–87

Endara MJ, Coley PD (2011) The resource availability hypothesis revisited: a meta- analysis. Funct Ecol 25:389–398

Endara MJ, Coley PD, Ghabash G et al (2017) Coevolutionary arms race versus host defense chase in a tropical herbivore–plant system. Proc Natl Acad Sci U S A 114:E7499–E7505

Erb M (2018) Plant defenses against herbivory: closing the fitness gap. Trends Plant Sci 31:187–194

Euler M, Baldwin IT (1996) The chemistry of defense and apparency in the corollas of *Nicotiana attenuata*. Oecologia 107:102–112

Feeny P (1970) Seasonal changes in oak leaf tannins and nutrients as a cause of spring feeding by winter moth caterpillars. Ecology 51:565–581

Feeny P (1976) Plant apparency and chemical defense. Rec Adv Phytochem 10:1–40

Fogelström E, Olofsson M, Posledovich D et al (2017) Plant-herbivore synchrony and selection on plant flowering phenology. Ecology 98:703–711

Franks SJ (2011) Plasticity and evolution in drought avoidance and escape in the annual plant *Brassica rapa*. New Phytol 190:249–257

Gianoli E, Carrasco-Urra F (2014) Leaf mimicry in a climbing plant protects against herbivory. Curr Biol 24:984–987

Gilbert LE (1980) Ecological consequences of a coevolved mutualism between butterflies and plants. In: Gilbert LE, Raven PH (eds) Coevolution of animals and plants. University of Texas Press, Austin, pp 210–240

Gruner DS, Taylor AD (2006) Richness and species composition of arboreal arthropods affected by nutrients and predators: a press experiment. Oecologia 147:714–724

Hampton E, Koski C, Barsoian O et al (2014) Use of early ripening cultivars to avoid infestation and mass trapping to manage *Drosophila suzukii* (Diptera: Drosophilidae) in *Vaccinium corymbosum* (Ericales: Ericaceae). J Econ Entomol 107:1849–1857

Hay ME (1984) Predictable spatial escapes from herbivory: how do these affect the evolution of herbivore resistance in tropical marine communities? Oecologia 64:396–407

Heisenberg M, Buchner E (1977) The role of retinula cell types in visual behavior of *Drosophila melanogaster*. J Comput Phys 117:127–162

Hendrix SD (1979) Compensatory reproduction in a biennial herb following insect defloration. Oecologia 42:107–118

Heschel MS, Riginos C (2005) Mechanisms of selection for drought stress tolerance and avoidance in *Impatiens capensis* (Balsaminaceae). Am J Bot 92:37–44

Hilker M, Meiners T (2008) Chemoecology of insect eggs and egg deposition. John Wiley & Sons, New York

Howe HF, Smallwood J (1982) Ecology of seed dispersal. Annu Rev Ecol Syst 13:47–228

Hyatt LA, Rosenberg MS, Howard TG et al (2003) The distance dependence prediction of the Janzen-Connell hypothesis: a meta-analysis. Oikos 103:590–602

Kelly D, Sork VL (2002) Mast seeding in perennial plants: why, how, where? Annu Rev Ecol Syst 33:427–447

Kessler A, Heil M (2011) The multiple faces of indirect defences and their agents of natural selection. Funct Ecol 25:348–357

Kharouba HM, Vellend M, Sarfraz RM, Myers JH (2015) The effects of experimental warming on the timing of a plant-insect herbivore interaction. J Anim Ecol 84:785–796

Kim TN (2017) How plant neighborhood composition influences herbivory: testing four mechanisms of associational resistance and susceptibility. PLoS One 12:e0176499

Knudsen JT, Eriksson R, Gershenzon J, Ståhl B (2006) Diversity and distribution of floral scent. Bot Rev 72:1–120

Kogan M (1986) Plant defense strategies and host-plant resistance. In: Kogan M (ed) Ecological theory and integrated pest management practice. Wiley, NewYork, pp 83–134

Kursar TA, Coley PD (2003) Convergence in defense syndromes of young leaves in tropical rainforests. Biochem Syst Ecol 31:929–949

Lampert EC, Zangerl AR, Berenbaum MR, Ode PJ (2008) Tritrophic effects of xanthotoxin on the polyembryonic parasitoid *Copidosoma sosares* (Hymenoptera: Encyrtidae). J Chem Ecol 34:783–790

Lev-Yadun S (2016) Butterfly egg mimicry. In: Defensive (anti-herbivory) coloration in land plants. Springer, Cham. https://doi.org/10.1007/978-3-319-42096-7_55

LoPresti EF, Karban R (2016) Chewing sandpaper: grit, plant apparency, and plant defense in sand-entrapping plants. Ecology 97:826–833

LoPresti EF, Pan V, Goidell J, Weber MG, Karban R (2019) Mucilage-bound sand reduces seed predation by ants but not by reducing apparency: a field test of 53 plant species. Ecology 100:e02809

Louda SM, Potvin MA, Collinge SK (1990) Predispersal seed predation, postdispersal seed predation and competition in the recruitment of seedlings of a native thistle in sandhills prairie. Am Midl Nat 124:105–113

Marquis RJ (2005) Impacts of herbivores on tropical plant diversity. In: Burslem D, Pinard M, Hartley S (eds) Biotic interactions in the tropics. Cambridge University Press, Cambridge, pp 328–346

McMahan EE, Steffan SA, Guédot C (2017) Population densities of Lepidopteran pests in selected cranberry cultivars in Wisconsin. J Econ Entomol 110:1113–1119

Miñarro M, Dapena E (2014) Resistance of apple cultivars to *Dysaphis plantaginea* (Hemiptera: Aphididae): role of tree phenology in infestation avoidance. Environ Entomol 36:1206–1211

Moreira X, Abdala-Roberts L, Galmán A et al (2020) Ontogenetic consistency in oak defence syndromes. J Ecol 108:1822–1834

Murali KS, Sukumar R (1993) Leaf flushing phenology and herbivory in a tropical dry deciduous forest, southern India. Oecologia 94:114–111

Nitao JK, Zangerl AR (1987) Floral development and chemical defense allocation in wild parsnip (*Pastinaca sativa*). Ecology 68:521–529

Niu Y, Sun H, Stevens M (2018) Plant camouflage: ecology, evolution, and implications. Trends Ecol Eol 33:608–618

Norghauer JM, Glauser G, Newbery DM (2014) Seedling resistance, tolerance and escape from herbivores: insights from co-dominant canopy tree species in a resource-poor African rain forest. Funct Ecol 28:1426–1439

Nottingham SF (1988) Host-plant finding for oviposition by adult cabbage root fly, *Delia radicum*. J Insect Physiol 34:227–234

Papaj DR (1986) Conditioning of leaf-shape discrimination by chemical cues in the butterfly, *Battus philenor*. Anim Behav 34:1281–1288

Pearse IS, LoPresti E, Schaeffer RN et al (2020) Generalising indirect defence and resistance of plants. Ecol Lett. https://onlinelibrary.wiley.com/doi/full/10.1111/ele.13512

Pilson D (2000) The evolution of plant response to herbivory: simultaneously considering resistance and tolerance in *Brassica rapa*. Evol Ecol 14:457–489

Pilson D, Decker KL (2002) Compensation for herbivory in wild sunflower: response to simulated damage by the head-clipping weevil. Ecology 83:3097–3107

Poelman EH, van Loon JJ, Dicke M (2008) Consequences of variation in plant defense for biodiversity at higher trophic levels. Trends Plant Sci 13:534–541

Prokopy RJ, Owens ED (1983) Visual detection of plants by herbivorous insects. Ann Rev Entomol 28:337–364

Rausher MD (1981) The effect of native vegetation on the susceptibility of *Aristolochia reticulata* (Aristolochiaceae) to herbivore attack. Ecology 62:1187–1195

Rausher MD, Feeny P (1980) Herbivory, plant density, and plant reproductive success: the effect of *Battus philenor* on *Aristolochia reticulata*. Ecology 61:905–917

Rengefors K, Karlsson I, Hansson L-A (1998) Algal cyst dormancy: a temporal escape from herbivory. Proc Roy Soc Lond Ser B Biol Sci 265:1353–1358

Rhoades DF, Cates RG (1976) Toward a general theory of plant antiherbivore chemistry. Rec Adv Phytochem 10:168–213

Root RB (1973) Organization of a plant-arthropod association in simple and diverse habitats: the fauna of collards (*Brassica oleracea*). Ecol Monogr 43:95–124

Russell GE (2013) Plant breeding for pest and disease resistance: studies in the agricultural and food sciences. London: Butterworth-Heinemann

Salazar D, Kelm DH, Marquis RJ (2013) Directed seed dispersal of *Piper* by *Carollia perspicillata* and its effect on understory plant diversity and folivory. Ecology 94:2444–2453

Schultes RE (1977) The odyssey of the cultivated rubber tree. Endeavour 1:133–138

Schwartzberg EG, Jamieson MA, Raffa KF, Reich PB, Montgomery RA, Lindroth RL (2014) Simulated climate warming alters phenological synchrony between an outbreak insect herbivore and host trees. Oecologia 175:1041–1049

Shapiro AM (1981a) Egg-mimics of *Streptanthus* (Cruciferae) deter oviposition by *Pieris sisymbrii* (Lepidoptera: Pieridae). Oecologia 48:142–143

Shapiro AM (1981b) The pierid red-egg syndrome. Am Nat 117:276–294

Smilanich AM, Fincher RM, Dyer LA (2016) Does plant apparency matter? Thirty years of data provide limited support but reveal clear patterns of the effects of plant chemistry on herbivores. New Phytol 210:1044–1057

Soltau U, Dötterl S, Liede-Schumann S (2009) Leaf variegation in *Caladium steudneriifolium* (Araceae): a case of mimicry? Evol Ecol 23:503–512

Stamp N (2003) Out of the quagmire of plant defense hypotheses. Q Rev Biol 78:23–55

Stowe KA, Marquis RJ (2011) Costs of defense: correlated responses to divergent selection for foliar glucosinolate content in *Brassica rapa*. Evol Ecol 25:763–775

Stowe KA, Marquis RJ, Hochwender CG, Simms EL (2000) The evolutionary ecology of tolerance to consumer damage. Annu Rev Ecol Syst 31:565–595

Strauss SY, Agrawal AA (1999) The ecology and evolution of plant tolerance to herbivory. Trends Ecol Evol 14:179–185

Strauss SY, Cacho NI (2013) Nowhere to run, nowhere to hide: the importance of enemies and apparency in adaptation to harsh soil environments. Am Nat 182:E1–E14

Strauss SY, Cacho NI, Schwartz MW, Schwartz AC, Burns KC (2015) Apparency revisited. Entomol Exp Appl 157:74–85

Strong DR (1974) Rapid asymptotic species accumulation in phytophagous insect communities: the pests of Cacao. Science 185:1064–1066

Stutz RS, Croak BM, Proschogo N et al (2017) Olfactory and visual plant cues as drivers of selective herbivory. Oikos 126. https://doi.org/10.1111/oik.03422

Terborgh JW (2015) Toward a trophic theory of species diversity. Proc Natl Acad Sci U S A 112:11415–11422

Thompson JN (1978) Within-patch structure and dynamics in *Pastinaca sativa* and resource availability to a specialized herbivore. Ecology 59:443–448

Thompson JN, Price PW (1977) Plant plasticity, phenology, and herbivore dispersion: wild parsnip and the parsnip webworm. Ecology 58:1112–1119

Trenbath BR (1993) Intercropping for the management of pests and diseases. Field Crop Res 34:381–405

Underwood N, Inouye BD, Hambäck PA (2014) A conceptual framework for associational effects: when do neighbors matter and how would we know? Q Rev Biol 89:1–19

Valdés A, Ehrlén J (2017) Caterpillar seed predators mediate shifts in selection on flowering phenology in their host plant. Ecology 98:228–238

Vidal MC, Lill J, Marquis RJ, Murphy SM (2020) Geographic variation in performance of a widespread generalist insect herbivore. Ecol Ent 45:617–625

Wenny DG (2001) Advantages of seed dispersal: a re-evaluation of directed dispersal. Evol Ecol Res 3:37–50

Wenny DG, Levey DJ (1998) Directed seed dispersal by bellbirds in a tropical cloud forest. Proc Natl Acad Sci U S A 95:6204–6207

Wiens D (1978) Mimicry in plants. In: Evolutionary biology. Springer, Boston, pp 365–403

Wiklund C, Friberg M (2009) The evolutionary ecology of generalization: among-year variation in host plant use and offspring survival in a butterfly. Ecology 90:3406–3417

Williams KS, Gilbert LE (1981) Insects as selective agents on plant vegetative morphology: egg mimicry reduces egg laying by butterflies. Science 212:467–469

Wise MJ (2009) To duck or not to duck: resistance advantages and disadvantages of the candy-cane stem phenotype in tall goldenrod, *Solidago altissima*. New Phytol 183:900–907

Wise MJ, Abrahamson WG (2008) Ducking as a means of resistance to herbivory in tall goldenrod, *Solidago altissima*. Ecology 89:3275–3281

Wise MJ, Ceal GY, Abrahamson WG (2009) Associational resistance, gall-fly preferences, and a stem dimorphism in *Solidago altissima*. Acta Oecol 35:471–476

Yang XB, Feng F (2001) Ranges and diversity of soybean fungal diseases in North America. Phytopathology 91:769–775

Zangerl AR, Rutledge CE (1996) The probability of attack and patterns of constitutive and induced defense: a test of optimal defense theory. Am Nat 147:599–608

The Genetic Basis of Plant-Herbivore Interactions

Liza M. Holeski

Contents

© Springer Nature Switzerland AG 2021
K. Del-Claro, H. M. Torezan-Silingardi (eds.), *Plant-Animal Interactions*, https://doi.org/10.1007/978-3-030-66877-8_4

⊜ **Learning Objectives**
This chapter will help readers to understand the following:
1. How advances in genetic technology have affected the study of plant-herbivore interactions
2. The prevalence of genetic variation in herbivory resistance traits in natural plant populations
3. Assessment of natural selection by herbivores in plant populations
4. Ways in which hypotheses for the evolution of plant defenses can be tested at a genetic level
5. How studies of gene expression can inform our understanding of plant-herbivore interactions

4.1 Introduction

A large proportion of global biodiversity and biomass consists of plants and their herbivores. Invertebrate herbivores such as insects, and vertebrate plant browsers such as deer and other mammals, consume plant tissue and impose a strong selective pressure on plants that has been ongoing for millions of years. Fossil evidence suggests that insects, for example, have been feeding on plants for an estimated 400 million years (Labandeira 2013; Bruce 2015). The evolutionary relationship between insects and their host plants is discussed in Ehrlich and Raven (1964), a now classic work that has received more than 2500 citations to date. Plants have evolved to produce a great diversity of defenses to resist herbivory (Hanley et al. 2007; Erb et al. 2012; Rasmann and Agrawal 2009). Phytochemical defenses are key among these defenses and are present in all higher plants in a wide variety of form and function (Fraenkel 1959; Wink 2003). The importance of *co-evolutionary* relationships, or reciprocal evolutionary interactions between herbivores and plants, in the evolution of both plants and herbivores has since been highlighted many times, at both macroevolutionary (e.g., Becerra 1997) and microevolutionary (e.g., Mauricio and Rausher 1997) scales.

At a macroevolutionary scale, phytochemicals can play a key role in the evolution of host shifts by herbivorous insects. For example, a molecular phylogenetic study in the plant genus *Bursera* and the beetle genus *Blepharida* shows that the patterns of host shifts in *Blepharida* beetles are strongly associated with patterns of host phytochemical similarity in the *Bursera* genus (Becerra 1997). The interaction between these beetles and plant genus is specialized and is evolutionarily old. The plants produce a variety of terpenes that are present in resin canals in the plant leaves and stems and decrease *Blepharida* survival and growth rate (Becerra and Venable 1990).

At the microevolutionary scale, there is abundant evidence that herbivory reduces plant fitness, and that herbivores are agents of natural selection on plant resistance traits (Marquis 1992; Núñez-Farfán and Dirzo 1994; Sagers and Coley 1995; Fornoni et al. 2003). For example, in *Arabidopsis thaliana*, the elimination of herbivores in a field experiment altered the pattern of selection on two defense traits, glucosinolate concentrations and trichome density (Mauricio and Rausher 1997). Likewise, herbivore-mediated natural selection was detected on stereochemistry of the secondary metabolites, sesquiterpene lactones, of common cocklebur (*Xanthium strumarium*). In natural environments, plants with cis-fused lactone ring junctions received higher levels of herbivory than those with trans-fused lactone ring junctions; herbivore damage was negatively correlated with plant fitness (Ahern and Whitney 2014). Finally, an assessment of selection imposed by both *generalist* and *specialist* herbivores (those with greater versus lesser dietary host breadth) in *Datura stramonium* demonstrated that generalists and specialists can impose divergent selection pressures on host plant resistance traits (Castillo et al. 2014). Geographic variation in herbivore community composition can thus lead to differences in resistance among populations across a plant species range.

4.2 Types of Resistance Traits

Plant defenses against herbivory include resistance, tolerance, and temporal avoidance. Resistance traits reduce the performance and/or preference of herbivores, while tolerance is a measure the extent to which plant fitness is affected by herbivory, relative to fitness in the absence of damage (Strauss and Agrawal 1999). In this chapter, I focus on plant resistance traits, and use "resistance" and "defense" interchangeably.

Plant *secondary compounds* are metabolites that do not play a role in the growth and development of the plant (Fraenkel 1959; Berenbaum and Zangerl 2008; but see Erb and Kliebenstein 2020). Hundreds of thousands of secondary compound structures have been elucidated, with many others yet uncharacterized (Wink 1988; Pichersky and Lewinsohn 2011). Secondary compounds can be toxic, anti-nutritive or anti-digestive, and/or act to repel herbivores through low palatability (Mithöfer and Boland 2012). These phytochemicals are highly structurally diverse, and include classes such as phenolics, terpenoids, and alkaloids, among others (Harborne et al. 1999; Wink 2018). Some secondary compounds, such as lignin, are more generalized defenses that affect many types of herbivores (Franceschi et al. 2005). Others, such as many alkaloids, have specific targets- enzymes or nucleic acids, for example- that they interact with in an herbivore (Mithöfer and Boland 2012). Secondary compounds can act individually or interactively to deter herbivores (Mason and Singer 2015).

There is often consistency in broad patterns of classes of compounds across closely related taxa (Wink 2003; Liscombe et al. 2005), but this consistency, or phylogenetic signal, is not always strong. Divergence in phytochemical defenses can occur through the evolution of novel compounds and/or the evolution of novel combinations of compounds. For example, in the wild parsnip system, plants escape from adapted herbivores by producing ecologically novel compounds, often from the same chemical precursor (Berenbaum 1978, 1983). Alternatively, in the tropical plant genus *Inga*, closely related species produce different combinations (and presence/absence patterns) of commonly produced compounds (Coley et al. 2018).

While much of the literature focuses on phytochemical traits, there are numerous, complex plant defenses that have evolved in response to herbivory, including *physical defenses*. Physical defenses are structural deterrents that impede the ability of herbivores to feed on the plant, and include traits such as toughened leaves, spines, thorns, or trichomes (hairlike extensions from the plant epidermis), or the incorporation of hard materials such as silica into the foliar tissue (Hanley et al. 2007). Thorns, spines, and trichomes can be present in many forms, and some trichomes produce glandular exudates that are toxic or can trap or repel herbivores (Levin 1973; Elle and Hare 2000; Hauser 2014). Trichome glandular exudates often contain secondary compounds, thus merging physical and chemical resistance (Glas et al. 2012). An example of this latter phenomenon is with stinging nettle (*Urtica dioica*), where the trichomes contain secondary compounds that are released by contact and confer a stinging sensation to mammals (Pollard and Briggs 1984). While they are overlooked in the literature to a much greater extent than phytochemical resistance traits, physical resistance traits have been clearly shown to be effective against herbivory (Mauricio 1998; Hanley et al. 2007; Barton 2016).

Resistance to herbivory can occur through *direct defenses*, which make the plant a less suitable host due to changes in physical or phytochemical defense traits, or *indirect defenses*, through which plants reduce levels of herbivory by interacting with herbivore enemies (Heil 2008; Pearse et al. 2020). Indirect defenses include the induction of volatile compounds that attract parasitoids and predators (Dicke 1999; Dicke and Hilker 2003), and traits that provide shelter, food, or other incentives to predators (Heil et al. 2001; Heil 2008; Weber and Agrawal 2014; Weber et al. 2016).

4.3 Temporal and Spatial Variation in Resistance Traits

Perhaps in part because of the sessile nature of plants, plant defenses are not static over time or across plants. Plants may produce direct and indirect defenses *constitutively*, in the absence of herbivory or regardless of levels of herbivory; alternatively, plants also *induce* defenses through plastic changes in levels of defense following herbivory (Adler and Karban 1994; Agrawal 1998; Cipollini 1998; Karban et al. 1999). Induction of defenses can be selected for if past/current herbivory is a reliable predictor of future herbivory, and if herbivory decreases plant fitness (Karban and Baldwin 1997; Harvell and Tollrian 1999). Plants can also plastically alter the availability of essential amino acids and nutrients available for digestion by the herbivore (Chen et al. 2005; Felton 2005). In addition to induction of defense within a plant generation, defenses can also be *transgenerationally induced*, whereby offspring defense phenotypes are altered by environmental signal in the parental generation and expressed independently of changes in the offspring genotype (Holeski et al. 2012b). This transgenerational induction can occur via *epigenetic* or maternal effects (Richards 2006; Roach and Wulff 1987). Epigenetic effects are heritable changes in traits that are mediated by mechanisms other than alterations in the DNA sequence, such as DNA methylation and histone modification (Rapp and Wendel 2005; Hauser et al. 2011).

Both constitutive and induced resistance can change as plants develop (Boege and Marquis 2005; Barton and Koricheva 2010; Holeski et al. 2012a). True leaves from different developmental, or *ontogenetic*, stages (e.g., juvenile versus adult) are usually anatomically and biochemically different, with different patterns of cellular differentiation (Poethig 1997; Mauricio 2005). The direction of change in resistance traits across ontogenetic stages is variable; some species have higher levels of resistance in the juvenile developmental stage relative to the adult (Price et al. 1987; Kearsley and Whitham 1989; Cole et al. 2020), while others have increased resistance in the adult developmental stage relative to the juvenile (Karban and Thaler 1999).

Finally, levels of defense can change in a predictable manner across the course of a growing season. These temporal changes in (usually phytochemical) defense are likely in part due to shifting allocational priorities, for example between defense, growth, and reproduction) as leaves age, as well as dilution as leaves expand. Physiological changes across a season can also affect defense concentrations, and are caused by shifts in photoperiod, temperature, and water and nutrient availability (Darrow and Bowers 1997; Holeski et al. 2012a; Koricheva and Barton 2012).

4.4 Evolution of Plant Resistance Hypotheses

Many hypotheses have been developed to explain how patterns in defense production within and across populations or closely related species may have evolved. Prominent among these are the Resource Allocation Hypothesis (RAH; Coley et al. 1985) and Optimal Defense Theory (ODT; Rhoades 1979).

The Resource Availability Hypothesis (RAH) hypothesis was formulated specifically for inter-species differences in plant defenses, while Optimal Defense Theory (ODT) is typically used to describe intra-species differences. Both are testable hypotheses. The RAH posits that defense investment is dependent on growth rate; long-lived species invest more heavily in defenses than do short-lived species, due to the cost-benefit ratio of the defense investment (Coley et al. 1985; Endara and Coley 2011). This hypothesis assumes that shorter life cycles are synonymous with rapid growth rate, so that the negative impact of losing leaf area is low in these species (Endara and Coley 2011). The RAH has since been extrapolated to an intra-species context (Hahn and Maron 2016; López-Goldar et al. 2020).

Three basic predictions of ODT (Rhoades 1979; Herms and Mattson 1992; Koricheva 2002; Stamp 2003) are that, first, plants will evolve a level of defense that is positively related to rates of herbivory and negatively related to allocational or ecological cost. Second, plants will differentially allocate defense to different parts or tissues, with greater investment in tissues with high fitness values or where the cost of defense is lower. Third, plants will increase defense in response to attack, a form of plasticity that is often referred to as induction. ODT predicts that the capacity for induction should be negatively correlated with levels of constitutive defense.

While hypotheses about the processes governing allocation of resources to plant defense differ, one common thread in contemplating the *evolution* of defense production within or across natural plant populations is the genetic basis of defenses. *Genetic variation* in traits within a plant population, or differences among individuals in DNA sequence of genes that underlie focal traits, is a necessary prerequisite to evolution. Historically, the study of plant defense traits has focused on phenotypic variation. As molecular genetic tools and knowledge gained from use of these tools have continued to develop, an increasing number of studies are directly assessing genetic-based patterns of trait production and trade-offs between traits. This work has provided insight into the genetic mechanisms behind the phenotypic patterns of defense trait evolution that we observe, as well as information about the evolutionary potential for plant resistance traits. Evolutionary/ecological hypotheses for patterns of defense trait production, among them the Research Allocation Hypothesis and Optimal Defense Theory, were developed when understanding of the genetic underpinnings of traits was not well understood. Studies of genetic variation during that era were at the level of protein electrophoresis. Testing these hypotheses at the level of genes or genetic correlations, rather than phenotypes and phenotypic correlations, was unprecedented 30–40 years ago (Fig. 4.1).

4.5 Microevolution of Plant Resistance

While decades of research have provided us with valuable information about defense phenotypes, microevolutionary inferences from these studies were limited until studies with the power to elucidate differences in plant defense among genotypes began in the 1980s (e.g., Berenbaum et al. 1986). In total, the relatively large body of work investigating genetic variation in resistance traits indicates it is widespread across both herbaceous and woody plant species (Stowe 1998; Moore et al. 2014).

4.5.1 Direct Defenses

While many studies demonstrating genotypic or genetic variation are done in herbaceous plants, due to ease of experimentation, woody plants also show substantial variation among genotypes in resistance traits, as showcased in multiple studies of *Populus* species (e.g., Havill and Raffa 1999; Lindroth and Hwang 1996; Holeski et al. 2012a; Cope et al. 2019). This work has also highlighted the interaction between genetics and the environment in influencing defense phenotypes. For example, in quaking aspen (*Populus angustifolia*), concentrations of phenolic glycosides, a phytochemical defense, vary substantially with genotype. Environmental factors such as light and nutrient availability also (significantly) affect phenolic glycoside concentrations, and do so differently among genotypes, but genotype is the dominant influence on variation in this trait (Osier and Lindroth 2001, 2004, 2006). In contrast, another phytochemical resistance trait, condensed tannin concentrations, is quite plastic, with variation in the trait typically influenced primarily

4

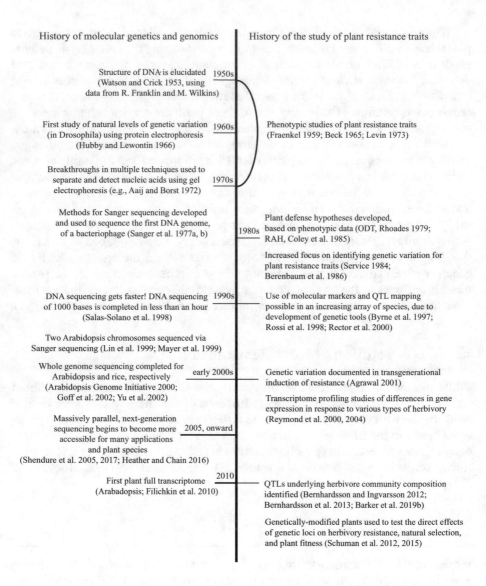

History of molecular genetics and genomics | History of the study of plant resistance traits

Structure of DNA is elucidated **1950s**
(Watson and Crick 1953, using
data from R. Franklin and M. Wilkins)

First study of natural levels of genetic variation **1960s**
(in Drosophila) using protein electrophoresis
(Hubby and Lewontin 1966)

Phenotypic studies of plant resistance traits
(Fraenkel 1959; Beck 1965; Levin 1973)

Breakthroughs in multiple techniques used to
separate and detect nucleic acids using gel **1970s**
electrophoresis (e.g., Aaij and Borst 1972)

Methods for Sanger sequencing developed
and used to sequence the first DNA genome,
of a bacteriophage (Sanger et al. 1977a, b) **1980s**

Plant defense hypotheses developed,
based on phenotypic data (ODT, Rhoades 1979;
RAH, Coley et al. 1985)

Increased focus on identifying genetic variation for
plant resistance traits (Service 1984;
Berenbaum et al. 1986)

DNA sequencing gets faster! DNA sequencing **1990s**
of 1000 bases is completed in less than an hour
(Salas-Solano et al. 1998)

Use of molecular markers and QTL mapping
possible in an increasing array of species, due to
development of genetic tools (Byrne et al. 1997;
Rossi et al. 1998; Rector et al. 2000)

Two Arabidopsis chromosomes sequenced via
Sanger sequencing (Lin et al. 1999; Mayer et al. 1999)

Whole genome sequencing completed for
Arabidopsis and rice, respectively **early 2000s**
(Arabidopsis Genome Initiative 2000;
Goff et al. 2002; Yu et al. 2002)

Genetic variation documented in transgenerational
induction of resistance (Agrawal 2001)

Transcriptome profiling studies of differences in gene
expression in response to various types of herbivory
(Reymond et al. 2000, 2004)

Massively parallel, next-generation
sequencing begins to become more **2005, onward**
accessible for many applications
and plant species
(Shendure et al. 2005, 2017; Heather and Chain 2016)

First plant full transcriptome **2010**
(Arabadopsis; Filichkin et al. 2010)

QTLs underlying herbivore community composition
identified (Bernhardsson and Ingvarsson 2012;
Bernhardsson et al. 2013; Barker et al. 2019b)

Genetically-modified plants used to test the direct effects
of genetic loci on herbivory resistance, natural selection,
and plant fitness (Schuman et al. 2012, 2015)

▫ Fig. 4.1 Timeline of advances in development of genetic and genomic technology (left panel) and of genetic understanding of plant-herbivore interactions (right panel)

by environmental factors such as light availability, or tissue defoliation, rather than genotype (e.g., Hemming and Lindroth 1995; Osier and Lindroth 2004, 2006). *Genotype-by-environment interactions*, whereby individuals of the same genotype respond to an environmental factor differently in terms of plastic trait expression (e.g., Barker et al. 2019a; ▫ Fig. 4.2) also influence condensed tannin concentrations in aspen, albeit to a lesser extent than particular environmental factors (Osier and Lindroth 2006). Ontogenetic trajectories of some resistance traits in cottonwood (*Populus fremontii, P. angustifolia*, and their hybrids) have also been shown

Fig. 4.2 Genotype-environment interactions schematic. Genotype 2 has a higher trait value in Environment A, relative to Genotype 1. Genotype 1 has a higher trait value in Environment B, relative to Genotype 2

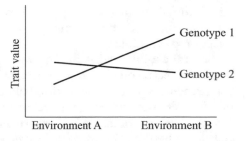

to have a genetic basis (Rehill et al. 2006; Holeski et al. 2012a; Cope et al. 2019). Research in another woody plant genus, *Eucalyptus,* has similarly found genotypic variation in resistance to mammalian browsers (O'Reilly-Wapstra et al. 2002, 2004, 2005). Genetic variation exists between different natural populations of *E. globulus* for resistance to browsing by a generalist marsupial, *Trichosurus vulpecula* (common brushtail possum). As in *Populus*, ontogenetic trajectories of at least some resistance traits are genetically-based in *E. globulus* (O'Reilly-Wapstra et al. 2007).

Research in multiple species illustrates that intra-specific genetic variation (Dungey et al. 2000; Gosney et al. 2014, 2017) as well as genotypic variation (Fritz and Price 1988; Underwood and Rausher 2000) can affect herbivore population dynamics and/or herbivore community composition. It should be noted, however that genetic variation in resistance traits is often not measured in these studies, and the mechanism behind the effects of host genetic variation on herbivores is not always known. Studies of the influence of genotypic variation on herbivore communities that do incorporate genotypic variation in resistance traits includes those in both herbaceous plants (*Arabidopsis thaliana*, Sato et al. 2019a; *Oenothera biennis*, Johnson and Agrawal 2005, 2007) and woody plants (*Populus sp.,* Wimp et al. 2007; Keith et al. 2010; Robinson et al. 2012; Barbour et al. 2016; Barker et al. 2018). Several of these experiments show that while genotypic variation in plant resistance traits does have some effect on arthropod community composition, community composition is most strongly associated with genotypic variation in other traits such as plant size, architecture, and phenology (e.g., Johnson and Agrawal 2005; Robinson et al. 2012; Barker et al. 2018).

4.5.2 Natural Selection on Herbivory Resistance

In addition to genetic variation for a trait, the trait must be acted upon by natural selection (or drift) in order to evolve within a population. Rausher (1996) succinctly described three necessary points for demonstrating that herbivores impose natural selection on resistance traits in their host plants. First, one must show that genetic variation for the focal resistance traits exists. Second, the resistance traits must be demonstrated to be under natural selection. This might be done by showing that genotypes that differ in resistance also differ in fitness. Third, natural herbivores should be manipulated in presence/absence or in density, so that selection on resistance traits can be assessed across treatments. This latter point allows the effects of herbivory to be disentangled from other environmental factors, and thus prevents selection on correlated traits from being confused with selection on resistance.

Several elegant studies in the 1980s–1990s completed these steps to demonstrate that herbivores do impose natural selection on plant resistance traits. One comprehensive set of manipulative field experiments showing both genetic variation in resistance and natural selection acting on resistance was in morning glory (*Ipomoea purpurea*). In one field experiment, genetic variation for resistance to specialist flea beetles (*Chaetocnema confinis*) was found under ambient levels of herbivory (Simms and Rausher 1987). In complementary experiments that manipulated levels of herbivory into ambient versus no herbivory via an insecticide spray, additive genetic variation for resistance to both specialist and generalist herbivores was detected, as well as natural selection acting on this resistance (Rausher and Simms 1989; Simms and Rausher 1989). Another study meeting the stringent criteria outlined by Rausher (1996) was done in jimson weed (*Datura stramonium*). The two major alkaloids in *D. stramonium* were found to be under negative directional selection and stabilizing selection, respectively, with insect herbivores as the agents of selection (Shonle and Bergelson 2000).

4.5.3 Indirect Defenses

While it is often inferred that indirect defense traits increase plant fitness if they reduce herbivore damage, the effects of indirect defenses on plant fitness are rarely directly demonstrated. Exceptions include an experimental study of leaf domatia, small hair-tufts or pockets, in cotton (*Gossypium hirsutum*; Agrawal and Karban 1997), and work on extrafloral nectaries and ant-plant associations in wild cotton (*Gossypium thurberi*; Rudgers 2004; Rudgers and Strauss 2004) and in partridge pea (*Chamaecrista fasciculata*; Rutter and Rausher 2004). In the latter example, there was evidence of costs of nectar production for plants in the absence of ants, and these costs were heightened by herbivory. When ants are present, however, natural selection favored high extrafloral nectar production (Rutter and Rausher 2004).

4.5.4 Transgenerational Defense

As with within-generation defense, the evolutionary relevance of transgenerational induction in defenses is contingent upon whether there is genetic variation, as well as the impacts of transgenerational induction on plant fitness (Kalisz and Purugganan 2004; Richards 2006; Day and Bonduriansky 2011). Genetic variation in transgenerational induction of defense has been shown in multiple species, including wild radish, monkeyflower, and dandelion (Agrawal 2001, 2002; Holeski 2007; Verhoeven et al. 2009; Colicchio 2017). Very few studies have incorporated an experimental design allowing both genetic variation for transgenerational plasticity of defense and the effects on traits related to organism fitness to be examined outside of a laboratory or greenhouse environment (Agrawal 2001, 2002; Holeski et al. 2013), and no published study has measured natural selection on transgenerational plasticity of defense.

Transgenerational plasticity in resistance has been demonstrated to affect plant fitness in natural conditions, although the rare studies investigating fitness have not also assessed genetic variation. In wild lima bean (*Phaseolus lunatus*), offspring of plants that experienced experimentally elevated levels of Chrysomelid beetle herbivory and offspring of control plants that experienced ambient herbivory were assessed in a field environment. Offspring of the elevated-herbivory plants showed higher levels of constitutive resistance traits in the seedlings and juvenile stages, as well as higher survival rates as seedlings (Ballhorn et al. 2016). Work in Carolina horsenettle (*Solanum carolinense*) demonstrates that the offspring of herbivore-damaged plants have decreased time to flowering, and/or produced more flowers than the offspring of control plants (Nihranz et al. 2020).

In summary, while research in both herbaceous and woody plant systems illustrates that genetic variation in resistance traits is widespread, comprehensive studies of the portion of genetic variation that selection acts upon, additive genetic variation, as well as investigation of natural selection on these traits is labor-intensive and is still relatively rare. In most cases, the ecological and evolutionary processes creating and maintaining the genetic variation and how they interact with the genome is still unclear. However, advances in genetic and genomic technology are allowing us to begin to elucidate these interactions, through identification of how genes underlying resistance traits are structured, as well as how genes affect the ecological interactions of plants.

4.6 Identification of Genes Underlying Resistance

4.6.1 Genetic Mapping

As the power to conduct genetic analyses developed, a number of studies in different plant species conducted genetic mapping experiments to identify genetic regions (*quantitative trait loci* or *QTL*) underlying defense trait variation; this technique can be followed by fine-scale mapping to identify specific genes within those regions (Doerge 2002). In plant species with short generation times that can be crossed with reasonable ease, genetic mapping can be done through controlled crosses of individuals divergent for the trait(s) of interest and QTL genetic mapping. Traditional QTL mapping is a labor-intensive process that involves phenotyping large numbers of individuals for traits of interest, in addition to molecular genetics work to genotype individuals at a number of *genetic markers*. Genetic markers are single nucleotides or small regions of the genome that are typically non-coding. Genetic markers are thus not directly involved with producing the trait of interest but may be linked to genes that do underlie these traits. QTL mapping is done through statistical techniques that associate presence of genetic variants (sequence variation, or different marker genotypes) to phenotypic variation in the traits of interest. Genetic markers that are linked to genes influencing the trait of interest will show non-random statistical associations between marker genotype and a particular phenotype. QTL mapping ultimately tells us the amount of observed phenotypic variation in a trait that can be explained by a particular genomic region (Falconer and Mackay 1996; Lynch and Walsh 1998).

In plants with long generation times or those are not amenable to controlled crosses, other techniques such as *genome-wide association mapping* (*GWAS*), also called linkage disequilibrium mapping, are often used (Stinchcombe and Hoekstra 2008; Hall et al. 2010). These studies require a study system in which application of next-generation sequencing technology is feasible (a rapidly increasing number of species) and have a number of logistical advantages over traditional QTL mapping (◘ Table 4.1; Nordborg and Weigel 2008; Hall et al. 2010; Ingvarsson and Street 2011).

Both traditional QTL mapping and GWAS studies have been used to gain understanding of the genetic basis of resistance traits. The initial goal of these studies is often to find out basic information about the genetic underpinnings of resistance traits. Genetic mapping studies of this nature have taken place in both agricultural and natural systems, to different ends. Genetic mapping for defense traits has been particularly common in agricultural plants, where this information can be used in breeding. Often in agriculture, the trait of interest is resistance itself, rather than phytochemical or physical traits conferring resistance. Extent of feeding, insect weight gain, and/or insect mortality are common attributes used to infer resistance against multiple different herbivores and across a variety of crop species such as soybean and tomato (Rector et al. 2000; Komatsu et al. 2005; Yesudas et al. 2010; Vargas-Ortiz et al. 2018). Genetic mapping of resistance in natural systems,

◘ **Table 4.1**   Positive and negative aspects of traditional QTL mapping analysis and genome-wide association mapping (GWAS)

Traditional QTL mapping		Genome-wide association mapping	
Pros	Cons	Pros	Cons
Relatively fewer genetic markers and genomic resources needed No statistical issues from population structure	Allelic variation is restricted to that of the wo parents in the initial cross Controlled crosses are not possible/ feasible for many species QTL identified typically encompass larger genomic regions than GWAS	Encompasses allelic variation within and across natural populations Linkage blocks typically smaller than in QTL mapping, results in more fine-scale mapping	Relatively more genetic markers and genomic resources needed for adequate coverage Population structure can lead to false positives

Both techniques require very large sample sizes. These techniques can also be used together to identify candidate genes. Both methods can be used in population and/or functional genomics studies while having some application to applied agricultural systems (Kloth et al. 2012), are most frequently done as part of fundamental evolutionary biology research, as part of the pursuit of understanding of the genetic basis of adaptation in natural plant populations (Orr and Coyne 1992; Rockman 2012).

4.6.2 QTL for Herbivory Resistance

Data accumulated from several decades of traditional QTL mapping and GWAS studies show that most resistance traits are quantitative traits that have multiple genes of both major and minor effects underlying phenotypes; as quantitative traits they may also be influenced by the environment. Model plant systems have had a head start in genetic mapping experiments due to the relative ease of obtaining genomic information and developing the genetic markers necessary for mapping. In the model system *Arabidopsis thaliana*, the first plant species to have a sequenced genome, informative work with QTL mapping of phytochemical defenses was developed two decades ago (Chan et al. 2010, 2011). These studies have contributed substantially to our understanding of the genetic underpinnings of resistance.

For example, an early study using genetic mapping in *A. thaliana* to investigate the genetic architecture of secondary compounds demonstrated that a relatively small number of genetic regions can underlie considerable variation in phytochemical resistance profiles. Glucosinolates are a large group of secondary metabolites in *Arabidopsis thaliana*, with an estimated 7–14 glucosinolates occurring in foliar tissue of a particular plant (Kliebenstein et al. 2001a). This allows *Arabidopsis* to generate a large number of possible combinations of glucosinolates in individual plants. In a study of the genetic basis of production of 34 different glucosinolates in *Arabidopsis* foliar tissue or seeds, variation at only five QTL resulted in 14 different foliar glucosinolate combinations (Kliebenstein et al. 2001a). Further work showed that a single QTL has a major effect in explaining variation in concentrations of a particular class of glucosinolates (Kliebenstein et al. 2001b).

Early genetic mapping studies exploring the overlap between genetic regions underlying resistance traits and those affecting herbivore performance also used *A. thaliana*. An investigation of QTL influencing feeding rates of generalist (cabbage looper, *Trichoplusia ni*) and specialist (diamondback moth, *Plutella xylostella*) herbivores demonstrated that five QTL for generalist insect feeding overlapped with those for glucosinolate resistance traits, while a relationship between loci underlying specialist herbivore feeding performance and glucosinolates was not found (Kliebenstein et al. 2002a). The diamondback moth can detoxify glucosinolates (Ratzka et al. 2002), potentially explaining the latter result. Similar results showing overlap between QTL underlying glucosinolate profile and those affecting *T. ni* performance were found in the Arabidopsis relative, *Boechera stricta* (Schranz et al. 2009).

While many specialists can detoxify or otherwise avoid the most detrimental effect of secondary compounds, their performance is frequently negatively affected by physical resistance traits such as trichomes (Rotter et al. 2018). Trichome density in *Arabidopsis* does influence oviposition success of the diamondback moth (Handley et al. 2005), and also provides resistance against herbivory by this specialist (Sletvold et al. 2010). In a field experiment, both glucosinolates and tri-chomes reduced levels of generalist and specialist herbivore damage (Mauricio and Rausher 1997; Mauricio 1998). Subsequently, a single gene was identified that influences both trichome density and resistance to chewing insects in a field envi-ronment (Sato et al. 2019a).

4.6.3 Linking Genes to Herbivore Communities

Research linking QTL to resistance traits and to insect performance has most fre-quently been done in cruciferous herbaceous plants. In contrast, investigations of overlap between QTL underlying resistance traits and QTL influencing herbivore community composition has most often been done in woody systems.

In European aspen (*Populus tremula*), geographic variation exists for multiple genes involved in defense against herbivory (Bernhardsson and Ingvarsson 2012; Bernhardsson et al. 2013). Several of these genes show evidence of undergoing *selective sweeps* (Bernhardsson and Ingvarsson 2011), in which beneficial mutations rise so rapidly in frequency in a population due to natural selection that alleles in nearby linked regions are "swept" along. In a GWAS study, multiple single nucleo-tide polymorphisms (SNPs) were identified that were directly associated with herbi-vore community metrics such as species abundances for specialist herbivores, species richness for generalist herbivores, and species abundances within the galling, min-ing, and leaf rolling feeding guilds (Bernhardsson et al. 2013). A GWAS study in a North American aspen species, quaking aspen (*Populus tremuloides*) similarly found multiple SNPs underlying aspects of insect community composition (Barker et al. 2019b). This study also found overlap between SNPs associated with variation in insect communities and those associated with variation in multiple plant traits, demonstrating a mechanistic link for the gene-insect associations. Plant traits included the phytochemical resistance traits phenolic glycosides and condensed tan-nins (Barker et al. 2019b). A previous QTL mapping study in hybrid *Populus* (*P. trichocarpa* x *P. deltoides*) also indicates a potential role for phenolic glycosides and condensed tannins in influencing herbivore community composition (DeWoody et al. 2013). In hybrid aspen, QTL underlying variation in different feeding guilds of herbivores contain genes in the shikimate-phenylpropanoid pathway, which pro-duces phenolic glycosides and condensed tannins (DeWoody et al. 2013).

4.7 Use of Genetics to Test Evolutionary Ecology Hypotheses

Techniques such as QTL mapping that identify genes or regions of the genome that underlie resistance traits can be used to test evolutionary hypotheses at previ-ously unprecedented mechanistic levels. Isolating the effects of single genes or genetic regions on a phenotype and/or on herbivores can be done through the use of traditional breeding designs (e.g., Lowry et al. 2019) or gene silencing (e.g., Kessler et al. 2004).

4.7.1 Inferring Ecological Consequences from Genetic Data

Experiments designed to link genetics with ecology and/or evolution became more common in the early twenty-first century with the advent of the field of *ecological genomics* and a growing realization of the lack of ecological context for model organisms in laboratory environments. The primary goal of ecological genomics is to identify the genetic and molecular mechanisms underlying natural trait variation and influencing organismal response to the environment (Feder and Mitchell-Olds 2003; McKay and Stinchcombe 2008; Ungerer et al. 2008). Genetic or genomic information provides a connection between ecology and phenotypic-based studies to the evolutionary trajectory of defenses and populations. Ecological genetics approaches can be used to elucidate the functional and ecological consequences of genes, with breadth ranging from looking at trade-offs influenced by individual QTL or transcriptome studies of patterns of gene expression.

While some studies of natural selection on defense traits were done in natural environments, most previous work related to resistance was done with a small number of herbivores in a laboratory environment. One early example of the gains in knowledge obtained from moving from a lab to an ecological context occurred in *Nicotiana attenuata* (wild tobacco), which has become a model system for the study of signaling pathways involved in induced resistance (Baldwin 1998a, b; Xu et al. 2018). Three genes playing a major role in plant wound recognition and signaling response were silenced, thus dampening induced response to herbivory (Kessler et al. 2004). In the lab, plants with these genes silenced were more susceptible to herbivory by the specialist tobacco hornworm, *Manduca sexta*. In the field, the community composition of the herbivores attacking the plants was highly altered, with some novel herbivores showing a preference for the plants, imposing heavy damage, and ovipositing (Kessler et al. 2004). This demonstrates that host plant selection is not determined only by the constitutive defenses of a plant, but also by a plant's induced response to herbivory. This unique result would have been difficult to discover without investigation of the effects of this genetic manipulation in a natural environment.

4.7.2 Genetic Correlations and QTL-Level Trade-Offs

QTL-level studies can be used to test predictions of trade-offs between multiple aspects of defense, the framework for which was developed based upon phenotypic information. Questions within this realm include whether genes for different aspects of defense, or genes influencing resistance vs. traits related to life history strategy, co-localize. For example, are functional trade-offs that are predicted by plant defense hypotheses such as the Resource Allocation Hypothesis and Optimal Defense Theory based upon *genetic correlations* and/or *co-localization* at the genomic level? The presence of genetic correlations implies that the genes that underlie the traits are inherited together (Lande 1979; Via and Hawthorne 2002). Genetic correlations can be due to *pleiotropy*, when one gene influences multiple traits, or to *linkage disequilibrium*, nonrandom associations of alleles at different genes affecting two traits (Falconer and Mackay 1996). Linkage disequilibrium is most often due to close physical linkage of the genes that underlie the traits (Lynch and Walsh 1998). Traditionally, genetic correlations have been estimated by assessing phenotypes of related individuals, and thus have not always been practical in studies of plant defense in non-model species. More recently, statistical methods have been developed for human data that allow genetic correlations to be calculated in very large groups of unrelated individuals based on genomic data (Lee et al. 2012; Sodini et al. 2018), although these methods are not often used in studies of plants. QTL mapping experiments, where QTL for multiple traits *co-localize*, or map to the same region of the genome, generally cannot distinguish whether the underlying mechanism for co-localization is pleiotrophy or physical linkage without being followed up by fine-mapping.

Knowledge of the extent of genetic correlations between traits can be more informative in studies of evolution than are phenotypic correlations. Analogous to studies of the evolution or evolutionary potential of a single trait, whereby genetic variation for a trait is necessary for evolution of the trait to occur within a population, genetic correlations are based upon genetic variance and covariances between traits and the rate and direction of their evolution depends on these parameters (Lynch and Walsh 1998). While phenotypic and genetic correlations sometimes correspond (Roff 1996), in many cases they do not (Willis et al. 1991), thus phenotypic correlations are not necessarily reliable substitutions when making evolutionary inferences for traits.

With genetic correlations, response to selection on one trait (i.e., changes in allele frequencies) will result in changes in other traits that are influenced by the same gene/group of genes. Such correlations can facilitate or constrain adaptation (Lande 1979; Via and Hawthorne 2005). QTL studies have been very informative in the exploration of the evolution of trade-offs in multiple aspects of defense, including between constitutive and induced resistance, between resistance and abiotic stress tolerance, and between resistance and other life history traits. Characterization of the strength and direction of genetic correlations between traits and/or the amount of phenotypic variation explained by QTL that co-localize and underlie traits allows inference about the evolutionary trajectory of these traits within populations (Conner and Hartl 2004).

Trade-offs between constitutive and induced defense are predicted to occur by Optimal Defense Theory, based upon allocation of resources given the probability of herbivory. Plant populations that experience consistent herbivory might invest more in constitutive defense, while populations that incur more sporadic herbivory might invest in induced defense only when necessary (Harvell 1990; Adler and Karban 1994; Cipollini et al. 2003). These patterns have been often, but not always, supported by phenotypic correlations (Karban and Baldwin 1997; Agrawal 1998). Investigation of genetic correlations and whether genetic underpinnings of these traits co-localize can help us to answer fundamental evolutionary questions regarding the genetic architecture of complex traits, and how trade-offs evolve.

Several studies of the genetic architecture of constitutive and induced defense have found QTL underlying these traits to co-localize. Kliebenstein et al. (2002b) found co-localization of QTLs influencing constitutive and induced glucosinolate levels in *Arabidopsis thaliana*. In contrast to the predictions of ODT, however, the genetic correlations between constitutive and induced defenses were positive. As predicted by ODT, negative genetic correlations were found between constitutive and induced trichome density in monkeyflower (*Mimulus guttatus*), with co-localization of some of the QTL underlying these traits (Holeski et al. 2010).

Trade-offs between resistance traits and abiotic stress tolerance have also been found to have a genetic basis. In a study of the mechanisms aiding or hindering range expansion in the ecological model species *Boechera stricta*, negative genetic correlations and co-localizing QTL were detected between glucosinolate production and drought stress tolerance. At the low elevation range boundary for the species, both increased levels of glucosinolates and increased drought stress tolerance were favored by selection. Thus, the genetic-based trade-off between these two attributes could contribute to limiting the range of the species by not allowing for simultaneous increase in resistance and increase in drought stress tolerance (Siemens et al. 2009; Olsen et al. 2019).

In monkeyflower, several studies have identified genetic-based trade-offs between resistance traits and traits related to life history strategy. In some perennial plants of *Mimulus guttatus*, an inverted portion of a chromosome (*DIV1*) underlies increases in both phytochemical resistance traits and traits related to long-term growth strategy such as plant height, adventitious root production, and number of stolons. The annual orientation of this region is associated with rapid development to reproduction and reduced phytochemical defense (Lowry et al. 2019). Similar trade-offs between rapid development to reproductive maturity and allocation to phytochemical defense is found within annual monkeyflower plants (*Mimulus guttatus*). A QTL of relatively large effect underlies a trade-off between phytochemical resistance and developmental rate; plants that have more rapid time to reproductive maturity have lower levels of phytochemical defense than plants with slower development times (Kooyers et al. 2020).

4.8 Gene Expression and Herbivore Resistance

Advances in genomic technology have allowed greater understanding of how plants respond to herbivory at the genomic level. While this work is still biased towards a relatively small number of model and agricultural species (◻ Table 4.2), research in species that are closely related to model species have been able to coopt genetic tools and molecular genetic resources. Species in the Brassicaceae family such as *Boechera* and *Brassica*, for example, are often used for study of the genetics of plant-herbivore interactions, and this is largely possible due to the use of genetic resources developed in *Arabidopsis* (Mitchell-Olds 2001; Anderson and Mitchell-Olds 2011). Likewise, experimental use of plants in the *Solanum* genus has taken advantage of the genomic tools developed in tomato and potato relatives (Schmidt

▫ Table 4.2   Genera used in genetic studies of plant resistance

Plant genus	Genome size of representative sequenced species	Year genome sequence made publicly available for a species in the genus
Arabidopsis	135 Mb	2000
Boechera	227 Mb	2017
Brassica	584 Mb	2011
Eucalyptus	640 Mb	2014
Glycine	1.1 Gb	2008
Ipomoea	750 Mb	2016
Mimulus	430 Mb	2008
Nicotiana	2.6–4.5 Gb	2012
Oryza	430 Mb	2002
Populus	500 Mb	2006
Solanum	840 Mb	2011
Triticum	17 Gb	2018
Zea (corn)	2.4 Gb	2009

This list represents many genera commonly used but is not comprehensive

et al. 2004). While a great deal of the work in gene expression has taken place in a greenhouse or laboratory environment, in the past decade multiple studies examining gene expression in plants grown in field-based common gardens have been published. These latter experiments have provided insight into the relationship between gene expression, ecologically-realistic abiotic and biotic environmental conditions, and population-level processes.

4.8.1 Transcriptome Profiling

Numerous studies have investigated patterns of gene induction and transcriptome patterns during and/or after herbivore feeding, along with the ecological effects of the induced changes in resistance. Expression profiling, also called *transcription profiling*, tracks the expression of hundreds to thousands of genes on *DNA microarrays*, whereby specific sequences are attached to a surface of a DNA chip, and act as probes to detect gene expression in samples (Bumgarner 2013). The expression profiles can then be compared in plants replicated across different environmental conditions of interest. In plant-herbivore interactions, this method has been used to compare gene expression in response to herbivores from different feeding guilds (Broekgaarden et al. 2010), generalist versus specialist herbivores (Reymond et al. 2004), and plant response to different natural herbivore communities (Broekgaarden et al. 2010). Plant response to herbivores of different genotypes has even been assessed (Zytynska et al. 2016). Transcription profiling has been conducted in a number of species, including *Arabidopsis* and relatives, sorghum (*Sorghum bicolor*), tobacco (*Nicotiana attenuate*), rice (*Oryza spp.*), and tomato (*Lycopersicon esculentum*); this profiling has highlighted the complexity of plant response to damage and has provided insight into damage-induced signaling pathways. Large-scale differences in results across studies have also demonstrated the need for use of more standardized experimental designs (Thompson and Goggin 2006). Unless genes previously characterized in function have been identified and are used, gene expression profiling represents a whole plant response to herbivory and thus can include changes in expression in genes underlying direct defenses, indirect defenses, and a myriad of physiological changes.

4.8.2 Gene Expression Following Herbivore Activity

Transcription profiling has been used to investigate gene expression in response to mechanical versus specialist herbivore (*Pieres rapae*) damage (Reymond et al. 2000) and damage by generalist (*Spodoptera littoralis*) versus specialist (*P. rapae*) chewing herbivores in *Arabidopsis thaliana* (Reymond et al. 2004). While gene expression was quite different between plants with mechanical vs. specialist damage (Reymond et al. 2000), substantial overlap occurred between genes expressed in response to the generalist and specialist caterpillars (Reymond et al. 2004). Another study comparing gene expression in response to two generalist and a specialist chewing herbivore in tobacco (*Nicotiana attenuata*) found that the most overlap occurred between the two generalists (*Heliothis virescens* and *Spodoptera exigua*), although over 60% of the genes up- or down-regulated by the specialist (*Manduca sexta*) were similarly expressed following herbivory by the generalists (Voelckel and Baldwin 2004).

Studies of gene expression have been used to address evolutionary ecological hypotheses regarding induction of plant defenses by generalist versus specialist herbivores. For example, a long-standing paradigm in plant-herbivore interactions predicts that phytochemical plant defenses will have less of an effect on specialist herbivores than on generalists, and that when damaged by generalists versus specialist herbivores, plant responses will differ (Ali and Agrawal 2012). In the latter point, plant responses are both dictated by the plant and manipulated by the herbivore (Felton and Eichenseer 1999; Felton and Tumlinson 2008; Erb et al. 2012). Phenotypic evidence for differential plant response to generalists versus specialist is not consistent (Bowers and Stamp 1993; Agrawal 2000; Ali and Agrawal 2012), while evidence for differential plant response to feeding by herbivores of different feeding guilds is clearer (Ali and Agrawal 2012).

Patterns of gene expression have generally supported trait-based results in tests of this paradigm, showing that feeding guild may have a stronger effect on overall differences in gene expression than does the diet breadth or degree of specialization of the herbivore. Analyses of gene expression following feeding by different feeding guilds in *Arabidopsis* showed limited overlap in the transcriptional response to feeding by chewing, cell-content feeding, and phloem-feeding herbivores (the caterpillar *Pieris rapae*, thrip *Frankliniella occidentalis*, and aphid *Myzus persicae*, respectively; De Vos et al. 2005). Broekgaarden et al. (2011) show similar results from a study of specialist caterpillars (*P. rapae*) and aphids (*Brevicoryne brassicae*) feeding on wild black mustard (*Brassica nigra*). Finally, in a study of four chewing lepidopteran herbivores and two aphid species, with generalists and specialists within each feeding guild, the effects of insect taxon or feeding guild had a substantially larger effect on patterns of gene expression than did insect diet breadth (Bidart-Bouzat and Kliebenstein 2011).

Gene expression data has highlighted at least some of the mechanistic basis for genotypic variation in resistance. In the *Arabidopsis* relative, white cabbage (*Brassica oleracea*), transcription responses to damage by a specialist (*Pieres rapae*) in two cultivars (genotypes) were compared using microarrays developed for *Arabidopsis* (Broekgaarden et al. 2007). The two genotypes differed in resistance as measured by *P. rapae* performance, although with this metric constitutive versus induced resistance cannot be disentangled. The two cultivars also differed fairly dramatically in transcriptional response, with 44% or more of the genes induced in one cultivar not induced in the other (Broekgaarden et al. 2007). Similarly, in a field experiment with 19 *Arabidopsis* accessions and ambient herbivory, more than half of the differences in gene expression in glucosinolate biosynthetic genes was among-accession variation (Sato et al. 2019b).

While much gene expression work has been done in a laboratory environment with one to three herbivores rather than in a natural context, some investigation of the relevance of transcriptional profiling in natural environments has been done. One such study elegantly ties herbivore community metrics to patterns of gene expression across plant genotypes. In a field study with two *Brassica oleracea* genotypes, nine herbivore species were present in similar abundances across the genotypes (Broekgaarden et al. 2010). Later in the season, one genotype hosted lower richness and abundances of both generalist and specialist herbivores than the other genotype. Levels of gene expression also differed more substantially between the genotypes relative to earlier in the season, with differences in expression levels in over 20 genes, including several genes known to play a role in herbivory defense (Broekgaarden et al. 2010).

When transcriptional profiling is used in combination with genes with a characterized function, a more comprehensive picture of plant response to herbivory can be obtained (He et al. 2020). Transcriptional profiling has been used for genes with previously characterized roles in direct and indirect defenses. In cucumber (*Cucumis sativus*), genes related to some direct defenses including phenylpropanoids and terpenoids were upregulated in response to spider mites (*Tetranychus urticae*), while genes underlying other direct defense phytochemicals were downregulated. Genes involved in the production of terpenoid emissions as an indirect defense were upregulated (He et al. 2020).

4.9 Genetic Basis of Indirect Defense

To date, study of the genetic basis of indirect defenses focuses almost exclusively on production of plant volatile compounds (Pearse et al. 2020). Levels, as well as the composition, of volatile blends emitted after herbivory are different than those emitted before herbivory; post-herbivory emissions of herbivory-induced plant volatiles (HIPVs) attract predators and parasitoids from multiple insect orders, as well as mites, nematodes, and birds (Dicke et al. 2003; McCormick et al. 2012). Plant volatiles can be induced by herbivore feeding and/or oviposition on a plant, and tend to attract primarily herbivore enemies (Kessler and Baldwin 2001, 2002; Poelman et al. 2008; Hilker and Meiners 2010), or egg parasitoids (Hilker et al. 2002; Hilker and Meiners 2002, 2006), accordingly.

Feeding by herbivores of different types and ages, feeding guilds, and abundances can affect the specific blend of volatiles that are released, and thus the enemies that are cued (McCormick et al. 2012). Volatiles induced by mechanical damage are different than those induced by herbivore feeding (Turlings et al. 1990; Baldwin et al. 2001). This could be a result of the differences in rate of tissue lost by herbivory vs. mechanical damage (Mithöfer et al. 2005), and/or to the lack of salivary cues by the herbivores (Turlings et al. 1990; Felton and Tumlinson 2008).

Research of indirect defenses has focused primarily on elucidating the signal-transduction pathways underlying plant response. This mechanistic research has taken place in model plant species, with genes underlying HIPV response identified in *Arabidopsis* (Van Poecke et al. 2001; Kappers et al. 2005), *Populus* (Irmisch et al. 2013; McCormick et al. 2019), tomato (Zhang et al. 2020), lima bean (Arimura et al. 2000), and maize (Erb et al. 2015), among others.

The genes underlying multiple aspects of direct and indirect defense have been identified in wild tobacco (*Nicotiana attenuata*; e.g., Dinh et al. 2013; Xu et al. 2020), and many aspects of the ecology of the species are known (Adam et al. 2018). However, extensions from the mechanistic genetic basis of HIPVs to plant fitness in natural environments to characterize the function of genes through generation of genetically altered plants are somewhat rare. One field-based study of HIPVs in wild tobacco estimated that they reduced the number of herbivores present by 90%, indicating the potential for substantial effects of HIPVs on plant fitness (Kessler and Baldwin 2001). Later studies have investigated the evolutionary and ecological effects of HIPVs in wild tobacco more directly through the use of genetically modified plants and have shown substantial effects of HIPVs on plant fitness.

One investigation of the effects of HIPVs on plant fitness in a field environment included plants that were genetically modified to have the genes underlying HIPVs silenced (Schuman et al. 2012). The specialist tobacco hornworm *Manduca sexta* had a large effect on flower production in the field environment; predation on *Manduca* was increased two-fold in plants that produced HIPVs, and these plants had twice as many buds and flowers as those in which HIPVs were silenced (Schuman et al. 2012). Another field experiment used wild tobacco plants that were genetically altered to produce reduced or enhanced levels of herbivore-induced volatiles, with some also having reduced levels of direct defenses (Schuman et al.

2015). Herbivore abundance was lowest on plants with enhanced levels of herbivore-induced volatiles, and these plants also had the lowest mortality rates. Enhanced levels of volatile production did not entirely compensate for reduced levels of direct defenses, in terms of plant mortality, but plants with enhanced levels of volatile emissions did improve the fitness of plants of other genotypes, when planted together (Schuman et al. 2015).

Work in wild tobacco using genetically modified plants has thus shown clear effects of variation in HIPV on plant fitness. Future studies investigating the effects of genetic variation in HIPV production within natural populations would provide more insight into the evolutionary potential of these indirect plant defenses.

Conclusions

Research in plant-herbivore interactions in the past half-century has shifted from a focus on phenotypic variation to an increasingly mechanistic genetic scale. We now know that genetic variation in resistance traits is widespread in natural populations. Characterizing patterns of genetic variation in traits and how natural selection by herbivores acts on these traits has given us a better understanding of evolutionary trajectories of resistance in natural populations.

Identification of genes or QTL underlying resistance traits has allowed us to test evolutionary and ecological hypotheses regarding the evolution of plant resistance at the level of individual genes or genetic regions, rather than at the level of phenotypes. Functional characterization of genes underlying herbivory resistance has led to a better understanding of the mechanistic pathways between genes and phenotypes.

As the variety of plant species in which genetic-based hypotheses can be tested increases, this will enable us to build a more comprehensive view of commonalities and differences in the genetic control of resistance. Investigations of gene expression and function across a broader array of species and in ecologically realistic environments will increase our understanding of how the genetic architecture of resistance functions and evolves in natural populations.

Key Points
Genetic and genomic studies of plant-herbivore interactions have contributed significantly to our understanding of:
- the genetic basis of plant resistance to herbivory
- evolutionary trajectories of resistance traits in natural populations
- evolutionary and ecological hypotheses for the evolution of resistance
- signaling pathways that underly plant response to herbivory

❓ Question
We have extensive knowledge of the natural history of plant-herbivore interactions in some systems, based on a century or more of research. How can genetic work be used to complement natural history to inform our understanding of plant-herbivore interactions?

Acknowledgments Thanks to Rick Lindroth for comments on a draft of this chapter. LMH was partially supported by an NAU RBS award.

References

Aaij C, Borst P (1972) The gel electrophoresis of DNA. Biochim Biophys Acta 269:192–200

Adam N, Kallenbach M, Meldau S, Veit D, van Dam NM, Baldwin IT, Schuman MC (2018) Functional variation in a key defense gene structures herbivore communities and alters plant performance. PLoS One 13:e0197221

Adler FR, Karban R (1994) Defended fortresses or moving targets? Another model of inducible defenses inspired by military metaphors. Am Nat 144:813–832

Agrawal AA (1998) Induced responses to herbivory and increased plant performance. Science 279:1201–1202

Agrawal AA (2000) Specificity of induced resistance in wild radish: causes and consequences for two specialist and two generalist caterpillars. Oikos 89:493–500

Agrawal AA (2001) Transgenerational consequences of plant responses to herbivory: an adaptive maternal effect? Am Nat 157:555–569

Agrawal AA (2002) Herbivory and maternal effects: Mechanisms and consequences of transgenerational induced plant resistance. Ecology 83:3408–3415

Agrawal AA, Karban R (1997) Domatia mediate plant-arthropod mutualism. Nature 387:562–563

Ahern JR, Whitney KD (2014) Sesquiterpene lactone stereochemistry influences herbivore resistance and plant fitness in the field. Ann Bot 113:731–740

Ali JG, Agrawal AA (2012) Specialist versus generalist insect herbivores and plant defense. Trends Plant Sci 17:293–302

Anderson JT, Mitchell-Olds (2011) Ecological genetics and genomics of plant defences: evidence and approaches. Funct Ecol 25:312–324

Arabidopsis Genome Initiative (2000) Analysis of the genome sequence of the flowering plant *Arabidopsis thaliana*. Nature 408:796–815

Arimura G, Ozawa R, Shimoda T, Nishioka T, Boland W, Takabayashi J (2000) Herbivory-induced volatiles elicit defence genes in lima bean leaves. Nature 406:512–515

Baldwin IT (1998a) Jasmonate-induced responses are costly but benefit plants under attack in native populations. Proc Natl Acad Sci U S A 95:8113–8118

Baldwin IT (1998b) Jasmonate-induced responses are costly but benefit plants under attack in native populations. Proc Natl Acad Sci 95:8113–8118

Baldwin IT, Halitschke R, Kessler A, Schittko U (2001) Merging molecular and ecological approaches in plant-insect interactions. Curr Opin Plant Biol 4:351–358

Ballhorn DJ, Kautz S, Laumann JM (2016) Herbivore damage induces a transgenerational increase of cyanogenesis in wild lima bean (*Phaseolus lunatus*). Chemoecology 26:1–5

Barbour MA, Fortuna MA, Bascompte J, Nicholson JR, Julkunen-Tiitto R, Jules ES, Crutsinger GM (2016) Genetic specificity of a plant-insect food web: implications for linking genetic variation to network complexity. Proc Natl Acad Sci U S A 113:2128–2133

Barker HL, Holeski LM, Lindroth RL (2018) Genotypic variation in plant traits shapes herbivorous insect and ant communities on a foundation tree species. PLoS One 13:e0200954

Barker HL, Holeski LM, Lindroth RL (2019a) Independent and interactive effects of plant genotype and environment on plant traits and insect herbivore performance: a meta-analysis with Salicaceae. Funct Ecol 33:122–135

Barker HL, Riehl JF, Bernhardsson C, Rubert-Nason KF, Holeski LM, Ingvarsson PK, Lindroth RL (2019b) Linking plant genes to insect communities: identifying the genetic bases of plant traits and community composition. Mol Ecol 28:4404–4421

Barton KE (2016) Tougher and thornier: general patterns in the induction of physical defense traits. Funct Ecol 30:181–187

Barton KE, Koricheva J (2010) The ontogeny of plant defense and herbivory: characterizing general patterns using meta-analysis. Am Nat 175:481–493

Becerra JX (1997) Insects on plants: macroevolutionary chemical trends in host use. Science 27:253–256

Becerra JX, Venable DL (1990) Rapid-terpene-bath and squirt-gun defense in *Bursera schlechtendalii* and the counterploy of chrysomelid beetles. Biotropica 22:320–323

Beck SD (1965) Resistance of plants to insects. Annu Rev Entomol 10:205–232

Berenbaum M (1978) Toxicity of a furanocoumarin to armyworms- case of biosynthetic escape from insect herbivores. Science 201:532–534

Berenbaum M (1983) Coumarins and caterpillars – a case for coevolution. Evolution 37:163–179

Berenbaum MR, Zangerl AR (2008) Facing the future of plant-insect interaction research: le retour à la "raison d'être". Plant Physiol 146:804–811

Berenbaum MR, Zangerl AR, Nitao JK (1986) Contraints on chemical coevolution- wild parsnips and the parsnip webworm. Evolution 40:1215–1228

Bernhardsson C, Ingvarsson PK (2011) Molecular population genetics of elicitor-induced resistance genes in European aspen (*Populus tremula* L., Salicaceae). PLoS One 6:e24867

Bernhardsson C, Ingvarsson PK (2012) Geographic structure and adaptive population differentiation in herbivore defense genes in European aspen (*Populus tremula* L., Salicaceae). Mol Ecol 21:2197–2207

Bernhardsson C, Robinson KM, Abreu IN, Jansson S, Albrectsen BR, Ingvarsson PK (2013) Geographic structure in metabolome and herbivore community co-occurs with genetic structure in plant defence genes. Ecol Lett 16:791–798

Bidart-Bouzat MG, Kliebenstein D (2011) An ecological genomic approach challenging the paradigm of differential plant responses to specialist versus generalist insect herbivores. Oecologia 167:677–689

Bidart-Bouzat MG, Kliebenstein D (2001) An ecological genomic approach challenging the paradigm of differential plant responses to specialist versus generalist insect herbivores. Oecologia 167:677–689

Boege K, Marquis RJ (2005) Facing herbivory as you grow up: the ontogeny of resistance in plants. Trends Ecol Evol 20:441–448

Bowers MD, Stamp N (1993) Effects of plant-age, genotype, and herbivory on *Plantago* performance and chemistry. Ecology 74:1778–1791

Broekgaarden C, Poelman EH, Steenhuis G, Voorrips RE, Dicke M, Vosman B (2007) Genotpic variation in genome-wide transcription profiles induced by insect feeding: *Brassica oleracea-Pieris rapae* interactions. BMC Genomics 8:239

Broekgaarden C, Poelman EH, Voorrips RE, Dicke M, Vosman B (2010) Intraspecific variation in herbivore community composition and transcriptional profiles in field-grown *Brassica oleracea* cultivars. J Exp Bot 61:807–819

Broekgaarden C, Voorrips RE, Dicke M, Vosman B (2011) Transcriptional responses of *Brassica nigra* to feeding by specialist insects of different feeding guilds. Insect Sci 18:259–272

Bruce TJA (2015) Interplay between insects and plants: dynamic and complex interactions that have coevolved over millions of years but act in milliseconds. J Exp Bot 66:455–465

Bumgarner R (2013) DNA microarrays: types, applications and their future. Curr Protocols Mol Biol 22.1.1–22.1-11

Byrne PF, McMullen MD, Wiseman BR, Snook ME, Musket TA, Theuri JM, Widstrom NW, Coe FH (1997) Identification of maize chromosome regions associated with antibiosis to corn earworm (Lepidoptera: Noctuidae) larvae. J Econ Entomol 90:1039–1045

Castillo G, Cruz LL, Tapia-López R, Olmedo-Vicente E, Carmona D, Anaya-Lang AL, Fornoni J, Andraca-Gómez G, Valverde PL, Núñez-Farfán J (2014) Selection mosaic exerted by specialist and generalist herbivores on chemical and physical defense of *Datura stramonium*. PLoS One 9:e102478

Chan EKF, Rowe HC, Kliebenstein DJ (2010) Understanding the evolution of defense metabolites in *Arabidopsis thaliana* using genome-wide association mapping. Genetics 185:991–1007

Chan EKF, Rowe HC, Corwin JA, Joseph B, Kliebenstein DJ (2011) Combining genome-wide association mapping and transcriptional networks to identify novel genes controlling glucosinolates in *Arabidopsis thaliana*. PLoS Biol 9:e1001125

Chen H, Wilkerson CG, Kuchar JA, Phinney BS, Howe GA (2005) Jasmonate-inducible plant enzymes degrade essential amino acids in the herbivore midgut. Proc Natl Acad Sci U S A 102:19237–19242

Cipollini DF (1998) Induced defenses and phenotypic plasticity. Trends Ecol Evol 13:200

Cipollini D, Purrington CB, Bergelson J (2003) Costs of induced responses. Basic Appl Ecol 4:79–89

Cole CT, Morrow CJ, Barker HL, Rubert-Nason KF, Riehl JFL, Kollner TG, Lackus ND, Lindroth RL (2020) Growing up aspen: ontogeny and trade-offs shape growth, defence, and reproduction in a foundation species. Am J Bot. https://doi.org/10.1093/aob/mcaa070

Coley PD, Bryant JP, Chapin FS III (1985) Resource availability and plant antiherbivore defense. Science 230:895–899

Coley PD, Endara M-J, Kursar TA (2018) Consequences of interspecific variation in defenses and herbivore host choice for the ecology and evolution of *Inga*, a speciose rainforest tree. Oecologia 187:361–376

Colicchio J (2017) Transgenerational effects alter plant defence and resistance in nature. J Evol Biol 30:664–680

Conner JK, Hartl DL (2004) A primer of ecological genetics. Sinauer, Sunderland

Cope OL, Kruger EL, Rubert-Nason KF, Lindroth RL (2019) Chemical defense over decadal scales: Ontogenetic allocation trajectories and consequences for fitness in a foundation tree species. Funct Ecol 33:2105–2115

Darrow K, Bowers MD (1997) Phenological and population variation in iridoid glycosides of *Plantago lanceolata*. Biochem Syst Ecol 25:1–11

Day T, Bonduriansky R (2011) A unified approach to the evolutionary consequences of genetic and nongenetic inheritance. Am Nat 178:E18–E36

De Vos M, Van Oosten VR, Van Poecke RMP, Van Pelt JA, Pozo MJ, Mueller MJ, Buchala AJ, Métraux JP, Van Loon LC, Dicke M, Pieterse CMJ (2005) Signal signature and transcriptome changes of *Arabidopsis* during pathogen and insect attack. Mol Plant-Microbe Interact 18:923–937

De-la-Cruz IM, Velázquez-Márquez S, Núñez-Farfán J (2020) What do we know about the genetic basis of plant defensive responses to herbivores? A mini-review. In: Núñez-Farfán J, Valverde P (eds) Evolutionary ecology of plant-herbivore interaction. Springer, Cham, pp 295–314

Dewoody J, Viger M, Lakatos F, Tuba K, Taylor G, Smulders MJM (2013) Insight into the genetic components of community genetics: QTL mapping of insect association in a fast-growing forest tree. PLoS One 8:e79925

Dicke M (1999) Evolution of induced indirect defence of plants. In: Tollrian R, Harvell CD (eds) The ecology and evolution of inducible defenses. Princeton University Press, Princeton, pp 62–88

Dicke M, Hilker M (2003) Induced plant defenses: from molecular biology to evolutionary biology. Basic Appl Ecol 4:3–14

Dicke M, van Poecke RMP, de Boer JG (2003) Inducible indirect defense of plants: from mechanisms to ecological functions. Basic Appl Ecol 4:27–42

Dinh ST, Baldwin IT, Galis I (2013) The *HERBIVORE ELICIOTR-REGULATED1* gene enhances abscisic acid levels and defenses against herbivores in *Nicotiana attenuata* plants. Plant Physiol 162:2106–2124

Doerge RW (2002) Mapping and analysis of quantitative trait loci in experimental populations. Nat Rev Genet 3:43–52

Dungey HS, Potts BM, Whitham TG, Li HF (2000) Plant genetics affects arthropod community richness and composition: evidence from a synthetic eucalypt hybrid population. Evolution 54:1936–1946

Ehrlich PR, Raven PH (1964) Butterflies and plants: a study in coevolution. Evolution 18:586–608

Elle E, Hare JD (2000) No benefit of glandular trichome production in natural populations of *Datura wrightii*? Oecologia 123:57–63

Endara M-J, Coley PD (2011) The resource-availability hypothesis: a meta-analysis. Funct Ecol 25:389–398

Erb M, Kliebenstein DJ (2020) Plant secondary metabolites as defenses, regulators, and primary metabolites: the blurred functional trichotomy. Plant Physiol. https://doi.org/10.1104/pp.20.00433

Erb M, Meldau S, Howe GA (2012) Role of phytohormones in insect-specific plant reactions. Trends Plant Sci 17:250–259

Erb M, Veyrat N, Rober CAM, Xu H, Frey M, Ton J, Turlings TCJ (2015) Indole is essential herbivore-induced volatile priming signal in maize. Nat Commun 6:6273

Falconer DS, Mackay TFC (1996) Introduction to quantitative genetics, 4th edn. Longmans Green, Harlow

Feder ME, Mitchell-Olds T (2003) Evolutionary and ecological functional genomics. Nat Rev Genet 4:649–655

Felton GW (2005) Indigestion is a plant's best defense. Proc Natl Acad Sci U S A 102:18771–18772

Felton GW, Eichenseer H (1999) Herbivore saliva and its effect on plant defense against herbivores and pathogens. In: Agrawal AA, Tuzun S, Bent E (eds) Induced plant defenses against pathogens and herbivores. Biochemistry, ecology and agriculture. APS Press, St. Paul, pp 19–36

Felton GW, Tumlinson JH (2008) Plant-insect dialogs: complex interactions at the plant-insect interface. Curr Opin Plant Biol 11:457–463

Filichkin SA, Priest HD, Givan SA, Shen R, Bryant DW, Fox SE, Wong WK, Mockler TC (2010) Genome-wide mapping of alternative splicing in *Arabidopsis thaliana*. Genome Res 20:45–58

Fornoni J, Valverde PL, Núñez-Farfán J (2003) Quantitative genetics of plant tolerance and resistance against natural enemies of two natural populations of *Datura stramonium*. Evol Ecol Res 5:1049–1065

Fraenkel GS (1959) The raison d'etre of secondary plant substances. Science 129:1466–1470

Franceschi VR, Krokene P, Christiansen E, Krekling T (2005) Anatomical and chemical defenses of conifer bark against bark beetles and other pests. New Phytol 167:353–375

Fritz RS, Price PW (1988) Genetic variation among plants and insect community structure: willows and saw-flies. Ecology 69:845–856

Glas JJ, Schimmel BCJ, Alba JM, Escobar-Bravo R, Schuurink RC, Kant MR (2012) Plant glandular trichomes as targets for breeding or engineering of resistance to herbivores. Int J Mol Sci 13:17077–17103

Goff SA, Ricke D, Lan TH, Presting G, Wang RL, Dunn M, Glazebrook J, Sessions A, Oeller P, Varma H et al (2002) A draft sequence of the rice genome (*Oryza sativa* L. ssp. *japonica*). Science 296:92–100

Gosney BJ, O Reilly-Wapstra JM, Forster LG, Barbour RC, Iason GR, Potts BM (2014) Genetic and ontogenetic variation in an endangered tree structures dependent arthropod and fungal communities. PLoS One 9:e114132

Gosney B, O'Reilly-Wapstra J, Forster L, Whiteley C, Potts B (2017) The extended community-level effects of genetic variation in foliar wax chemistry in the forest tree *Eucalyptus globus*. J Chem Ecol 43:532–542

Hahn PG, Maron JL (2016) A framework for predicting intraspecific variation in plant defense. Trends Ecol Evol 31:646–656

Hall D, Tegström C, Ingvarsson PK (2010) Using association mapping to dissect the genetic basis of complex traits in plants. Briefings Funct Genomics 9:157–165

Handley R, Ekbom B, Agren J (2005) Variation in trichome density and resistance against a specialist insect herbivore in natural populations of *Arabidopsis thaliana*. Ecol Entomol 30:284–292

Hanley ME, Lamont BB, Fairbanks MM, Rafferty CM (2007) Plant structural traits and their role in anti-herbivore defence. Perspect Plant Ecol Evol Syst 8:157–158

Harborne JB, Baxter H, Moss GP (1999) Phytochemical dictionary: a handbook of bioactive compounds from plants, 2nd edn. CRC Press, Boca Raton

Harvell CD (1990) The ecology and evolution of inducible defenses. Q Rev Biol 65:323–340

Harvell CD, Tollrian R (1999) Why inducible defenses? In: Tollrian R, Harvell CD (eds) The ecology and evolution of inducible defenses. Princeton University Press, Princeton, pp 3–9

Hauser M-T (2014) Molecular basis of natural variation and environmental control of trichome patterning. Front Plant Sci 5:320

Hauser M-T, Aufsatz W, Jonak C, Luschnig C (2011) Transgenerational epigenetic inheritance in plants. Biochimica et Biophysica Acta Gene Regul Mech 1809:459–468

Havill NP, Raffa KF (1999) Effects of elicitation treatment and genotypic variation on induced resistance in *Populus*: impacts on gypsy moth (Lepidoptera: Lymantriidae) development and feeding behavior. Oecologia 120:295–303

He J, Bouwmeester HJ, Dicke M, Kappers IF (2020) Transcriptional and metabolite analysis reveal a shift in direct and indirect defenses in response to spider-mite infestation in cucumber (*Cucumis sativus*). Plant Mol Biol 103:489–505

Heather JM, Chain B (2016) The sequence of sequencers: the history of sequencing DNA. Genomics 107:1–8

Heidel-Fischer HM, Musser RO, Vogel H (2014) Plant transcriptome responses to herbivory. In: Voelckel C, Jander G (eds) Insect-plant interactions. Wiley Blackwell, Chichester, pp 155–196

Heil M (2008) Indirect defense via tritrophic interactions. New Phytol 178:41–61

Heil M, Koch T, Hilpert A, Fiala B, Boland W, Linsenmair KE (2001) Extrafloral nectar production of the ant-associated plant, *Macaranga tanarius*, is an induced, indirect, defensive response elicited by jasmonic acid. Proc Natl Acad Sci U S A 98:1083–1088

Hemming JDC, Lindroth RL (1995) Intraspecific variation in aspen phytochemistry: effects on performance of gypsy moths and forest tent caterpillars. Oecologia 103:79–88

Herms DA, Mattson WJ (1992) The dilemma of plants: to grow or defend. Q Rev Biol 67:283–335

Hilker M, Meiners T (2002) Induction of plant responses towards oviposition and feeding of herbivorous arthropods: a comparison. Entomologia Experimentalis et Applicata 104:181–192

Hilker M, Meiners T (2006) Early herbivore alert: insect eggs induce plant defense. J Chem Ecol 32:1379–1397

Hilker M, Meiners T (2010) How do plants "notice" attack by herbivorous arthropods? Biol Rev 85:267–280

Hilker M, Kobs C, Varama M, Schrank K (2002) Insect egg deposition induces *Pinus sylvestris* to attract egg parasitoids. J Exp Biol 205:455–461

Holeski LM (2007) Within and between generation phenotypic plasticity in trichome density of *Mimulus guttatus*. J Evol Biol 20:2092–2100

Holeski LM, Chase-Alone R, Kelly JK (2010) The genetics of phenotypic plasticity in plant defense: Trichome production in *Mimulus guttatus*. Am Nat 175:391–400

Holeski LM, Hillstrom ML, Whitham TG, Lindroth RL (2012a) Relative importance of genetic, ontogenetic, induction, and seasonal variation in producing a multivariate defense phenotype in a foundation tree species. Oecologia 170:695–707

Holeski LM, Jander G, Agrawal AA (2012b) Transgenerational defense induction and epigenetic inheritance in plants. Trends Ecol Evol 27:618–626

Holeski LM, Zinkgraf MS, Whitham TG, Lindroth RL (2013) Maternal herbivory reduces offspring allocation to resistance traits, but not growth, in cottonwood. J Ecol 101:1062–1073

Hubby JL, Lewontin RC (1966) A molecular approach to the study of genetic heterozygosity in natural populations. I. The number of alleles at different loci in *Drosophila pseudoobscura*. Genetics 54:577–594

Ingvarsson PK, Street NR (2011) Association genetics of complex traits in plants. New Phytol 189:909–922

Irmisch S, McCormick AC, Boeckler GA, Schmidt A, Reichelt M, Schneider B, Block K, Schnitzler JP, Gershenzon J, Unsicker SB, Kollner TG (2013) Two herbivore-induced cytochrome P450 enzymes CYP79D6 and CYP79D7 catalyze the formation of volatile aldoximes involved in poplar defense. Plant Cell 25:4737–4754

Johnson MTJ, Agrawal AA (2005) Plant genotype and environment interact to shape a diverse arthropod community on evening primrose (*Oenothera biennis*). Ecology 86:874–885

Johnson MTJ, Agrawal AA (2007) Covariation and composition of arthropod species across plant genotypes of evening primrose, *Oenothera biennis*. Oikos 116:941–956

Kalisz S, Purugganan MD (2004) Epialleles via DNA methylation: consequences for plant evolution. Trends Ecol Evol 19:309–314

Kappers IF, Aharoni A, van Herpen TWJM, Luckerhoff LLP, Dicke M, Bouwmeester HJ (2005) Genetic engineering of terpenoid metabolism attracts bodyguards to *Arabidopsis*. Science 309:2070–2072

Karban R, Baldwin IT (1997) Induced responses to herbivory. University of Chicago Press, Chicago

Karban R, Thaler JS (1999) Plant phase change and resistance to herbivory. Ecology 80:510–517

Karban R, Agrawal AA, Thaler JS, Adler LS (1999) Induced plant responses and information content about risk of herbivory. Trends Ecol Evol 14:443–447

Kearsley MJ, Whitham TG (1989) Developmental changes in resistance to herbivory: implications for individuals and populations. Ecology 70:422–434

Keith AR, Bailey JK, Whitham TG (2010) A genetic basis to community repeatability and stability. Ecology 91:3398–3406

Kessler A, Baldwin IT (2001) Defensive function of herbivore-induced plant volatile emissions in nature. Science 291:2141–2144

Kessler A, Baldwin IT (2002) Plant responses to insect herbivory: the emerging molecular analysis. Annu Rev Plant Biol 53:299–328

Kessler A, Halitschke R, Baldwin IT (2004) Silencing the jasmonate cascade: Induced plant defenses and insect populations. Science 305:665–668

Kliebenstein DJ (2014) Quantitative genetics and genomics of plant resistance to insects. In: Voelckel C, Jander G (eds) Insect-plant interactions. Wiley Blackwell, Chichester, pp 235–262

Kliebenstein DJ, Gershenzon J, Mitchell-Olds T (2001a) Comparative quantitative trait loci mapping of aliphatic, indolic, and benzylic glucosinolate production in *Arabidopsis thaliana* leaves and seeds. Genetics 159:359–370

Kliebenstein DJ, Kroymann J, Brown P, Figuth A, Pedersen D, Gershenzon J, Mitchell-Olds T (2001b) Genetic control of natural variation in *Arabidopsis* glucosinolate accumulation. Plant Physiol 126:811–825

Kliebenstein DJ, Pedersen D, Barker B, Mitchell-Olds T (2002a) Comparative analysis of quantitative trait loci controlling glucosinolates, myrosinase and insect resistance in *Arabidopsis thaliana*. Genetics 161:325–332

Kliebenstein DJ, Figuth A, Mitchell-Olds T (2002b) Genetic architecture of plastic methyl jasmonate responses in *Arabidopsis thaliana*. Genetics 161:1685–1696

Kloth KJ, Thoen MPM, Bouwmeester HJ, Jongsma MA, Dicke M (2012) Association mapping of plant resistance to insects. Trends Plant Sci 17:311–319

Komatsu K, Okuda S, Takahashi M, Matsunaga R, Nakazawa Y (2005) QTL mapping of antibiosis resistance to common cutworm (*Spodoptera litura* Fabricius) in soybean. Crop Sci 45:2044–2048

Kooyers NJ, Donofrio A, Blackman BK, Holeski LM (2020) The genetic architecture of plant defense tradeoffs in a common monkeyflower. J Hered. https://doi.org/10.1093/jhered/esaa015

Koricheva J (2002) Meta-analysis of sources of variation in fitness costs of plant antiherbivore defenses. Ecology 83:176–190

Koricheva J, Barton KE (2012) Temporal changes in plant secondary metabolite production. In: Iason GR, Dicke M, Hartley SE (eds) The ecology of plant secondary metabolites: from genes to global processes. Ecological reviews. Cambridge University Press, Cambridge

Labandeira CC (2013) A paleobiologic perspective on plant–insect interactions. Curr Opin Plant Biol 16:414–421

Lande R (1979) Quantitative genetic analysis of multivariate evolution, applied to brain: body size allometry. Evolution 33:402–416

Lee SH, Yang J, Goddard ME, Visscher PM, Wray NR (2012) Estimation of pleiotropy between complex diseases using single-nucleotide polymorphism-derived genomic relationships and restricted maximum likelihood. Bioinformatics 28:2540–2542

Levin DA (1973) The role of trichomes in plant defense. Q Rev Biol 48:3–15

Lin XY, Kaul S, Rounsley S, Shea TP, Benito MI, Town CD, Fujii CY, Mason T, Bowman CL, Barnstead M et al (1999) Sequence and analysis of chromosome 2 of the plant *Arabidopsis thaliana*. Nature 402:761–768

Lindroth RL, Hwang S-Y (1996) Clonal variation in foliar chemistry of quaking aspen (*Populus tremuloides* Michx.). Biochem Syst Ecol 24:357–364

Liscombe DK, MacLeod BP, Loukanina N, Nand OI, Facchini PJ (2005) Evidence for the monophyletic evolution of benzylisoquinoline alkaloid biosynthesis in angiosperms. Phytochemistry 66:1374–1393

López-Goldar X, Zaz R, Sampedro L (2020) Resource availability drives microevolutionary patterns of plant defences. Funct Ecol. https://doi.org/10.1111/1365-2435.13610

4

Lowry DB, Popovic D, Brennan DJ, Holeski LM (2019) Mechanisms of a locally adaptive shift in allocation among growth, reproduction, and herbivore resistance in *Mimulus guttatus*. Evolution 73:1168–1181

Lynch M, Walsh B (1998) Genetics and analysis of quantitative traits. Sinauer, Sunderland

Marquis RJ (1992) The selective impact of herbivores. In: Fritz RS, Simms EL (eds) Plant resistance to herbivores and pathogens. Univ. of Chicago Press, Chicago, pp 301–325

Mason PA, Singer MS (2015) Defensive mixology: Combining acquired chemicals toward defense. Funct Ecol 29:441–450

Mauricio R (1998) Costs of resistance to natural enemies in field populations of the annual plant *Arabidopsis thaliana*. Am Nat 151:20–28

Mauricio R (2005) Ontogenetics of QTL: the genetic architecture of trichome density over time in *Arabidopsis thaliana*. Genetica 123:75–85

Mauricio R, Rausher MD (1997) Experimental manipulation of putative selective agents provides evidence for the role of natural enemies in the evolution of plant defense. Evolution 51:1435–1444

Mayer K, Schuller C, Wambutt R, Murphy G, Volckaert G, Pohl T, Dusterhoft A, Stiekema W, Entian KD, Terryn N et al (1999) Sequence and analysis of chromosome 4 of the plant *Arabidopsis thaliana*. Nature 402:769–777

McCormick AC, Unsicker SB, Gershenzon J (2012) The specificity of herbivore-induced plant volatiles in attracting herbivore enemies. Trends Plant Sci 17:303–310

McCormick AC, Irmisch S, Boeckler GA, Gershenzon J, Köllner TG, Unsicker SB (2019) Herbivore-induced volatile emission from old-growth black poplar trees under field conditions. Sci Rep 9:77

McKay JK, Stinchcombe JR (2008) Ecological genomics of model eukaryotes. Evolution 62:2953–2957

Mitchell-Olds T (2001) *Arabidopsis thaliana* and its wild relatives: a model system for ecology and evolution. Trends Ecol Evol 16:693–700

Mithöfer A, Boland W (2012) Plant defense against herbivores: chemical aspects. Annu Rev Plant Biol 63:431–450

Mithöfer A, Wanner G, Boland W (2005) Effects of feeding *Spodoptera littoralis* on lima bean leaves. II. Continuous mechanical wounding resembling insect feeding is sufficient to elicit herbivory-related volatile emission. Plant Physiol 137:1160–1168

Moore BD, Andrew RL, Külheim C, Foley JT (2014) Explaining intraspecific diversity in secondary metabolites in an ecological context. New Phytol 201:733–750

Nihranz CT, Walker WS, Brown SJ, Mescher MC, De Moraes CM, Stephenson AG (2020) Transgenerational impacts of herbivory and inbreeding on reproductive output in Solanum caro-linense. Am J Bot 107:286–297

Nordborg M, Weigel D (2008) Next-generation genetics in plants. Nature 456:720–723

Núñez-Farfán J, Dirzo R (1994) Evolutionary ecology of *Datura stramonium* L. in central Mexico: natural selection for resistance to herbivorous insects. Evolution 48:423–436

O'Reilly-Wapstra JM, McArthur C, Potts BM (2002) Genetic variation in resistance of *Eucalyptus globulus* to marsupial browsers. Oecologia 130:289–296

O'Reilly-Wapstra JM, McArthur C, Potts BM (2004) Linking plant genotype, plant defensive chemistry and mammal browsing in a *Eucalyptus* species. Funct Ecol 18:677–684

O'Reilly-Wapstra JM, McArthur C, Potts BM, Davies NW (2005) Effects of nutrient variability on the genetic-based resistance of *Eucalyptus globulus* to a mammalian herbivore and on plant defensive chemistry. Oecologia 142:597–605

O'Reilly-Wapstra JM, Humphreys JR, Potts BM (2007) Stability of genetic-based defensive chemistry across life stages in a Eucalyptus species. J Chem Ecol 33:1876–1884

Olsen J, Gill GS, Haugen R, Matzner SL, Alsdurf J, Siemens DH (2019) Evolutionary constraint on low elevation range expansion: defense-abiotic stress-tolerance trade-offs in crosses of the ecological model *Boechera stricta*. Ecol Evol 9:11532–11544

Orr HA, Coyne JA (1992) The genetics of adaptation: a reassessment. Am Nat 140:725

Osier TL, Lindroth RL (2001) Effects of genotype, nutrient availability, and defoliation on aspen phytochemistry and insect performance. J Chem Ecol 27:1289–1313

Osier TL, Lindroth RL (2004) Long-term effects of defoliation on quaking aspen in relation to geno-type and nutrient availability: plant growth, phytochemistry and insect performance. Oecologia 139:55–65

Osier TL, Lindroth RL (2006) Genotype and environment determine allocation to and costs of resistance in quaking aspen. Oecologia 148:293–303

Pearse IS, LoPresti E, Schaeffer RN, Mooney KA, Ali JG, Ode PJ, Eubanks MD, Bronstein JL, Weber MG (2020) Generalising indirect defense and resistance of plants. Ecol Lett 23:1137–1152

Pichersky E, Lewinsohn E (2011) Convergent evolution in plant specialized metabolism. Annu Rev Plant Biol 62:549–566

Poelman EH, van Loon JJA, Dicke M (2008) Consequences of variation in plant defense for biodiversity at higher trophic levels. Trends Plant Sci 13:534–541

Poethig RS (1997) Leaf morphogenesis in flowering plants. Plant Cell 9:1077–1087

Pollard AJ, Briggs D (1984) Genecological studies of *Urtica dioica* L. III. Stinging hairs and plant-herbivore interactions. New Phytol 97:507–522

Price PW, Roinenen H, Tahvanainen J (1987) Plant age and attack by the bud galler, *Euura mucronata*. Oecologia 73:334–337

Rapp RA, Wendel JF (2005) Epigenetics and plant evolution. New Phytol 168:81–91

Rasmann S, Agrawal AA (2009) Plant defense against herbivory: progress in identifying synergism, redundancy, and antagonism between resistance traits. Curr Opin Biol 12:473–478

Ratzka A, Vogel H, Kliebenstein DJ, Mitchell-Olds T, Kroymann J (2002) Disarming the mustard oil bomb. Proc Natl Acad Sci U S A 99:11223–11228

Rausher MD (1996) Genetic analysis of coevolution between plants and their natural enemies. Trends Genet 12:212–217

Rausher MD, Simms EL (1989) The evolution of resistance to herbivory in *Ipomoea purpurea*. I. Attempts to detect selection. Evolution 43:563–572

Rector BG, All JN, Parrott WA, Boerma HR (2000) Quantitative trait loci for antibiosis resistance to corn earworm in soybean. Crop Sci 40:233–238

Rehill B, Whitham T, Martinsen G, Schweitzer J, Bailey J, Lindroth R (2006) Developmental Trajectories in Cottonwood Phytochemistry. J Chem Ecol 32:2269–2285

Reymond P, Weber H, Damond M, Farmer EE (2000) Differential gene expression in response to mechanical wounding and insect feeding in *Arabidopsis*. Plant Cell 12:707–719

Reymond P, Bodenhausen N, Van Poecke RMP, Krishnamurthy V, Dicke M, Farmer EE (2004) A conserved transcript pattern in response to a specialist and a generalist herbivore. Plant Cell 16:3132–3147

Rhoades DF (1979) Evolution of plant chemical defense against herbivores. In: Rosenthal GA, Janzen DH (eds) Herbivores: their interaction with secondary plant metabolites. Academic Press, New York, pp 3–54

Richards EJ (2006) Inherited epigenetic variation: revisiting soft inheritance. Nat Rev Genet 7:395–401

Roach DA, Wulff RD (1987) Maternal effects in plants. Annu Rev Ecol Syst 18:209–235

Robinson KM, Ingvarsson PK, Jansson S, Albrectsen BR (2012) Genetic variation in functional traits influences arthropod community composition in aspen (*Populus tremula* L.). PLoS One 7:e37679

Rockman MV (2012) The QTN program and the alleles that matter for evolution: all that's gold does not glitter. Evolution 66:1–17

Roff DA (1996) The evolution of genetic correlations: an analysis of patterns. Evolution 50:1392–1403

Rossi M, Goggin F, Milligan S, Kaloshian I, Ullman D, Williams V (1998) The nematode resistance gene Mi of tomato confers resistance against the potato aphid. Proc Natl Acad Sci U S A 95:9750–9754

Rotter MC, Couture JJ, Rothwell EM, Garcia J, Holeski LM (2018) Evolutionary ecology of plant resistance traits across the herbivore diet spectrum: a test in the model plant *Mimulus guttatus*. Evol Ecol Res 19:423–440

Rudgers JA (2004) Enemies of herbivores can shape plant traits: selection in a facultative ant-plant mutualism. Ecology 85:192–205

Rudgers JA, Strauss SY (2004) A selection mosaic in the facultative mutualism between ants and wild cotton. Proc R Soc B Biol Sci 271:2481–2488

Rutter MT, Rausher MD (2004) Natural selection on extrafloral nectar production in *Chamaecrista fasciculata*: the costs and benefits of a mutualism trait. Evolution 58:2657–2668

Sagers CL, Coley PD (1995) Benefits and costs of defense in a neotropical shrub. Ecology 76:1835–1843

Salas-Solano O, Carrilho E, Kotler L, Miller AW, Goetzinger W, Sosic Z, Karger BL (1998) Routine DNA sequencing of 1000 bases in less than one hour by capillary electrophoresis with replaceable linear polyacrylamide solutions. Anal Chem 70:3996–4003

Sanger F, Air GM, Barrell BG, Brown NL, Coulson AR, Fiddes JC, Hutchinson CA, Slocombe PM, Smith M (1977a) Nucleotide sequence of bacteriophage phi X174 DNA. Nature 265:687–695

Sanger F, Nicklen S, Coulson AR (1977b) DNA sequencing with chain-terminating inhibitors. Proc Natl Acad Sci U S A 74:5463–5467

Sato Y, Shimizu-Inatsugi R, Yamazaki M, Shimizu KK, Nagano AJ (2019a) Plant trichomes and a single gene *GLABRA1* contribute to insect community composition on field-grown *Arabidopsis thaliana*. BMC Plant Biol 19:163

Sato Y, Tezuka A, Kashima M, Deguchi A, Shimizu-Inatsugi R, Yamazaki M, Shimizu KK, Nagano AJ (2019b) Transcriptional variation in glucosinolate biosynthetic genes and inducible responses to aphid herbivory on field grown *Arabidopsis thaliana*. Front Genet 10:787

Schmidt DD, Kessler A, Kessler D, Schmidt S, Lim M, Gase K, Baldwin IT (2004) *Solanum nigrum*: a model ecological expression system and its tools. Mol Ecol 13:981–995

Schranz ME, Manzaneda AJ, Windsor AJ, Clauss MJ, Mitchell-Olds T (2009) Ecological genomics of *Boechera stricta*: identification of a QTL controlling the allocation of methionine- vs branched-chain amino acid derived glucosinolates and levels of insect herbivory. Heredity 102:465–474

Schuman MC, Barthel K, Baldwin IT (2012) Herbivory-induced volatiles function as defenses increasing fitness of the native plant *Nicotiana attenuata* in nature. elife 1:e00007

Schuman M, Allmann S, Baldwin IT (2015) Plant defense phenotypes determine the consequences of volatile emission for individuals and neighbors. elife 4:e04490

Service P (1984) Genotypic interaction in an aphid host plant relationship, *Uroleucon rudbeckiae* and *Rudbeckia laciniata*. Oecologia 61:271–276

Shendure J, Porreca GJ, Reppas NB, Lin XX, McCutcheon JP, Rosenbaum AM, Wang MD, Zhang K, Mitra RD, Church GM (2005) Accurate multiplex polony sequencing of an evolved bacterial genome. Science 309:1728–1732

Shendure J, Balasubramanian S, Church GM, Gilbert W, Rogers J, Schloss JA, Waterston RH (2017) DNA sequencing at 40: past, present, and future. Nature 550:345–353

Shonle I, Bergelson J (2000) Evolutionary ecology of the tropane alkaloids of *Datura stramonium* L. (Solanaceae). Evolution 54:778–788

Siemens DH, Haugen R, Matzner S, VanAsma N (2009) Plant chemical defense allocation constrains evolution of local range. Mol Ecol 18:4974–4983

Simms EL, Rausher MD (1987) Costs and benefits of plant resistance to herbivory. Am Nat 130:570–581

Simms EL, Rausher MD (1989) The evolution of resistance to herbivory in *Ipomoea purpurea*. II. Natural selection by insects and cost of resistance. Evolution 43:573–585

Sletvold N, Huttunen P, Handley R, Kärkkäinen K, Ägren A (2010) Cost of trichome production and resistance to a specialist insect herbivore in *Arabidobsis lyrata*. Evol Ecol 24:1307–1319

Sodini SM, Kemper KE, Wray NR, Trzaskowski (2018) Comparison of genotypic and phenotypic correlations: Cheverud's conjecture in humans. Genetics 209:941–948

Stamp N (2003) Out of the quagmire of plant defense hypotheses. Q Rev Biol 78:23–55

Stinchcombe JR, Hoekstra HE (2008) Combining population genomics and quantitative genetics: finding the genes underlying ecologically important traits. Heredity 100:158–170

Stowe K (1998) Experimental evolution of resistance in *Brassica rapa*: correlated response of tolerance in lines selected for glucosinolate content. Evolution 52:703–712

Strauss SY, Agrawal AA (1999) The ecology and evolution of plant tolerance to herbivory. Trends Ecol Evol 14:179–185

Thompson GA, Goggin FL (2006) Transcriptomics and functional genomics of plant defence induction by phloem-feeding insects. J Exp Bot 57:755–766

Turlings TCJ, Tumlinson JH, Lewis WJ (1990) Exploitation of herbivore-induced plant odors by host-seeking parasitic wasps. Science 250:1251–1253

Underwood N, Rausher MD (2000) The effects of host-plant genotype on herbivore population dynamics. Ecology 81:1565–1576

Ungerer MC, Johnson LC, Herman MA (2008) Ecological genomics: Understanding gene and genome function in the natural environment. Heredity 100:178–183

Van Poecke RMP, Posthumus MA, Dicke M (2001) Herbivore-induced volatile production by *Arabidopsis thaliana* leads to attraction of the parasitoid *Cotesia rubecula*: chemical, behavioral, and gene-expression analysis. J Chem Ecol 27:1911–1928

Vargas-Ortiz E, Gonda I, Smeda JR, Mutschler MA, Giovannoni JJ, Jander G (2018) Genetic mapping identifies loci that influence tomato resistance against Colorado potato beetles. Sci Rep 8:7429

Verhoeven KJF, Jansen JJ, van Dijk PJ, Biere A (2009) Stress-induced DNA methylation changes and their heritability in asexual dandelions. New Phytol 185:1108–1118

Via S, Hawthorne D (2002) The genetic architecture of ecological specialization: Correlated gene effects on host use and habitat choice in pea aphids. Am Nat 159:S76–S88

Via S, Hawthorne D (2005) Back to the future: genetic correlations, adaptation, and speciation. Genetica 123:147–156

Voelckel C, Baldwin IT (2004) Generalist and specialist lepidopteran larvae elicit different transcriptional responses in *Nicotiana attenuata*, which correlate with larval FAC profiles. Ecol Lett 7:770–775

Watson J, Crick F (1953) Molecular structure of nucleic acids. Nature 171:709–756

Weber MG, Agrawal AA (2014) Defense mutualisms enhance plant diversification. Proc Natl Acad Sci U S A 111:16442–16447

Weber MG, Porturas LD, Taylor SA (2016) Foliar nectar enhances plant-mite mutualisms: The effect of leaf sugar on the control of powdery mildew by domatia-inhabiting mites. Ann Bot. https://doi.org/10.1093/aob/mcw118

Willis JH, Coyne JA, Kirkpatrick M (1991) Can one predict the evolution of quantitative characters without genetics? Evolution 45:441–444

Wimp GM, Wooley S, Bangert RK, Young WP, Martinsen GD, Keim P, Rehill B, Lindroth RL, Whitham TG (2007) Plant genetics predicts intra-annual variation in phytochemistry and arthropod community structure. Mol Ecol 16:5057–5069

Wink M (1988) Plant breeding- Importance of plant secondary metabolites for protection against pathogens and herbivores. Theor Appl Genet 75:225–233

Wink M (2003) Evolution of secondary metabolites from an ecological and molecular phylogenetic perspective. Phytochemistry 64:3–19

Wink M (2018) Plant secondary metabolites modulate insect behavior- steps toward addiction? Front Physiol 9:364

Xu S, Brockmöller T, Navarro-Quezada A, Kuhl H, Gase H, Ling Z, Zhou W, Kreitzer C, Stanke M, Tang H, Lyons E, Pandey P, Pandey SP, Timmermann B, Gaquerel E, Baldwin IT (2018) Wild tobacco genomes reveal the evolution of nicotine biosynthesis. Proc Natl Acad Sci U S A 114:6133–6138

Xu S, Kreitzer C, McGale E, Lackus ND, Guo H, Köllner TG, Schuman MC, Baldwin IT, Zhou W (2020) Allelic differences of clustered terpene synthases contribute to correlated intraspecific variation of floral and herbivory-induced volatiles in a wild tobacco. New Phytol. https://doi.org/10.1111/nph.16739

Yesudas CR, Sharma H, Lightfoot DA (2010) Identification of QTL in soybean underlying resistance to herbivory by Japanese beetles (*Popillia japonica*, Newman). Theor Appl Genet 121:353–362

Yu J, Hu S, Wang J, Wong GKS, Li SG, Liu B, Deng YJ, Dai L, Zhou Y, Zhang XQ et al (2002) A draft sequence of the rice genome (*Oryza sativa* L. ssp. *indica*). Science 296:79–92

Zhang PJ, Zhao C, Ye ZH, Yu XP (2020) Trade-off between defense priming by herbivore-induced plant volatiles and constitutive defense in tomato. Pest Manag Sci 76:1893–1901

Zytynska SE, Jourdie V, Naseeb S, Delneri D, Preziosi RF (2016) Induced expression of defense-related genes in barley is specific to aphid genotype. Biol J Linn Soc 117:672–685

Further Reading/Additional Resources

Anderson and Mitchell-Olds (2011) review ecological genomics and plant-herbivore interactions

A recent book chapter by Mijail De-la-Cruz, Sabina Velázquez-Márquez, and Juan Núñez-Farfán (2020) provides a complementary review of genomics work in plant-herbivore interactions

A book chapter by Heidel-Fischer et al. (2014) reviews plant transcriptomic responses to herbivory

A book chapter by Kliebenstein (2014) reviews quantitative genetic studies of plant resistance

Pearse et al. (2020) review the state of the field in the evolutionary ecology of indirect defenses

Biotic Defenses Against Herbivory

*Renan F. Moura, Eva Colberg,
Estevão Alves-Silva, Isamara Mendes-Silva,
Roberth Fagundes, Vanessa Stefani,
and Kleber Del-Claro*

Contents

© Springer Nature Switzerland AG 2021
K. Del-Claro, H. M. Torezan-Silingardi (eds.), *Plant-Animal
Interactions*, https://doi.org/10.1007/978-3-030-66877-8_5

🎯 Learning Objectives

After completing this chapter, you should have an understanding of the following:
1. The terms biotic defense and indirect defense.
2. Several examples of biotic defense relationships between different organisms.
3. The potential benefits and costs to each organism involved in a biotic defense relationship.
4. The variation in generality and specificity of biotic defense partnerships.

5.1 Introduction

Herbivory is a strong negative pressure known to shape the diversity and distribution of numerous plant species (Marquis and Braker 1994). As an evolutionary response to the negative effects of herbivory, plants have developed a myriad of defensive strategies that can be further categorized as direct or indirect defenses (Price et al. 1980; Dicke and Sabelis 1988). Direct defenses are traits that reduce herbivory by acting directly upon herbivores without any additional mediators; this includes the production and presence of chemical compounds that function as repellents and toxins, and physical structures that harm or deter herbivores, such as thorns and spines. These defenses are mostly constitutive and are continuously expressed in plants across ontogeny (Boege and Marquis 2005). The other category, indirect defenses, are traits that do not directly affect herbivores, but rather enhance plant fitness by altering the behavior or presence of natural enemies of herbivores (Pearse et al. 2020), and are considered inducible defenses (Zangerl and Rutledge 1996). Although indirect defenses can still involve plant chemistry (in the form of volatile attractants or nutrient-rich extrafloral nectar and food bodies), their proximal mechanism is the predators they attract, which act as biotic agents of defense, or "biotic defenses". While many studies address the impacts of direct defenses in plants, biotic defenses are still overlooked, especially if the definition of biotic defense is expanded to include interactions beyond classic plant-herbivore-predator relationships (e.g., Heil 2014).

5.1.1 A Classic Example of Biotic Defense: Myrmecophily and Extrafloral Nectar

Perhaps the most classic example of biotic defense involves the interaction between plants bearing extrafloral nectaries (EFNs) and ants, spiders, and other predators (◻ Figs. 5.1 and 5.2). EFNs are nectar-secreting structures not associated with the pollination of mature flowers, but rather are found on other above ground plant parts, such as leaves, stems, stipules, and flower buds (e.g., Machado et al. 2008; Schoereder et al. 2010; Marazzi et al. 2013). EFNs produce liquid comprising carbohydrate-rich compounds but also small amounts of other organic compounds such as amino acids and lipids (González-Teuber and Heil 2009). Plants with EFNs are characterized as myrmecophilous ("ant-loving") organisms due to their ability

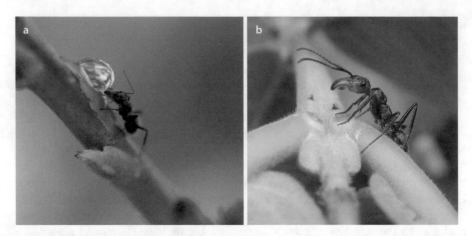

◘ Fig. 5.1 Myrmecophilous ants feeding on extrafloral nectaries. **a** *Camponotus crassus* ant using nectar from an extrafloral nectary (EFN) of a bract on *Ouratea spectabilis* (Ochnaceae); **b** *Ectatomma tuberculatum* with an extrafloral nectar droplet in its mandibles from a *Qualea multiflora* (Vochysiaceae) petiolar EFN (D. Lange picture)

to attract and interact with ants. Numerous studies have demonstrated that EFNs may specifically attract aggressive omnivorous ants (Koptur 1994; Blüthgen et al. 2004; Bixenmann et al. 2011), which can then increase plant performance by preying upon and decreasing the damage caused by herbivores (Cuautle et al. 2005; Baker-Méio and Marquis 2012). A meta-analysis conducted by Rosumek et al. (2009) concluded that plant fitness decreased by almost 60% when ants were excluded from plants. Furthermore, Trager et al. (2010) demonstrated that the presence of ants enhanced plant reproductive output by 49% and decreased herbivory by 62%. From this perspective, recent studies strongly suggest that extrafloral nectaries may act as inducible defenses decreasing costs and maximizing benefits against herbivores (Calixto et al. 2020).

Relationships between plants and ants are widespread and diverse (Rico-Gray and Oliveira 2007; Del-Claro et al. 2016; Calixto et al. 2018), with EFNs occurring in at least 100 plant families around the world (Weber and Keeler 2013). Consequently, these relationships are usually generalized and highly facultative (Rosumek et al. 2009; Chamberlain and Holland 2009; Del-Claro et al. 2018; Calixto et al. 2018). Nectaries of different shapes and locations, secreting nectar at different times, composed of different micronutrients, evolved in parallel in several plant species, including Pteridophytes, Gymnosperms, and Angiosperms (Marazzi et al. 2013; Weber and Keeler 2013). This wide diversity of EFNs is matched with an equally diverse amount of interactions with various ant species (Buckley 1982; Rico-Gray and Oliveira 2007; Dáttilo et al. 2014; Del-Claro et al. 2018; Nelsen et al. 2018). As more studies are published, more species are included in the list of plant-protecting ants (Rico-Gray and Oliveira 2007; Trager et al. 2010; Calixto et al. 2018). Additionally, in tropical environments such as the Brazilian Cerrado, a single plant species can provide nectar for dozens of ants

⬛ Fig. 5.2 Spiders on plants. **a** *Zuniga magna* (Salticidae) is a spider that mimics ants to easily access extrafloral nectaries and feed on them unnoticed; **b** *Peucetia flava* (Oxyopidae) preying upon a bug nymph (Hemiptera) that was trapped on the glandular trichomes of a *Chamaecrista* shrub; **c** a crab spider (Thomisidae) preying upon a bee (*Trigona* sp.) pollinator in an EFN-bearing Legumino-sae; **d** *Tmarus* sp. (Thomisidae) spider preying upon a *Pseudomyrmex* ant on an EFN-bearing plant

(Lange and Del-Claro 2014; Costa et al. 2016; Sendoya et al. 2016; Fagundes et al. 2018) and spider species (Nahas et al. 2017).

Although EFN-mediated defensive interactions are widespread in land environments, the mechanisms and effects of biotic defense can easily become complicated and conditional on different biotic and abiotic factors (Del-Claro et al. 2016, 2018). For one, EFNs may not be the only resource incentivizing ants to spend time on plants. Shelter in the form of domatia (Letourneau 1983; Fiala and Maschwitz 1991) and other food sources such as food bodies and fruit (Fiala and Maschwitz 1992; Dutra et al. 2006) can also serve as rewards to ants that then patrol and protect the plant from herbivores. For example, although some *Macaranga* plants possess many EFNs that incentivize ant protectors, myrmecophytic (ant-inhabited)

Macaranga plants with hollow stems and food bodies have reduced or even absent EFNs (Fiala and Maschwitz 1991). In other shelter-based ant-plant systems, hollow stems or petioles provide shelter and nesting space for ant defenders, and the plants might provide food bodies instead of EFNs as an additional resource for the ants (Yu and Davidson 1997).

As Pearse et al. (2020) have argued, indirect defense can apply to relationships beyond the plant-herbivore-predator canon. Indeed, ant-plant relationships do not always involve trophic defense. Ants can also protect their host plants from competitors, removing vines (Schupp 1986; Fiala et al. 1989) and other nearby or encroaching vegetation (Frederickson et al. 2005) that might otherwise pose a competitive threat. However, biotic defenses do not always yield complete or even net benefits to the partner being defended. In many cases, the defender species can determine the effectiveness of defense (Young et al. 1996; Fagundes et al. 2017). In other cases, defenders might also deter other plant mutualists, such as pollinators (Assunção et al. 2014), and species-specific combinations of defense efficacy and pollinator deterrence can lead to variability in the net costs and benefits of biotic defense (Ohm and Miller 2014).

In the rest of this chapter, we will call attention to the diversity of biotic defenses beyond the classic ant-plant mutualism mediated by extrafloral nectaries, a diversity in large part due to the lack of specificity in the mechanisms that induce biotic defense. Ants are not the only predators attracted to EFNs and other plant rewards, but rather other insects and even spiders can also use these resources and then act as biotic defenders (◘ Fig. 5.2). Investigating these non-ant defenders can lead to a broader understanding of the partner generality of biotic defense; we will use spiders as an example of this. Additionally, plants are not the only organisms that secrete sugary solutions in return for ant defense—we will use myrmecophilous hemipterans and lepidopterans as examples of how an animal's biotic defense can compete for, directly counteract, or otherwise interact with a plant's biotic defense. Throughout these examples, we will point out commonalities in how these defense interactions are selected for and maintained. Considering these diverse types of biotic defense relationships can lead to a deeper understanding of the various costs and benefits of biotic defense, and ultimately help us predict and understand the evolutionary trajectories of these relationships and their contributions to overall biodiversity.

5.2 Spiders as Plant Bodyguards and the Role of Extrafloral Nectaries

Although the interactions between EFN-bearing plants and ants are relatively well-studied, interactions involving other species of nectar-consuming animals (Heil 2015) such as wasps (Cuautle and Rico-Gray 2003), beetles (Agarwal and Rastogi 2010), flies (Agarwal and Rastogi 2010), bees (Thorp and Sugden 1990), neuropterans (Limburg and Rosenheim 2001), and spiders (Taylor and Foster 1996; Nahas et al. 2017) are underexplored. The latter is particularly surprising, considering that spiders are among the most common arthropods inhabiting vegetation (Wise 1993; Foelix 2011) and have been observed feeding upon EFNs (Ruhren and Handel 1999; Cross and Jackson 2009; Nahas et al. 2017, ◙ Fig. 5.2a). Spiders inhabiting vegetation have adaptations that facilitate their relationship with host plants. These are generally sensory adaptations, which allow spiders to find and discern specific plant species through visual, olfactory, and tactile stimuli (Vasconcellos-Neto et al. 2017). Further evidence that spiders regularly consume EFN can be demonstrated by using cold-anthrone tests for the presence and concentration of fructose in the bodies of spiders found on EFN-bearing plants (Nyffeler et al. 2016). In the Brazilian Cerrado, such methods have provided evidence that 39 species across seven spider families consume EFN from at least eight different host plant species (Nahas et al. 2017). Thus, EFN-feeding behavior is common in spiders, and sugar-rich solutions like extrafloral nectar can be a beneficial energy source for spiders (e.g., Taylor and Bradley 2009; Nahas et al. 2012; Patt et al. 2012), supplementing the animal-based aspects of their diet.

Similar to ants, spiders are generalist predators that can reduce the number of herbivores on a plant while also feeding upon nectar from EFNs (Stefani et al. 2015; Del-Claro et al. 2017). This can work in the plant's favor when such predation is enough to reduce foliar herbivory and increase productivity (Nahas et al. 2012; Nelson and Jackson 2013; Stefani et al. 2015). In some cases, plants receive even more defense from herbivores, so the fitness increases when spiders co-occur with ants (Nahas et al. 2012), although spiders and ants can also compete for the same resources (Halaj et al. 1997, ◙ Fig. 5.2d). For instance, Stefani et al. (2015) showed that visitation by both spiders and ants had strong synergistic effects on seed production of *Eriotheca gracilipes*, an EFN-bearing plant. Plants visited by both spiders and ants produced an average of 13 viable seeds, which was higher than the average seed production of plants visited by either spiders (3) or ants (5) alone. However, just as with ants, spiders can also have negative effects on plants. Spiders can reduce plant fitness by preying on or deterring pollinators (◙ Fig. 5.2c), thereby reducing pollination rates and consequently the average flower fertility rate (Romero and Koricheva 2011).

However, EFNs are not the only plant resource available to spiders. Spiders can also eat other plant-derived food, including floral nectar, food bodies, and pollen (Nyffeler et al. 2016). Additionally, plants can provide spiders with resources such as refuge, favorable microclimatic conditions, anchorage points for webs, and places for nuptial encounters, oviposition, and hunting (Greenstone 1984; Uetz 1991; Dennis et al. 1998; Silva et al. 2020).

Spiders can also serve as biotic defenders in plants that have glandular trichomes on the surfaces of their leaves and stems (◙ Fig. 5.2b). These structures produce sticky substances that trap and kill small insects, a direct defense against herbivores and pathogens (Duffey 1986; Vasconcellos-Neto et al. 2017). However, these trichomes can also act as indirect defenses, as some families of spiders (e.g., Thomisidae and Oxyopidae) are often found foraging on insects trapped by glandular trichomes, and can positively contribute to plant defenses in the process (Romero et al. 2008; Krimmel and Pearse 2013). This protection was verified by

Silva et al. (2020), who demonstrated that spiders reduce the abundance of herbivores on plants with glandular trichomes (*Chamaecrista neesiana*, Fabaceae), resulting in lower herbivory and increased seed production.

5.2.1 Future Directions in the Study of Spiders as Biotic Defenders

Although studies on the interactions between spiders and plants have increased in the last 15 years, research is still incipient, and anecdotal studies show both positive and negative effects of spiders on plants. A major step is to identify all nectar-consuming spiders and how they interact with herbivores and ants (Del-Claro et al. 2017). If spiders, specifically nectar-feeding species, play a positive role in defending the plant against herbivory in conjunction with ants (Nahas et al. 2012), then the occasional negative interactions between ants and spiders should not be enough to reduce the benefits of spiders, ants, and the host plant. However, spiders are more aggressive and attack more herbivores when they are consuming nectar (Stefani et al. 2015; Del-Claro et al. 2017), therefore the influence of the host plant rewards on spider behavior and predation needs to be more thoroughly studied. Additionally, spiders have strong chemical senses, which may help them detect herbivore attacks on plants through plant release of volatile organic compounds, but the extent to which chemicals mediate spider behavior as a biotic defense remains to be explored.

5.3 Ant-Hemipteran Associations: Animal-Animal Biotic Defense and Plant Impacts

Aphids and membracids (Hemiptera) are important trophobiont (organisms that provide food rewards) herbivores that employ ants as biotic defenses. Even though the terms biotic and indirect defenses are usually applied to plants, trophobiont hemipterans can also recruit ants as bodyguards by releasing a sugar-rich solution (honeydew) (Way 1963; Stadler and Dixon 1999; Shingleton et al. 2005; Styrsky and Eubanks 2007). The consequences of this interaction can extend up beyond the local trophic chain, not only influencing the fitness of the insects involved, but also that of the aphid host plants and their associated herbivores (Styrsky and Eubanks 2007; Yao 2014).

Honeydew is a by-product of phloem ingestion by hemipterans, composed of a series of sugars (and also organic acids, amino acids, lipids, and proteins), and is a key mediator of the reciprocal interaction between ants and aphids (Völkl et al. 1999; Leroy et al. 2011a; Sabri et al. 2013). Ants will tend honeydew-producing hemipterans, reducing the abundance of hemipterans predators on the plant and increasing hemipterans fitness (Flatt and Weisser 2000; Renault et al. 2005, Vilela and Del-Claro 2018, ◘ Fig. 5.3a). Ants may also benefit aphids by protecting their eggs from fungal infections without receiving any apparent benefit (Matsuura and

☐ **Fig. 5.3** Ants tending lycaenid caterpillars and a membracid. **a** *Camponotus* ant visiting extrafloral nectaries of a Malpighiaceae and simultaneously tending a trophobiont hemipteran (*Enchenopa* sp., Membracidae); **b** *Camponotus* ant tending a larval Lycaenidae (*Rekoa marius*; A. Bächtold picture); **c** *Ectatomma tuberculatum* approaching a lycaenid caterpillar on a flower bud (E. Alves-Silva picture); **d** *Camponotus* sp. tending a caterpillar (E. Calixto picture)

Yashiro 2006). Without ant bodyguards, aphids are vulnerable to predation: during feeding, aphids insert their stylets (long mouthparts) into the leaves, preventing them from any rapid escape from natural enemies. In addition, these insects are gregarious (and thus conspicuous to predators) and have few direct defenses against natural enemies (Stadler and Dixon 2005; Suzuki and Ide 2008).

Despite the benefits provided by ants, most hemipterans are not tended by ants, and for those species that are actually tended, they can also be attacked and even preyed upon by ants (Billick et al. 2007). Aside from harassing and preying upon aphids, ants can also prevent hyperparasitoids from attacking parasitized aphids (Völkl 1992). Additionally, ants can fail to detect and defend aphids from specialist predators that mimic their chemical profile and then remain unnoticed by ants (Lohman et al. 2006). Furthermore, ant-tending behavior may be related to the presence of certain sugars, so aphids that do not release or metabolize specific sugars may go unattended by ants (Fischer and Shingleton 2001). Competition for ant attention exists when more than one species of aphid inhabits the same plant, and this too directly affects aphid fitness and persistence (Addicott 1978; Cushman and Addicott 1989).

In addition to the ways in which ants can reduce aphid fitness, honeydew may also cue parasitoids to the presence of aphid colonies on host plants (Budenberg 1990). Aphid feeding also prompts the release of volatile allelochemicals (herbivore-induced plant volatiles, or HIPVs), which natural enemies use to locate aphids (Hatano et al. 2008). In other instances, parasitoids of aphids feed on honeydew, although this is not their preferred food source (Lee et al. 2004). More recently, studies have shown that bacteria from aphid guts produce volatiles that effectively attract natural enemies as well (Leroy et al. 2011b). This shows that honeydew, although a necessary component of the ant-mediated biotic defense of aphids, can also have important and sometimes detrimental side effects for the aphids.

5.3.1 Aphid-Ant Effects on Plants

If ants can serve as biotic defenders of plants, but also as biotic defenders of aphids, what is the overall impact of ants on aphid host plants? After all, when ants protect aphids, they are protecting an herbivore of that plant—many of the 4000 described aphid species are most known for their direct negative effects on plants. By constantly feeding on phloem, aphids often cause severe damage to plants, particularly crops (Ragsdale et al. 2011). In addition, their high reproductive rates due to parthenogenesis, efficient settlement, and resistance to insecticides make aphids one of the most abundant pests (Stadler and Dixon 2005; Giordanengo et al. 2010; Jaouannet et al. 2014). Aphid damage to plants can range from microscopic (pathogen transmission, changes in nutrient flow, especially nitrogen and sugar allocation) to macroscopic (wilting of leaves and aesthetic damage on fruit surface), with devastating results (Jaouannet et al. 2014).

Hundreds of plant species consumed by aphids, coccids and also membracids possess EFNs (Del-Claro and Marquis 2015) which scientists have hypothesized can distract ants from trophobiont hemipterans (and other myrmecophilous insects), leaving the honeydew-producing insects vulnerable to natural enemies (distraction hypothesis – reviewed in Del-Claro et al. 2016). In fact, aphid species that are not attended by ants may suppress the production of EFN, thus weakening the plant's defense and deterring plant-bodyguards (ants, wasps) from visiting the plants (Yoshida et al. 2018). Historically, the distraction hypothesis has had good support because other forms of plant defense are less effective against aphids, as excreted honeydew incurs fungal infection on plants and also suppresses the JA pathway, which is used by plants as a form of chemical defense (Stadler and Müller 1996; Schwartzberg and Tumlinson 2014). For example, Engel et al. (2001) showed that the EFNs of *Vicia faba* (Leguminosae) can displace ants from aphids by producing nectar with more sugar than honeydew.

However, despite being herbivores, the effect of some aphids on EFN-bearing plants is negligible in terms of plant fitness loss (Suzuki et al. 2004), causing no severe damage (Rico-Gray and Castro 1996). In some cases, aphids might even indirectly benefit the plants—evidence shows that ants might shift their patrolling behavior from EFNs towards aphids and their honeydew, and that the presence of aphids can increase the number of ants on the plant (Sakata and Hashimoto 2000; Katayama and Suzuki 2003, reviewed in Styrsky and Eubanks 2007). A review by Zhang et al. (2012) also demonstrated that the mutualism between ants and honeydew-releasing hemipterans has a wide range of ecological effects on the host plant, such as reduced herbivory, low abundance of herbivores, and low fruit removal. In fact, the abundance of ants on some EFN plants is only enhanced in the presence of aphids. In *Vicia angustifolia* L. (Leguminosae) the number of *Lasius japonicus* ants visiting EFNs was 10% of the total of ants foraging on the plant; however, when aphids were present, the number of foraging ants on plants increased two-fold and 50% of ants foraged on honeydew (Suzuki et al. 2004). This shows the extent to which aphids influence the natural history of the host plants, and can be an important facilitator of biotic defense for the plant. Nonetheless, the herbivory of trophobiont insects might outweigh the suppression of other herbivores. Plants will only benefit if the cost of feeding the trophobiont is lower than the cost of suppressing other herbivores, thus a trade-off is expected (Styrsky and Eubanks 2007).

The herbivory-reducing effects of ant-hemipteran associations are higher compared to plants that rely solely on EFNs; however, plants that provide direct benefit to ants (e.g., EFNs, domatia, and food bodies) generally exhibit increased growth and reproduction whereas ant-hemipteran presence does not significantly affect plant performance (Chamberlain and Holland 2009, but see Fiala and Maschwitz 1991). This shows that the presence of direct rewards to ants is more beneficial to one aspect of plant life history, while the presence of trophobiont insects benefits other aspects.

It is evident that tritrophic interactions involving ants, aphids, and plants are extremely complex and conditional, thus it is hard to make generalizations (Stadler and Dixon 1999, 2005; Stadler et al. 2002). The role of honeydew goes far beyond simply mediating the aphid-ant mutualism. It is both a blessing and a curse for the aphids themselves as well as their host plants, and many questions remain unanswered about honeydew-mediated interactions. For instance, how do physiological changes incurred by aphid herbivory in host plants influence other insects, such as herbivores and pollinators? How do aphid-plant interactions affect the biodiversity of the entomofauna associated with plants? Aphid feeding behavior elicits plant antiherbivore defenses, some of which are constitutive and general, while others are local or systemic (Jaouannet et al. 2014), but further consideration of the fitness effects of such defenses is needed. In addition, some other insects may avoid aphid-infested plants. By investigating the oviposition patterns of Alcon Blue butterflies (*Maculinea alcon*), researchers found that aphid presence on the host plant strongly inhibited the oviposition of butterflies (Arnyas et al. 2009). If we are to understand biodiversity and multitrophic interactions in their full magnitude, such topics deserve further evaluation and detailed investigations.

5.4 Myrmecophilous Lepidoptera

Plants and trophobiont hemipterans are not unique in their myrmecophily. Some butterfly larvae also secrete sugary liquids that attract ant bodyguards (Pierce et al. 2002, ▪ Fig. 5.3b–d). This lepidopteran myrmecophily is broadly known and documented for two butterfly families: Lycaenidae and Riodinidae (DeVries 1991;

Pierce et al. 2002), which combined compose 30% of all butterfly species (Shields 1989). In these families, 75% of species interact with ants during their larval stages, exhibiting a variety of associations varying from obligatory to facultative and parasitic to mutualistic (Fiedler and Hölldobler 1992; Stadler et al. 2001).

Just as with EFN-bearing plants, the ants involved in these interactions protect their food source against natural enemies, and therefore these associations are known as defense or protective mutualisms (Agrawal and Fordyce 2000). Butterfly larvae possess specialized glands that produce nutritive secretions, which serve as ant food (Stadler et al. 2003; Daniels et al. 2005). In counterpart, the butterflies receive ant protection against predators and parasitoids during their most vulnerable life stage (pupal and larval instars), representing another case of animal-animal biotic defense (Pierce et al. 2002).

In addition to providing liquid rewards, myrmecophilous butterflies have also developed a series of morphological, behavioral, acoustic, and chemical characteristics to attract, pacify, alarm, or even trick ants into exhibiting defensive behavior (see review by Casacci et al. 2019). The resulting interaction is complex, with varied costs and benefits for the organisms involved (butterflies, ants, and plants), depending on the mechanisms (▶ Box 5.1) and the interaction degree, as discussed in the following sections.

Box 5.1   Mechanisms of Butterfly Myrmecophily

Ant-associated organs: Among the different chemical mechanisms responsible for attracting and maintaining ant's attendance of butterfly larvae, the nectary organs in Lycaenidae and Riodinidae deserve special attention. In Lycaenidae, the exudate is produced by the dorsal nectar organ (DNO) (Newcomer 1912) and in Riodinidae by the tentacle nectar organs (TNOs) (DeVries 1988). These secretions are composed mostly of amino acids and carbohydrates and are even richer in amino acids than extrafloral nectar and hemipteran honeydew (Yao and Akimoto 2002; Blüthgen et al. 2004; Daniels et al. 2005). It is important to consider exudates when evaluating this system's trade-off since the secretion quality and quantity influences ant persistence in attending the larvae (Wada et al. 2001; Hojo et al. 2015) and consequently ant permanence on the plant.

The eversible tentacle organs (TOs) in some lycaenid larvae and the anterior tentacle organs (ATOs) in riodinids, as well as the perforated cupola organs (PCOs) present in almost all species of both families, facilitate chemical communication between butterflies and ants (Malicky 1970). Authors have reported that eversion of the TOs is related to the liberation of volatile composites that incite alert and aggressive behavior in patrolling ants (Axén et al. 1996).

Cuticular hydrocarbons (CHCs): To avoid ant attacks, some myrmecophilous butterfly species mimic ant CHCs in order to be recognized as a colony member (Hojo et al. 2009). Beyond these, other CHC strategies used by butterflies include chemical insignificance, in which the larvae become "invisible" to the ants by reducing their cuticular chemical components, and chemical camouflage, in which the larvae take on the cuticular profile of the host plant via diet or contact (see review by Barbero 2016). The important role of CHCs is evident in insect-plant interactions, but the benefits to the emitter or the receiver are little understood.

Vibroacoustic signals: Both Lycaenidae and Riodinidae can produce sound via stridulation, a mechanism that may play an important role in the maintenance of their interactions with ants (DeVries 1990; Mathew et al. 2008). Stridulation organs are also present in non-myrmecophilous lycaenids (Álvarez et al. 2014), but can nonetheless influence the ant protective behavior in myrmecophiles. Travassos and Pierce (2000) show how the larvae and pupae of *Jalmenus evagoras* use a complex repertoire of sounds to improve the attendance of associated *Iridomyrmex* ants. Furthermore, pupae of the lycaenid *Spindasis lohita* do not possess secretory organs but still communicate with *Crematogaster rogenhoferi* ant attendants via vibration (Lin et al. 2019).

5.4.1 Benefits of the Ant-Butterfly Biotic Defense Relationship

The benefits of attracting ants as biotic defenses are widely studied from the butterfly perspective (Pierce et al. 1987; Baylis and Pierce 1991; Stadler et al. 2001; Kaminski et al. 2013; Bächtold et al. 2014; Mizuno et al. 2019). Experimental manipulations have demonstrated that butterfly eggs and larvae are less parasitized, larvae show better development, and survive at significantly higher rates when attending ants are present (Pierce et al. 1987; Nakabayashi et al. 2020). From the ant's point of view, however, the costs and benefits are less studied. Ants presumably benefit from feeding on the nutritive secretions, and lab experiments have shown higher colony survivorship in the presence of butterfly larvae (Cushman et al. 1994; ◘ Fig. 5.1). Nonetheless, the energetic and opportunity costs of tending butterfly larvae should also be included when evaluating the costs and benefits for ants. For example, the costs for ants could outweigh the benefits when butterfly larvae manipulate the ant behavior by means other than liquid rewards, such as CHC volatile emissions and stridulation.

5.4.2 How Does Lepidopteran Myrmecophily Affect Plants?

At first glance, the butterfly-ant interaction is detrimental to host plants, since Lycaenidae and Riodinidae larvae may directly reduce leaf area through herbivory (DeVries 1989) and directly reduce plant reproductive success by consuming floral buds and flowers (Bächtold et al. 2013). In this sense, the protection provided by ants to caterpillars might reduce plant fitness by allowing the larvae to damage host plants. On the other hand, the ants involved in this interaction are generally not specialists on butterfly secretions, but also attack and consume other arthropods on the plant (Davidson 1997). The aggressive predatory behavior of ants may reduce the abundance and feeding activity of other herbivores, which may have a positive effect on the plant. The net result for the host plant thus depends on whether the reduction of non-butterfly herbivory is greater than the extent of the myrmecophilous butterfly larvae's herbivory (Oliveira 1997; Rico-Gray and Oliveira 2007). Therefore, butterfly-ant associations may confer biotic defense value to their host plants, since the constant ant presence induced by the myrmecophilous larvae may negatively impact other herbivores and limit the areas free of natural enemies for herbivores (◘ Fig. 5.1).

Although the results of these interactions depend on factors that vary in space and time such as plant phenology and density (Rodrigues et al. 2010), Rico-Gray and Oliveira (2007) proposed three criteria that the butterfly-ant association must follow in order to benefit the plant: (1) the butterfly must not be the main plant herbivore; (2) ant attendance cannot increase the population density of the myrmecophile; and (3) associated ants must deter other plant herbivores (Rico-Gray and Oliveira 2007).

Unfortunately, the effects of the butterfly-ant interaction on plant reproductive success remain largely unknown. DeVries (1991) suggested butterfly myrmecophily evolved as a consequence of lasting interactions between ants and angiosperms, in a complex system that also involves EFN-bearing plants and those with trophobiont hemipterans. Despite the notable importance of vegetation in the evolutionary and ecological context of these interactions (Moreau et al. 2006), their impacts on and interactions with plants have received little attention. For instance, several studies have shown that ant presence mediates host plant selection by adult butterflies (Atsatt 1981; Fraser et al. 2002; Bächtold et al. 2016). The role of plant chemistry (CHCs or volatile emissions) in host plant choice by myrmecophilous butterflies is unknown. Also, the possibility that myrmecophilous organisms may manipulate the ant's behavior via multiple strategies and how that affects plant fitness is still little understood (Mannino et al. 2018; Casacci et al. 2019). By being a multitrophic interaction, the interaction between ants and butterflies have multiple ecological effects, since the presence of myrmecophilous larvae alters the abundance and behavior of ants on plants. Exploring the effects of this butterfly-ant association on other trophic levels (Elgar et al. 2016), such as host plants, constitutes the key to the ecological and evolutionary comprehension of the mutualism between myrmecophilous Lepidoptera and ants.

5.5   Methodological Approaches and Biotic Defense Networks

As we have observed throughout this chapter, biotic defense interactions are complex, context-dependent, and involve many distinct species. Faced with this complexity, an equally complex methodology is necessary to synthesize this plethora of interactions into patterns that can be analyzed and generalized (Huxley and Cutler 1991; Beattie and Hughes 2002; Del-Claro et al. 2018). Graph Theory, which studies three-dimensional geometrical shapes transforming vertices and edges into three-dimensional networks of points connected by lines (Barnes and Harary 1983; Biggs et al. 1986; Bascompte 2010), provided one such approach to describe and model complex biological relationships in interaction networks (Pascual and

Dunne 2006; Dale 2017; Dáttilo and Rico-Gray 2018). The Theory of Ecological Networks (see references in Ings and Hawes 2018) emerged from this integrative approach, providing a way to study the multitrophic and multispecific nature of complex plant-ant-herbivore interactions (Bascompte 2010; Dehling 2018; Dáttilo and Rico-Gray 2018).

According to the Theory of Ecological Networks, ecological interactions are multispecific, that is, a given species can simultaneously interact with many other species. This creates a three-dimensional network, in which the tangle of different interactions regulates each other and together contribute to the overall structure and properties of the network (Bascompte 2010). This network can be visualized through a network graph, a three-dimensional geometric figure in which the species are represented as points, and the interactions between the species are represented by lines that connect these points (Dehling 2018). Several parameters can be calculated from this network structure, including form, degree of connection, diversity and specificity of connections, and the position of species in the network, so changes in these parameters can be used to model changes in the patterns of interactions (Pascual and Dunne 2006; Bascompte 2010; Dehling 2018). These parameters quantify interactions between species, often measured by the number of encounters between a set of coexisting species (Bascompte and Jordano 2007). For instance, an interaction between ants and plants can be conceived when individual ants collecting nectar from different plants are used to establish a network of species interactions (Dáttilo et al. 2014; see also Luna and Dáttilo, ▶ Chap. 10).

Ecological networks can assume different arrangements, ranging from a more specialized, modular pattern (Grilli et al. 2016), to a more generalized, nested pattern (Bascompte et al. 2003; Fortuna et al. 2010). Interactions between ants and plants (Bascompte et al. 2003) are typically nested, as described for several ant-plant networks in Brazil (Guimarães et al. 2006; Dáttilo et al. 2013, 2014; Lange and Del-Claro 2014; Costa et al. 2016; Fagundes et al. 2016; Del-Claro et al. 2018) and worldwide (Nielsen and Bascompte 2007; Vázquez et al. 2009; Chamberlain et al. 2010; Díaz-Castelazo et al. 2010; Rico-Gray et al. 2012). In nested networks, there are unequal interactions between species, where a few (central) species interact with many others, but most species are less interactive (peripheral, Dáttilo et al. 2014). In these nested networks, the low specialization of interactions and high sharing of partners leads to a high redundancy of interactions, as peripheral species interact with the same species as central ones (Dehling 2018).

Ant-plant interactions are consistently nested within different communities (Dáttilo et al. 2013, 2014; Costa et al. 2016; Sendoya et al. 2016; Fagundes et al. 2018), under different environmental conditions (Rico-Gray et al. 2012; Dáttilo et al. 2013), at different periods of the day (Dáttilo et al. 2014) and seasons (Díaz-Castelazo et al. 2013; Lange et al. 2013; Santos et al. 2014), and even after severe disturbances (Falcão et al. 2014; Costa et al. 2018; Fagundes et al. 2018). Nested structures resist change due to a high overlap of interacting partners, which allows flexibility in the replacement of species without necessarily disrupting the network (Dáttilo 2012; Passmore et al. 2012). Therefore, there is evidence of evolutionary selection for interactions that strengthen the nested structure of the ant-plant networks (Del-Claro et al. 2018). However, few studies have tested whether community-level benefits to biotic defense remain after changes in species composition (Piovia-Scott 2011) or network structure (Plowman et al. 2017), because it is difficult to measure these benefits in the field.

Network approaches can also be used for biotic defense relationships beyond ant and EFN-bearing plants, and comparing the network structure of different biotic defense types can elucidate broader patterns or sources of variation. For example, when comparing different ant-mediated defense relationships, Cagnolo and Tavella (2015) found ant-lepidopteran and ant-myrmecophyte (domatia-bearing plants) networks to be highly modular (specialized), unlike more nested ant-EFN relationships. Furthermore, biotic defenses by definition involve at least three layers of interaction: the defender, the defended, and the offender. As network approaches become more widely used, they are being expanded to accommodate more complex interactions such as multiple trophic levels (Cagnolo 2018), which could allow more accurate descriptions of biotic defense relationships.

Ecological networks are not the only approach to studying biotic defenses, and indeed, other approaches are also warranted. Experimental tests of the immediate costs and benefits to both the defenders and defended organisms, the longer-term fitness ramifications, and the conditions that affect these costs and benefits will all provide insight to the intricacies of these relationships. For instance, there are trade-off effects between ant aggressiveness and plant protection. Although highly aggressive ants provide increased protection against herbivores, they may drive away potential pollinators, reducing the reproductive output of plants (Melati and Leal 2018). Resource quality also affects the outcomes of ant-defended plants. Nectar and honeydew rich in sugars and amino acids attract more ants and increase plant defense (Blüthgen et al. 2004; Alves-Silva and Del-Claro 2013; Fagundes et al. 2017; Pacelhe et al. 2019). However, as a facultative and non-specialized interaction, plants with EFNs are susceptible to robbers and visiting ants that provide little protection (Dáttilo et al. 2014). Further observation of natural history will also help expand the known diversity of biotic defense interactions and is an important key step in plant-animal interactions in general (Del-Claro et al. 2013).

5.6 Reconsidering Biotic Defenses: Implications for Biodiversity and Future Directions

In the past centuries, researchers have investigated and proposed explanations for why communities and ecosystems differ in terms of biodiversity. Although climate and geography have been regularly used to explain species distribution since the time of Humboldt (Norder 2019), only after the 1960s did studies start recognizing the importance of ecological interactions for patterns of biodiversity (e.g., Janzen 1966; Paine 1966; Breedlove and Ehrlich 1968). Currently, we recognize that some species are so dependent on interactions that their extinction may cause cascading effects that ultimately lead to the extinction of their associated species (Forup et al. 2008). Although these effects are usually demonstrated with key predators and pollinators, there is also evidence that biotic defenses may contribute to the persistence of species in several systems (see Bronstein, ▶ Chap. 11).

Biotic defenses may contribute to biodiversity by increasing the performance and fitness of associated organisms. As we have seen in this chapter, plants with EFNs are not the only ones benefited by the presence of natural enemies such as ants, spiders, and parasitoids, since plants contribute to their defenders' survival by offering valuable resources. For example, Byk and Del-Claro (2011) showed that the offer of extrafloral nectar increased the survival and growth rate of myrmecophilous ants. Biotic defenses also affect the herbivore competition and host recognition with enough potential to shape entire populations and communities of arthropods (Agrawal and Sherriffs 2001; González-Megías and Gómez 2003; Ohgushi 2005; Kessler and Heil 2011). This was demonstrated in studies involving the experimental removal of herbivores of distinct guilds. As an example, Waltz and Whitham (1997) revealed that responses of cottonwoods attacked by a leaf-galling aphid (*Pemphigus betae*) maintain high levels of species richness and diversity through the attraction of natural enemies and other herbivores. When the aphids were removed, the species richness and relative abundance of arthropods decreased by 32% and 55%, respectively.

Given that biotic defense relationships play integral roles in the development and maintenance of biodiversity, understanding how anthropogenic and natural disturbances impact these relationships is an important part of predicting, managing, and when possible preventing the effects of future disturbances. Climatic changes are impacting all biotic interactions by disrupting some relationships and altering the outcomes of others. Many studies have shown that increasing temperatures and changing precipitation are shifting plant and animal phenology (Munson and Long 2017; reviewed by Cohen et al. 2018), with strong impacts on pollination and biotic defenses (Vilela et al. 2018). Furthermore, deforestation and habitat fragmentation may disrupt and weaken ecological interactions (Sabatino et al. 2010). However, modeling network techniques can be used to detect keystone species and predict the possible consequences of their extinction (Messeder et al. 2020). This knowledge can be used to set management and conservation strategies to mitigate ecological disturbances.

We have selected these example systems here to illustrate some of the main aspects of biotic defense, but we encourage readers to be open-minded in what they categorize as biotic defense. We and several authors before us have made the case that indirect defense encompasses more than simple plant-herbivore-natural enemy interactions—other organisms also manipulate natural enemies to reduce the effects of predation, and considering these alternate systems can lead to broader conclusions about defense as a whole (Pearse et al. 2020). Not all antagonistic threats are herbivores, and not all natural enemies are other animals—some fungal microbes might reduce more pathogenic fungi in fruit (Cipollini and Stiles 1993), although this hypothesis seems to have been left untested.

Conclusion

In this chapter, we discussed four examples of biotic defense systems and their mechanisms. We showed that plants protected by natural enemies such as ants and spiders exhibit less herbivore damage, produce more fruits and seeds, and may ultimately increase their fitness. The benefits provided by the biotic agents, however, are not pervasive and depend on the involved species and the ecological context. Additionally, plants are not the only organisms employing biotic defenses—honeydew-producing aphids and caterpillars also provide rewards to ants which can then provide protection from other natural enemies. These interactions can still benefit host plants by attracting natural enemies, but such benefits are tempered by the herbivory of the honeydew-producing insects. Such complex biotic interactions convey methodological challenges, but ecological networks provide a useful approach to studying biotic interactions and assessing how deforestation and climatic changes are extinguishing and disrupting biotic interactions across the globe. Despite the fact that biotic interactions contribute to the maintenance of biodiversity, most conservation efforts focus on species loss. However, biotic defense interactions are a major piece of biodiversity, and their conservation is necessary for maintaining healthy environments and essential ecosystem services.

Key Points
- The stability of biotic defense systems will depend on the costs and benefits to both partners, as well as the generality or specificity of the defense and rewards.
- Although the most widely-studied biotic defense interactions involve plants, extrafloral nectar, and natural enemies of herbivores, other organisms can also engage in biotic defense relationships.
- Honeydew-producing herbivores and extrafloral nectar-producing plants can attract the same biotic defenders, with synergistic or competitive effects depending on the fidelity of natural enemies to one or both rewards, the extent of herbivory, and the effectiveness of defense.

❓ Questions
- Which conditions determine the stability of biotic defense interactions?
- What other organisms participate in biotic defense systems, particularly understudied interactions?
- How will disturbance affect the outcomes of biotic defenses?

References

Addicott JF (1978) Competition for mutualists: aphids and ants. Can J Zool 56:2093–2096

Agarwal VM, Rastogi N (2010) Ants as dominant insect visitors of the extrafloral nectaries of sponge gourd plant, *Luffa cylindrica* (L.)(Cucurbitaceae). Asian Myrmecol 3:45–54

Agrawal AA, Fordyce JA (2000) Induced indirect defence in a lycaenid-ant association: the regulation of a resource in a mutualism. Proc R Soc London Ser B Biol Sci 267:1857–1861

Agrawal AA, Sherriffs MF (2001) Induced plant resistance and susceptibility to late-season herbivores of wild radish. Ann Entomol Soc Am 94:71–75

Álvarez M, Munguira ML, Martínez-Ibáñez MD (2014) Comparative study of the morphology of stridulatory organs of the Iberian lycaenid butterfly pupae (Lepidoptera). J Morphol 275:414–430

Alves-Silva E, Del-Claro K (2013) Effect of post-fire resprouting on leaf fluctuating asymmetry, extrafloral nectar quality, and ant–plant–herbivore interactions. Naturwissenschaften 100:525–532

Arnyas E, Bereczki J, Toth A et al (2009) Oviposition preferences of *Maculinea alcon* as influenced by aphid (*Aphis gentianae*) and fungal (*Puccinia gentianae*) infestation of larval host plants. Ecol Entomol 34:90–97

Assunção MA, Torezan-Silingardi HM, Del-Claro K (2014) Do ant visitors to extrafloral nectaries of plants repel pollinators and cause an indirect cost of mutualism? Flora-Morphology, Distrib Funct Ecol Plants 209:244–249

Atsatt PR (1981) Lycaenid butterflies and ants: selection for enemy-free space. Am Nat 118:638–654

Axén AH, Leimar O, Hoffman V (1996) Signalling in a mutualistic interaction. Anim Behav 52:321–333

Bächtold A, Alves-Silva E, Del-Claro K (2013) Lycaenidae larvae feeding on *Peixotoa parviflora* (Malpighiaceae) in a semi-deciduous forest in Southeastern Brazil. J Lepid Soc 67:65–67

Bächtold A, Alves-Silva E, Kaminski LA, Del-Claro K (2014) The role of tending ants in host plant selection and egg parasitism of two facultative myrmecophilous butterflies. Naturwissenschaften 101:913–919

Bächtold A, Silva EA, Del-Claro K (2016) Ants, plant characteristics and habitat conservation status affect the occurrence of myrmecophilous butterflies on an extrafloral nectaried Malpighiaceae. Stud Neotrop Fauna Environ 51:112–120

Baker-Méio B, Marquis RJ (2012) Context dependent benefits from ant–plant mutualism in three sympatric varieties of *Chamaecrista desvauxii*. J Ecol 100:242–252

Barbero F (2016) Cuticular lipids as a cross-talk among ants, plants and butterflies. Int J Mol Sci 17:1966

Barnes JA, Harary F (1983) Graph theory in network analysis. Soc Networks 5:235–244

Bascompte J (2010) Structure and dynamics of ecological networks. Science 329:765–766

Bascompte J, Jordano P (2007) Plant-animal mutualistic networks: the architecture of biodiversity. Annu Rev Ecol Evol Syst 38:567–593

Bascompte J, Jordano P, Melián CJ, Olesen JM (2003) The nested assembly of plant–animal mutualistic networks. Proc Natl Acad Sci 100:9383–9387

Baylis M, Pierce NE (1991) The effect of host-plant quality on the survival of larvae and oviposition by adults of an ant-tended lycaenid butterfly, *Jalmenus evagoras*. Ecol Entomol 16:1–9

Beattie A, Hughes L (2002) Ant-plant interactions. In: Herrera CM, Pellmyr O (eds) Plant animal interactions: an evolutionary approach. Blackwell Science, Malden, pp 211–236

Biggs N, Lloyd EK, Wilson RJ (1986) Graph theory, 1736–1936. Oxford University Press, New York

Billick I, Hammer S, Reithel JS, Abbot P (2007) Ant-aphid interactions: are ants friends, enemies, or both? Ann Entomol Soc Am 100:887–892

Bixenmann RJ, Coley PD, Kursar TA (2011) Is extrafloral nectar production induced by herbivores or ants in a tropical facultative ant–plant mutualism? Oecologia 165:417–425

Blüthgen N, E Stork N, Fiedler K (2004) Bottom-up control and co-occurrence in complex communities: honeydew and nectar determine a rainforest ant mosaic. Oikos 106:344–358

Boege K, Marquis RJ (2005) Facing herbivory as you grow up: the ontogeny of resistance in plants. Trends Ecol Evol 20:441–448

Breedlove DE, Ehrlich PR (1968) Plant-herbivore coevolution: lupines and lycaenids. Science 162:671–672

Buckley RC (1982) Ant-plant interactions: a world review. In: Ant-plant interactions in Australia. Springer, Dordrecht, pp 111–141

Budenberg WJ (1990) Honeydew as a contact kairomone for aphid parasitoids. Entomol Exp Appl 55:139–148

Byk J, Del-Claro K (2011) Ant–plant interaction in the Neotropical savanna: direct beneficial effects of extrafloral nectar on ant colony fitness. Popul Ecol 53:327–332

Cagnolo L (2018) The future of ecological networks in the tropics. In: Ecological networks in the tropics. Springer, Cham, pp 171–183

Cagnolo L, Tavella J (2015) The network structure of myrmecophilic interactions. Ecol Entomol 40:553–561

Calixto ES, Lange D, Del-Claro K (2018) Protection mutualism: an overview of ant-plant interactions mediated by extrafloral nectaries. Oecologia Aust 22:410

Calixto ES, Lange D, Bronstein J et al (2020) Optimal Defense Theory in an ant–plant mutualism: extrafloral nectar as an induced defence is maximized in the most valuable plant structures. J Ecol 109:167–178

Casacci LP, Bonelli S, Balletto E, Barbero F (2019) Multimodal signaling in myrmecophilous butterflies. Front Ecol Evol 7:454

Chamberlain SA, Holland JN (2009) Quantitative synthesis of context dependency in ant–plant protection mutualisms. Ecology 90:2384–2392

Chamberlain SA, Kilpatrick JR, Holland JN (2010) Do extrafloral nectar resources, species abundances, and body sizes contribute to the structure of ant–plant mutualistic networks? Oecologia 164:741–750

Cipollini ML, Stiles EW (1993) Fungi as biotic defense agents of fleshy fruits: alternative hypotheses, predictions, and evidence. Am Nat 141:663–673

Cohen JM, Lajeunesse MJ, Rohr JR (2018) A global synthesis of animal phenological responses to climate change. Nat Clim Change 8:224–228

Costa FV, Mello MAR, Bronstein JL et al (2016) Few ant species play a central role linking different plant resources in a network in rupestrian grasslands. PLoS One 11:e0167161

Costa FV, Blüthgen N, Viana-Junior AB et al (2018) Resilience to fire and climate seasonality drive the temporal dynamics of ant-plant interactions in a fire-prone ecosystem. Ecol Indic 93:247–255

Cross FR, Jackson RR (2009) Odour mediated response to plants by evarcha culicivora, a blood-feeding jumping spider from East Africa. New Zeal J Zool 36:75–80

Cuautle M, Rico-Gray V (2003) The effect of wasps and ants on the reproductive success of the extrafloral nectaried plant *Turnera ulmifolia* (Turneraceae). Funct Ecol 17:417–423

Cuautle M, Rico Gray V, Díaz Castelazo C (2005) Effects of ant behaviour and presence of extrafloral nectaries on seed dispersal of the Neotropical myrmecochore *Turnera ulmifolia* L.(Turneraceae). Biol J Linn Soc 86:67–77

Cushman JH, Addicott JF (1989) Intra-and interspecific competition for mutualists: ants as a limited and limiting resource for aphids. Oecologia 79:315–321

Cushman JH, Rashbrook VK, Beattie AJ (1994) Assessing benefits to both participants in a lycaenid ant association. Ecology 75:1031–1041

Dale MRT (2017) Applying graph theory in ecological research. Cambridge University Press, Cham

Daniels H, Gottsberger G, Fiedler K (2005) Nutrient composition of larval nectar secretions from three species of myrmecophilous butterflies. J Chem Ecol 31:2805–2821

Dáttilo W (2012) Different tolerances of symbiotic and nonsymbiotic ant-plant networks to species extinctions. Netw Biol 2:127

Dáttilo W, Rico-Gray V (2018) Ecological networks in the tropics. Cham, Springer

Dáttilo W, Guimarães PR Jr, Izzo TJ (2013) Spatial structure of ant–plant mutualistic networks. Oikos 122:1643–1648

Dáttilo W, Marquitti FMD, Guimarães PR Jr, Izzo TJ (2014) The structure of ant–plant ecological networks: Is abundance enough? Ecology 95:475–485

Davidson DW (1997) The role of resource imbalances in the evolutionary ecology of tropical arboreal ants. Biol J Linn Soc 61:153–181

Dehling DM (2018) The structure of ecological networks. In: Ecological networks in the tropics. Springer, Cham, pp 29–42

Del-Claro K, Marquis RJ (2015) Ant species identity has a greater effect than fire on the outcome of an ant protection system in Brazilian Cerrado. Biotropica 47:459–467

Del-Claro K, Stefani V, Lange D et al (2013) The importance of natural history studies for a better comprehension of animal-plant interaction networks. Biosci J 29:439–448

Del-Claro K, Rico-Gray V, Torezan-Silingardi HM et al (2016) Loss and gains in ant–plant interactions mediated by extrafloral nectar: fidelity, cheats, and lies. Insectes Soc 63:207–221

Del-Claro K, Stefani V, Nahas L, Torezan-Silingardi HM (2017) Spiders as plant partners: complementing ant services to plants with extrafloral nectaries. In: Behaviour and ecology of spiders. Springer, Cham, pp 215–226

Del-Claro K, Lange D, Torezan-Silingardi HM et al (2018) The complex ant–plant relationship within tropical ecological networks. In: Ecological networks in the tropics. Springer, Cham, pp 59–71

Dennis P, Young MR, Gordon IJ (1998) Distribution and abundance of small insects and arachnids in relation to structural heterogeneity of grazed, indigenous grasslands. Ecol Entomol 23:253–264

DeVries PJ (1988) The larval ant-organs of Thisbe irenea (Lepidoptera: Riodinidae) and their effects upon attending ants. Zool J Linn Soc 94:379–393

DeVries PJ (1989) Detecting and recording the calls produced by butterfly caterpillars and ants. J Res Lepid 28:258–262

DeVries PJ (1990) Enhancement of symbioses between butterfly caterpillars and ants by vibrational communication. Science 248:1104–1106

DeVries PJ (1991) Evolutionary and ecological patterns in myrmecophilous riodinid butterflies. In: Ant-plant interactions. Oxford University Press, Oxford, pp 143–156

Díaz-Castelazo C, Guimaraes PR Jr, Jordano P et al (2010) Changes of a mutualistic network over time: reanalysis over a 10-year period. Ecology 91:793–801

Díaz-Castelazo C, Sánchez-Galván IR, Guimarães PR Jr et al (2013) Long-term temporal variation in the organization of an ant–plant network. Ann Bot 111:1285–1293

Dicke M, Sabelis MW (1988) How plants obtain predatory mites as bodyguards. Netherlands J Zool 38:148–165

Duffey SS (1986) Plant glandular trichomes: their partial role in defence against insects. Insects and the plant surface. Edward Arnold, London, pp 151–172

Dutra HP, Freitas AVL, Oliveira PS (2006) Dual ant attraction in the Neotropical shrub Urera baccifera (Urticaceae): the role of ant visitation to pearl bodies and fruits in herbivore deterrence and leaf longevity. Funct Ecol 20:252–260

Elgar MA, Nash DR, Pierce NE (2016) Eavesdropping on cooperative communication within an ant-butterfly mutualism. Sci Nat 103:84

Engel V, Fischer MK, Wäckers FL, Völkl W (2001) Interactions between extrafloral nectaries, aphids and ants: are there competition effects between plant and homopteran sugar sources? Oecologia 129:577–584

Fagundes R, Dáttilo W, Ribeiro SP et al (2016) Food source availability and interspecific dominance as structural mechanisms of ant-plant-hemipteran multitrophic networks. Arthropod Plant Interact 10:207–220

Fagundes R, Dáttilo W, Ribeiro SP et al (2017) Differences among ant species in plant protection are related to production of extrafloral nectar and degree of leaf herbivory. Biol J Linn Soc 122:71–83

Fagundes R, Lange D, Anjos DV et al (2018) Limited effects of fire disturbances on the species diversity and structure of ant-plant interaction networks in Brazilian Cerrado. Acta Oecologica 93:65–73

Falcão JCF, Dáttilo W, Izzo TJ (2014) Temporal variation in extrafloral nectar secretion in different ontogenic stages of the fruits of Alibertia verrucosa S. Moore (Rubiaceae) in a Neotropical savanna. J Plant Interact 9:137–142

Fiala B, Maschwitz U (1991) Extrafloral nectaries in the genus Macaranga (Euphorbiaceae) in Malaysia: comparative studies of their possible significance as predispositions for myrmecophytism. Biol J Linn Soc 44:287–305

Fiala B, Maschwitz U (1992) Food bodies and their significance for obligate ant-association in the tree genus Macaranga (Euphorbiaceae). Bot J Linn Soc 110:61–75

Fiala B, Maschwitz U, Pong TY, Helbig AJ (1989) Studies of a South East Asian ant-plant association: protection of Macaranga trees by Crematogaster borneensis. Oecologia 79:463–470

Fiedler K, Hölldobler B (1992) Ants and *Polyommatus icarus* immatures (Lycaenidae)—sex-related developmental benefits and costs of ant attendance. Oecologia 91:468–473

Fischer MK, Shingleton AW (2001) Host plant and ants influence the honeydew sugar composition of aphids. Funct Ecol 15:544–550

Flatt T, Weisser WW (2000) The effects of mutualistic ants on aphid life history traits. Ecology 81:3522–3529

Foelix R (2011) Biology of spiders. Oxford University Press, Oxford

Fortuna MA, Stouffer DB, Olesen JM et al (2010) Nestedness versus modularity in ecological networks: two sides of the same coin? J Anim Ecol 79(4):811–817

Forup ML, Henson KSE, Craze PG, Memmott J (2008) The restoration of ecological interactions: plant–pollinator networks on ancient and restored heathlands. J Appl Ecol 45:742–752

Fraser AM, Tregenza T, Wedell N et al (2002) Oviposition tests of ant preference in a myrmecophilous butterfly. J Evol Biol 15:861–870

Frederickson ME, Greene MJ, Gordon DM (2005) 'Devil's gardens' bedevilled by ants. Nature 437:495–496

Giordanengo P, Brunissen L, Rusterucci C et al (2010) Compatible plant-aphid interactions: how aphids manipulate plant responses. C R Biol 333:516–523

González-Megías A, Gómez JM (2003) Consequences of removing a keystone herbivore for the abundance and diversity of arthropods associated with a cruciferous shrub. Ecol Entomol 28:299–308

González-Teuber M, Heil M (2009) Nectar chemistry is tailored for both attraction of mutualists and protection from exploiters. Plant Signal Behav 4:809–813

Greenstone MH (1984) Determinants of web spider species diversity: vegetation structural diversity vs. prey availability. Oecologia 62:299–304

Grilli J, Rogers T, Allesina S (2016) Modularity and stability in ecological communities. Nat Commun 7:1–10

Guimarães PR Jr, Rico-Gray V, Furtado dos Reis S, Thompson JN (2006) Asymmetries in specialization in ant–plant mutualistic networks. Proc R Soc B Biol Sci 273:2041–2047

Halaj J, Ross DW, Moldenke AR (1997) Negative effects of ant foraging on spiders in Douglas-fir canopies. Oecologia 109:313–322

Hatano E, Kunert G, Michaud JP, Weisser WW (2008) Chemical cues mediating aphid location by natural enemies. Eur J Entomol 105:797–806

Heil M (2014) Herbivore induced plant volatiles: targets, perception and unanswered questions. New Phytol 204:297–306

Heil M (2015) Extrafloral nectar at the plant-insect interface: a spotlight on chemical ecology, phenotypic plasticity, and food webs. Annu Rev Entomol 60:213–232

Hojo MK, Wada-Katsumata A, Akino T et al (2009) Chemical disguise as particular caste of host ants in the ant inquiline parasite Niphanda fusca (Lepidoptera: Lycaenidae). Proc R Soc B Biol Sci 276:551–558

Hojo MK, Pierce NE, Tsuji K (2015) Lycaenid caterpillar secretions manipulate attendant ant behavior. Curr Biol 25:2260–2264

Huxley CR, Cutler DF (1991) Ant-plant interactions. Oxford University Press, Oxford

Ings TC, Hawes JE (2018) The history of ecological networks. In: Ecological networks in the tropics. Springer, Cham, pp 15–28

Janzen DH (1966) Coevolution of mutualism between ants and acacias in Central America. Evolution 20:249–275

Jaouannet M, Rodriguez PA, Thorpe P et al (2014) Plant immunity in plant–aphid interactions. Front Plant Sci 5:663

Kaminski LA, Mota LL, Freitas AVL, Moreira GRP (2013) Two ways to be a myrmecophilous butterfly: natural history and comparative immature-stage morphology of two species of *Theope* (Lepidoptera: Riodinidae). Biol J Linn Soc 108:844–870

Katayama N, Suzuki N (2003) Changes in the use of extrafloral nectaries of *Vicia faba* (Leguminosae) and honeydew of aphids by ants with increasing aphid density. Ann Entomol Soc Am 96:579–584

Kessler A, Heil M (2011) The multiple faces of indirect defences and their agents of natural selection. Funct Ecol 25:348–357

Koptur S (1994) Floral and extrafloral nectars of Costa Rican Inga trees: a comparison of their constituents and composition. Biotropica 26(3):276–284

Krimmel BA, Pearse IS (2013) Sticky plant traps insects to enhance indirect defence. Ecol Lett 16:219–224

Lange D, Del-Claro K (2014) Ant-plant interaction in a tropical savanna: may the network structure vary over time and influence on the outcomes of associations? PLoS One 9:e105574

Lange D, Dattilo W, Del Claro K (2013) Influence of extrafloral nectary phenology on ant–plant mutualistic networks in a neotropical savanna. Ecol Entomol 38:463–469

Lee JC, Heimpel GE, Leibee GL (2004) Comparing floral nectar and aphid honeydew diets on the longevity and nutrient levels of a parasitoid wasp. Entomol Exp Appl 111:189–199

Leroy PD, Sabri A, Heuskin S et al (2011a) Microorganisms from aphid honeydew attract and enhance the efficacy of natural enemies. Nat Commun 2:1–7

Leroy PD, Wathelet B, Sabri A et al (2011b) Aphid-host plant interactions: does aphid honeydew exactly reflect the host plant amino acid composition? Arthropod Plant Interact 5:193–199

Letourneau DK (1983) Passive aggression: an alternative hypothesis for the *Piper-Pheidole* association. Oecologia 60:122–126

Limburg DD, Rosenheim JA (2001) Extrafloral nectar consumption and its influence on survival and development of an omnivorous predator, larval *Chrysoperla plorabunda* (Neuroptera: Chrysopidae). Environ Entomol 30:595–604

Lin Y-H, Liao Y-C, C-CS Y et al (2019) Vibrational communication between a myrmecophilous butterfly *Spindasis lohita* (Lepidoptera: Lycaenidae) and its host ant *Crematogaster rogenhoferi* (Hymenoptera: Formicidae). Sci Rep 9:1–10

Lohman DJ, Liao Q, Pierce NE (2006) Convergence of chemical mimicry in a guild of aphid predators. Ecol Entomol 31:41–51

Machado SR, Morellato LPC, Sajo MG, Oliveira PS (2008) Morphological patterns of extrafloral nectaries in woody plant species of the Brazilian cerrado. Plant Biol 10:660–673

Malicky H (1970) New aspects of the association between lycaenid larvae (Lycaenidae) and ants (Formicidae, Hymenoptera). J Lepid Soc 24:190–202

Mannino G, Abdi G, Maffei ME, Barbero F (2018) *Origanum vulgare* terpenoids modulate *Myrmica scabrinodis* brain biogenic amines and ant behaviour. PLoS One 13:e0209047

Marazzi B, Bronstein JL, Koptur S (2013) The diversity, ecology and evolution of extrafloral nectaries: current perspectives and future challenges. Ann Bot 111:1243–1250

Marquis RJ, Braker HE (1994) Plant-herbivore interactions: diversity, specificity and impact. In: La Selva: ecology and natural history of a neotropical rain forest. University of Chicago Press, Chicago, pp 261–281

Mathew J, Travassos MA, Canfield MR et al (2008) The singing reaper: diet, morphology and vibrational signaling in the nearctic species *Feniseca tarquinius* (Lepidoptera: Lycaenidae, Miletinae). Trop Lepid Res 18(1):24–29

Matsuura K, Yashiro T (2006) Aphid egg protection by ants: a novel aspect of the mutualism between the tree-feeding aphid *Stomaphis hirukawai* and its attendant ant *Lasius productus*. Naturwissenschaften 93:506–510

Melati BG, Leal LC (2018) Aggressive bodyguards are not always the best: preferential interaction with more aggressive ant species reduces reproductive success of plant bearing extrafloral nectaries. PLoS One 13:e0199764

Messeder JVS, Guerra TJ, Dáttilo W, Silveira FAO (2020) Searching for keystone plant resources in fruit-frugivore interaction networks across the Neotropics. Biotropica 52:857–870

Mizuno T, Hagiwara Y, Akino T (2019) Varied effects of tending ant species on the development of facultatively myrmecophilous lycaenid butterfly larvae. Insects 10:234

Moreau CS, Bell CD, Vila R et al (2006) Phylogeny of the ants: diversification in the age of angiosperms. Science (80) 312:101–104

Munson SM, Long AL (2017) Climate drives shifts in grass reproductive phenology across the western USA. New Phytol 213:1945–1955

Nahas L, Gonzaga MO, Del Claro K (2012) Emergent impacts of ant and spider interactions: herbivory reduction in a tropical savanna tree. Biotropica 44:498–505

Nahas L, Gonzaga MO, Del-Claro K (2017) Wandering and web spiders feeding on the nectar from extrafloral nectaries in neotropical savanna. J Zool 301:125–132

Nakabayashi Y, Mochioka Y, Tokuda M, Ohshima I (2020) Mutualistic ants and parasitoid communities associated with a facultative myrmecophilous lycaenid, *Arhopala japonica*, and the effects of ant attendance on the avoidance of parasitism. Entomol Sci 23:233–244

Nelsen MP, Ree RH, Moreau CS (2018) Ant–plant interactions evolved through increasing interdependence. Proc Natl Acad Sci 115:12253–12258

Nelson XJ, Jackson RR (2013) Hunger-driven response by a nectar-eating jumping spider to specific phytochemicals. Chemoecology 23:149–153

Newcomer EJ (1912) Some observations on the relations of ants and lycaenid caterpillars, and a description of the relational organs of the latter. J New York Entomol Soc 20:31–36

Nielsen A, Bascompte J (2007) Ecological networks, nestedness and sampling effort. J Ecol 95(5):1134–1141

Norder SJ (2019) Alexander von Humboldt (1769–1859): connecting geodiversity, biodiversity and society. J Biogeogr 46:1627–1630

Nyffeler M, Olson EJ, Symondson WOC (2016) Plant-eating by spiders. J Arachnol 44:15–27

Ohgushi T (2005) Indirect interaction webs: herbivore-induced effects through trait change in plants. Annu Rev Ecol Evol Syst 36:81–105

Ohm JR, Miller TEX (2014) Balancing anti herbivore benefits and anti-pollinator costs of defensive mutualists. Ecology 95:2924–2935

Oliveira PS (1997) The ecological function of extrafloral nectaries: herbivore deterrence by visiting ants and reproductive output in *Caryocar brasiliense* (Caryocaraceae). Funct Ecol 11:323–330

Pacelhe FT, Costa FV, Neves FS et al (2019) Nectar quality affects ant aggressiveness and biotic defense provided to plants. Biotropica 51:196–204

Paine RT (1966) Food web complexity and species diversity. Am Nat 100:65–75

Pascual M, Dunne JA (2006) Ecological networks: linking structure to dynamics in food webs. Oxford University Press, Oxford

Passmore HA, Bruna EM, Heredia SM, Vasconcelos HL (2012) Resilient networks of ant-plant mutualists in Amazonian forest fragments. PLoS One 7:e40803

Patt JM, Pfannenstiel RS, Meikle WG, Adamczyk JJ (2012) Supplemental diets containing yeast, sucrose, and soy powder enhance the survivorship, growth, and development of prey-limited cursorial spiders. Biol Control 63:237–245

Pearse IS, LoPresti E, Schaeffer RN et al (2020) Generalising indirect defence and resistance of plants. Ecol Lett 23:1137–1152

Pierce NE, Kitching RL, Buckley RC et al (1987) The costs and benefits of cooperation between the Australian lycaenid butterfly, *Jalmenus evagoras*, and its attendant ants. Behav Ecol Sociobiol 21:237–248

Pierce NE, Braby MF, Heath A et al (2002) The ecology and evolution of ant association in the Lycaenidae (Lepidoptera). Annu Rev Entomol 47:733–771

Piovia-Scott J (2011) The effect of disturbance on an ant–plant mutualism. Oecologia 166:411–420

Plowman NS, Hood ASC, Moses J et al (2017) Network reorganization and breakdown of an ant–plant protection mutualism with elevation. Proc R Soc B Biol Sci 284:20162564

Price PW, Bouton CE, Gross P et al (1980) Interactions among three trophic levels: influence of plants on interactions between insect herbivores and natural enemies. Annu Rev Ecol Syst 11:41–65

Ragsdale DW, Landis DA, Brodeur J et al (2011) Ecology and management of the soybean aphid in North America. Annu Rev Entomol 56:375–399

Renault CK, Buffa LM, Delfino MA (2005) An aphid-ant interaction: effects on different trophic levels. Ecol Res 20:71–74

Rico-Gray V, Castro G (1996) Effect of an ant-aphid interaction on the reproductive fitness of *Paullinia fuscecens* (Sapindaceae). Southwest Nat 41:434–440

Rico-Gray V, Oliveira PS (2007) The ecology and evolution of ant-plant interactions. University of Chicago Press, Chicago

Rico-Gray V, Díaz-Castelazo C, Ramírez-Hernández A et al (2012) Abiotic factors shape temporal variation in the structure of an ant–plant network. Arthropod Plant Interact 6:289–295

Rodrigues D, Kaminski LA, Freitas AVL, Oliveira PS (2010) Trade-offs underlying polyphagy in a facultative ant-tended florivorous butterfly: the role of host plant quality and enemy-free space. Oecologia 163:719–728

Romero GQ, Koricheva J (2011) Contrasting cascade effects of carnivores on plant fitness: a meta-analysis. J Anim Ecol 80:696–704

Romero GQ, Souza JC, Vasconcellos-Neto J (2008) Anti-herbivore protection by mutualistic spiders and the role of plant glandular trichomes. Ecology 89:3105–3115

Rosumek FB, Silveira FAO, Neves F de S et al (2009) Ants on plants: a meta-analysis of the role of ants as plant biotic defenses. Oecologia 160:537–549

Ruhren S, Handel SN (1999) Jumping spiders (Salticidae) enhance the seed production of a plant with extrafloral nectaries. Oecologia 119:227–230

Sabatino M, Maceira N, Aizen MA (2010) Direct effects of habitat area on interaction diversity in pollination webs. Ecol Appl 20:1491–1497

Sabri A, Vandermoten S, Leroy PD et al (2013) Proteomic investigation of aphid honeydew reveals an unexpected diversity of proteins. PLoS One 8:e74656

Sakata H, Hashimoto Y (2000) Should aphids attract or repel ants? Effect of rival aphids and extra-floral nectaries on ant–aphid interactions. Popul Ecol 42:171–178

Santos GMM, Dattilo W, Presley SJ (2014) The seasonal dynamic of ant-flower networks in a semi-arid tropical environment. Ecol Entomol 39:674–683

Schoereder JH, Sobrinho TG, Madureira MS et al (2010) The arboreal ant community visiting extra-floral nectaries in the Neotropical cerrado savanna. Terr Arthropod Rev 3:3–27

Schupp EW (1986) *Azteca* protection of *Cecropia*: ant occupation benefits juvenile trees. Oecologia 70:379–385

Schwartzberg EG, Tumlinson JH (2014) Aphid honeydew alters plant defence responses. Funct Ecol 28:386–394

Sendoya SF, Blüthgen N, Tamashiro JY et al (2016) Foliage-dwelling ants in a neotropical savanna: effects of plant and insect exudates on ant communities. Arthropod Plant Interact 10:183–195

Shields O (1989) World numbers of butterflies. J Lepid Soc 43:178–183

Shingleton AW, Stern DL, Foster WA (2005) The origin of a mutualism: a morphological trait promoting the evolution of ant aphid mutualisms. Evolution 59:921–926

Silva LA, Vasconcellos-Neto J, Del-Claro K, Stefani V (2020) Seasonally variable effects of spiders on herbivory and seed production of *Chamaecrista neesiana* (Leguminosae Caesalpinioideae). Ethol Ecol Evol 32(5):1–15

Stadler B, Dixon AFG (1999) Ant attendance in aphids: why different degrees of myrmecophily? Ecol Entomol 24:363–369

Stadler B, Dixon AFG (2005) Ecology and evolution of aphid-ant interactions. Annu Rev Ecol Evol Syst 36:345–372

Stadler B, Müller T (1996) Aphid honeydew and its effect on the phyllosphere microflora of *Picea abies* (L.) Karst. Oecologia 108:771–776

Stadler B, Fiedler K, Kawecki TJ, Weisser WW (2001) Costs and benefits for phytophagous myrmecophiles: when ants are not always available. Oikos 92:467–478

Stadler B, Dixon AFG, Kindlmann P (2002) Relative fitness of aphids: effects of plant quality and ants. Ecol Lett 5:216–222

Stadler B, Kindlmann P, Šmilauer P, Fiedler K (2003) A comparative analysis of morphological and ecological characters of European aphids and lycaenids in relation to ant attendance. Oecologia 135:422–430

Stefani V, Pires TL, Torezan-Silingardi HM, Del-Claro K (2015) Beneficial effects of ants and spiders on the reproductive value of *Eriotheca gracilipes* (Malvaceae) in a tropical savanna. PLoS One 10.e0131843

Styrsky JD, Eubanks MD (2007) Ecological consequences of interactions between ants and honeydew-producing insects. Proc R Soc B Biol Sci 274:151–164

Suzuki N, Ide T (2008) The foraging behaviors of larvae of the ladybird beetle, *Coccinella septempunctata* L.,(Coleoptera: Coccinellidae) towards ant-tended and non-ant-tended aphids. Ecol Res 23:371

Suzuki N, Ogura K, Katayama N (2004) Efficiency of herbivore exclusion by ants attracted to aphids on the vetch *Vicia angustifolia* L.(Leguminosae). Ecol Res 19:275–282

Taylor RM, Bradley RA (2009) Plant nectar increases survival, molting, and foraging in two foliage wandering spiders. J Arachnol 37:232–237

Taylor RM, Foster WA (1996) Spider nectarivory. Am Entomol 42:82–86

Thorp RW, Sugden EA (1990) Extrafloral nectaries producing rewards for pollinator attraction in *Acacia longifolia* (Andr.) Willd. Isr J Bot 39:177–186

Trager MD, Bhotika S, Hostetler JA et al (2010) Benefits for plants in ant-plant protective mutualisms: a meta-analysis. PLoS One 5:e14308

Travassos MA, Pierce NE (2000) Acoustics, context and function of vibrational signalling in a lycaenid butterfly–ant mutualism. Anim Behav 60:13–26

Uetz GW (1991) Habitat structure and spider foraging. In: Habitat structure. Springer, Cham, pp 325–348

Vasconcellos-Neto J, Messas YF, da Silva Souza H et al (2017) Spider–plant interactions: an ecological approach. In: Behaviour and ecology of spiders. Springer, Cham, pp 165–214

Vázquez DP, Blüthgen N, Cagnolo L, Chacoff NP (2009) Uniting pattern and process in plant–animal mutualistic networks: a review. Ann Bot 103:1445–1457

Vilela AA, Del-Claro K (2018) Effects of different ant species on the attendance of neighbouring hemipteran colonies and the outcomes for the host plant. J Nat Hist 52:415–428

Vilela AA, Del-Claro VTS, Torezan-Silingardi HM, Del-Claro K (2018) Climate changes affecting biotic interactions, phenology, and reproductive success in a savanna community over a 10-year period. Arthropod Plant Interact 12:215–227

Völkl W (1992) Aphids or their parasitoids: who actually benefits from ant-attendance? J Anim Ecol 61(2):273–281

Völkl W, Woodring J, Fischer M et al (1999) Ant-aphid mutualisms: the impact of honeydew production and honeydew sugar composition on ant preferences. Oecologia 118:483–491

Wada A, Isobe Y, Yamaguchi S et al (2001) Taste-enhancing effects of glycine on the sweetness of glucose: a gustatory aspect of symbiosis between the ant, *Camponotus japonicus*, and the larvae of the lycaenid butterfly, *Niphanda fusca*. Chem Senses 26:983–992

Waltz AM, Whitham TG (1997) Plant development affects arthropod communities: opposing impacts of species removal. Ecology 78:2133–2144

Way MJ (1963) Mutualism between ants and honeydew-producing Homoptera. Annu Rev Entomol 8:307–344

Weber MG, Keeler KH (2013) The phylogenetic distribution of extrafloral nectaries in plants. Ann Bot 111:1251–1261

Wise DH (1993) Spiders in ecological webs. Cambridge University Press, New York. ISBN-13, pp 521–978

Yao I (2014) Costs and constraints in aphid-ant mutualism. Ecol Res 29:383–391

Yao I, Akimoto S (2002) Flexibility in the composition and concentration of amino acids in honeydew of the drepanosiphid aphid *Tuberculatus quercicola*. Ecol Entomol 27:745–752

Yoshida T, Kakuta H, Choh Y (2018) Pea aphids (*Acyrthosiphon pisum* Harris) reduce secretion of extrafloral nectar in broad bean (*Vicia faba*). Ecol Entomol 43:134–136

Young TP, Stubblefield CH, Isbell LA (1996) Ants on swollen-thorn acacias: species coexistence in a simple system. Oecologia 109:98–107

Yu DW, Davidson DW (1997) Experimental studies of species-specificity in *Cecropia*–ant relationships. Ecol Monogr 67:273–294

Zangerl AR, Rutledge CE (1996) The probability of attack and patterns of constitutive and induced defense: a test of optimal defense theory. Am Nat 147:599–608

Zhang S, Zhang Y, Ma K (2012) The ecological effects of the ant–hemipteran mutualism: a meta-analysis. Basic Appl Ecol 13:116–124

Pollination Ecology: Natural History, Perspectives and Future Directions

Helena Maura Torezan-Silingardi,
Ilse Silberbauer-Gottsberger,
and Gerhard Gottsberger

Contents

© Springer Nature Switzerland AG 2021
K. Del-Claro, H. M. Torezan-Silingardi (eds.), *Plant-Animal
Interactions*, https://doi.org/10.1007/978-3-030-66877-8_6

⊜ Learning Objectives

After reading the text you should:

1. Comprehend the development of the flower through evolutionary time;
2. Realize the conditioned outcomes of flower development considering mutualistic interactions;
3. Understand the benefits of the crosspollination;
4. Have an overview of animal-mediated pollination relationships;
5. Realize the impact of species invasions on plant-pollinator interactions;
6. Understand the impacts of fragmentation, loss of natural areas and pesticides on pollinator populations and pollinating process in native and introduced species, in crops and natural areas.

▪ Foreword

Pollination and seed dispersion are possibly the most important mutualistic plant-animal interactions. By promoting plant reproduction, pollinators support the majority of the world's plant diversity, playing a crucial role in most ecological networks. Hence, pollinators are concomitantly vital to both natural ecosystems and human food security, which is a singular position for such a group of organisms. The past three decades have seen unparalleled interest in pollination ecology, plant-pollinator's biology and natural history, in some way stimulated by worries about the decline of pollinator abundance and diversity in several natural and agricultural ecosystems. In this chapter we will present a brief history of the origins and evolution of pollination, followed by a characterization of the main pollinator groups, what is basic to open ways to whom could be interested in to follow this important research issue. Additionally, we will briefly discuss human impact on pollination systems and point out some future direction.

6.1 A Brief Historical Approach of Pollination and Flower Origin

6.1.1 Definition

Pollination is the transport of pollen grains to the stigmatic surface of a flower. Pollen is the special structure that contains the plant's male gametophyte. A successful pollination event needs to happen between flowers of the same species, and preferably from distinct individuals (Wilcock and Neiland 2002). Post-pollination outcomes depend on many factors. A pollen grain hydrates on the stigma before it's germination and production of the pollen tube, which goes through the stigmatic surface and the transmitting tissue into the style, then down to the ovary and the ovule inside. One of the gametic nuclei (sperm nuclei) inside the pollen tube reaches and fuses with the oocyte, producing an embryo, while the second gametic nuclei fuses with the two polar nucleus in the central cell, resulting in the endosperm. This process is known as double-fertilization and is found in almost all Angiosperms, configuring an important characteristic in this plant clade (but see also Williams and Friedman 2002). These processes can be interrupted at any time due to incompatibility reactions, which may prevent self-pollination and pollination between flowers from distinct species. In the latter case, heterospecific pollen can physically block the stigmatic surface and disturb conspecific pollen performance in several ways extensively discussed by Streher et al. (2020) and references therein. Specialized floral morphology and physiology limit the visits to the few pollinators able to perform the best pollen transfer. However generalist flowers are much more common in nature and they usually have to support heterospecific pollination consequences (Wilcock and Neiland 2002; Ollerton et al. 2007).

Pollination happens in gymnosperm and angiosperm species, as they are able to produce pollen grains. Pollen can be transferred abiotically by wind or water, biotically by animals or even automatically by the flower itself. Most plants depend on animal pollination for their reproduction (Klein et al. 2007). As an example, possibly 94% of plants from tropical communities are pollinated by animals (Ollerton et al. 2011).

From the animal point of view, pollination is often an accident or a by-product of floral resource collection (Vitali and Machado 1994). A generalist pollinator is able to collect floral resources from a wide array of plant species (Waser et al. 1996), while a specialist pollinator has close evolutionary adaptations that limits its interactions to a few plant taxa (Dormann 2011).

6.1.2 Flower Evolutionary History and Morphology: What Have We Learnt So Far?

Flowers are present on Angiosperm species and are responsible for fruit and seed production, ensuring sexual reproduction. The definition of a flower has changed a little bit over time (Moseley 1898; Bateman et al. 2006). Today we can say that a flower is a very compressed and determinate bisexual or unisexual reproductive axis, with its structures placed in concentric whorls with radial (actinomorphic) or bilateral (zygomorphic) symmetry. The central whorls consist of carpels (megasporangia) and/or stamens (microsporangia), surrounded by a sterile perianth with at least one laminar organ usually composed by distinct petals and sepals, or by similar structures called tepals (Bateman et al. 2006; Theissen and Melzer 2007; Smyth 2018). The carpel presents the stigmatic surface where pollen may be deposited, the style and the ovary containing the ovules. The stamen is composed of the filament and the anther, which is responsible for pollen grain production. The fossil history of stamens is well explained by Crepet and Nixon (1996). Floral morphology plays a central role in the pollination process, allowing legitimate pollinators to contact the anthers and transfer pollen grains to the stigma. Hence, it is important to understand the origin of flowers, their formation and the factors affecting floral development and morphology.

Due to the evolutionary importance of the flower in defining the clade of the flowering plants (Angiosperms), one could expect a general agreement about the concept of the flower, but no perfect consensus has been achieved until now. A central problem in understanding the flower's evolutionary history is the difficulty in homology assignments among flowering plants and between the most recent common ancestor of Angiosperms and Gymnosperms (Fröhlich 2003; Theissen and Melzer 2007).

In order to try to fill this gap, diverse tentatives were made to elucidate the evolution of flower morphology. Traditionally, the ancestral flower was considered to be similar to modern Magnoliidae flowers (Crepet 1984; Crepet et al. 1991), what is partially supported by recent data based on phylogenetic ancestral reconstruction (Saunquet et al. 2017). Based on this last approach, the structure of the basal Angiosperm flower had a radially symmetric perianth composed of tepals surrounding a bisexual structure, with carpels organized in a spiral and an androecium organized in whorls. From this basal floral structure and counting on an elegant genetic control and pollinator pressure, a huge diversity of flower shapes can be found among the extant Angiosperms.

6.1.3 The Enigmatic Origin of Flowering Plants and Animal Pollination

Charles Darwin was fascinated by angiosperm emergence and its intriguing fast radiation which he called the "abominable mystery", as it initially seemed improbable. Darwin suggested angiosperm-insect associations could have increased the angiosperm diversification (Friedman 2009). According to the fossil record, angiosperms appear in the Mesozoic era, Triassic period, about 200 million years after the Gymnosperms (Doyle 2012; Li et al. 2019). However, recent DNA sequencing studies and molecular dating methods suggest that the first angiosperms originated in the upper Triassic, about 209 million years ago (Li et al. 2019) or a little after that in the Jurassic (Doyle 2012). The time of angiosperm appearance is still nowadays a question to be solved (van der Kooi and Ollerton 2020).

6.1.3.1 Herbivory and Pollination

Since the early development of the first land plants by the Ordovician period (Paleozoic era), spores were spread by water and wind. Those spores belonged to very simple plants, possibly the bryophyte ancestors (Wellman et al. 2003). Sometime after, in the Devonian period, land was full of moss-like vegetation (Bryophytes), ferns (Pteridophytes) and conifers (Gymnosperms). Those plants were damaged by primitive arthropods looking for food such as spores, pollen grains and plant juices, which offered a very good nutrient source for arthropods, even for small ones unable to masticate (Kevan et al. 1975; Ren et al. 2009). Herbivores used to feeding on those other plant resources gradually started to consume the newly emerged angiosperms (Nyman et al. 2019). Herbivory is the consumption of any living vegetal tissue (Abrahamson 1989, Marquis and Moura: ▶ Chap. 3).

Antagonistic plant-animal interactions, such as pollen consumption by animals, could have been the first step that led to the emergence of animal pollination (Ren et al. 2009, see review by Del-Claro and Torezan-Sililingardi: ▶ Chap. 1). Still now herbivores and pollinators can use the same cues to find a specific flower, such as floral scent (Nunes et al. 2016). Liquid floral resources such as ovular secretions, stigmatic exudate and nectar of angiosperm flowers were (and still are) commonly used as food by animals (Lloyd and Wells 1992; Kato and Inouye 1994). Plant reproductive parts show rapid development, which makes it difficult to form rigid fibers, so the reproductive parts are usually soft and rich in water, mineral nutrients and carbohydrates, which make them very interesting for insect herbivores (Varanda et al. 2005; Mccall and Irwin 2006), both to those that feed inside (endophytic) or outside (exophytic) the structure (Labandeira 1998; Torezan-Silingardi 2011).

Evidence from the earliest seed plants indicates animal (Kato and Inouye 1994; Labandeira 1998; Ollerton and Coulthard 2009) and also wind pollination (Crepet and Feldman 1991; Niklas 1997). The distinct insect mouthparts from the Devonian period indicate their ability to collect and use many floral resources. For example, there are many fossil records of insects covered by gymnosperm pollen (Grimaldi 1999; Hu et al. 2008), as Mecoptera (scorpionflies) (Ren et al. 2009), Thysanoptera (thrips) (Peñalver et al. 2012), Neuroptera (lacewings) (Labandeira et al. 2016), Diptera (flies) (Peñalver et al. 2015) and Coleoptera (beetles) (Peris et al. 2017a). There is also coprolite evidence of cycad pollen grains consumed by beetles (Klavins et al. 2005). These two types of fossil evidence indicate gymnosperm-insect pollination. Some other fossil evidence points to the development of angiosperm species in the same space and time as when gymnosperms were insect pollinated (Ollerton 2017). Many of those Cretaceous gymnosperm pollinator insects are nowadays exclusively angiosperm pollinators (Peris et al. 2017b).

6.1.3.2 The Ecological Dominance of Angiosperms

The emergence of the flower promoted additional chances for highly-faithful mutualisms and coevolution (Crepet and Nixon 1996; Hu et al. 2008). Hu et al. (2008) prepared a reconstruction of the evolution of pollination modes in angiosperm species. Their conclusions support the hypothesis that insect pollination was the initial pollination mode, so more specialized animal pollination modes should be derived. They present fossil evidence implicating insect pollination in approximately 86% of the basal angiosperms, many of which produced clumped sticky pollen and pollen ornamented surfaces adequate for insect pollination. Several other fossil records also support insect pollination, with insects feeding on just one type of pollen grains and in a big quantity, which suggests a food preference (Grimaldi 1999) or specialized pollination (Crepet and Nixon 1996). Those records indicating insects feeding on many flowers from the same species may suggest a situation where the animal was a regular flower visitor due to its feeding on flower resources and, consequently it could act as a constant pollen transfer, characterizing a pollination mutualisms.

A certain degree of dependence evolved in some of those flower-insect interactions, suggesting coevolution. In these cases, species develop a series of morphological and physiological characteristics that make them more apt to establish mutualistic interactions over evolutionary time (Lomáscolo et al. 2019). The associations between angiosperms and animals have benefited both pollination processes and seed dispersal and are an important part of the explanation for the great diversification of angiosperms (see Corlett: ► Chap. 7).

The fossil record shows evidence of insect pollination of angiosperm species by Coleoptera (Crepet 2008; Poinar Jr 2016; Nabozhenko 2019), Diptera (Grimaldi 1999 and references therein), Lepidoptera (Crepet 1984) and Apoidea: bees (Crepet 1984; Crepet 2008). These associations facilitated the rapid increase in richness of species, abundance in number of individuals and angiosperm radiation worldwide (e.g. Thompson 2014, see also Del-Claro and Torezan-Silingardi: ► Chap. 1).

6.2 Floral Rewards

A floral reward is any resource produced by floral tissue that is used by animals and leads to an increase in visitation rates helping the pollination process (Simpson and Neff 1981). Pollen, nectar, oils, resins, fragrances and floral tissue are floral rewards used by animals mainly for their nutritious value. Many animals can pollinate, but not all visitors are efficient pollen vectors, some are just looking for an interesting floral resource and do not pollinate adequately. For example, spiders feed on stigma exudates (Marquínez et al. 2010) but they usually just use one of the several floral products without pollinating. Occasional pollination is possible due to insects looking for the flower as a sleeping or mating place, heat source and fiber or exudate sources for nest construction (Neff and Simpson 1981). Evolution has provided flowers with plenty of possibilities.

6.2.1 Pollen

Pollen is a nutritious resource, its composition depends on the plant family and can be very variable (Campos et al. 2008). Pollen grains are composed of proteins and amino acids, lipids, carbohydrates, vitamins, carotenoids, sterols, fibers, minerals and salts (Day et al. 1990; Roulston et al. 2000; Campos et al. 2008). Pollen protein and lipid ratios are a factor able to induce selection on plant-pollinator interactions (Vaudo et al. 2020). Depending on the taxonomic plant group, pollen presents secondary compounds as flavonoids, alkaloids, terpenoids, and phenolic compounds (Palmer-Young et al. 2018). Bees can selectively forage on distinct pollen species based on their quality, which influences larval development and colony fitness (Vanderplanck et al. 2014; Cane 2016).

6.2.2 Floral Nectar

Nectar is the most frequent floral resource in South American communities (Silberbauer-Gottsberger and Gottsberger 1988; Quirino and Machado 2014). Floral nectar is produced in many angiosperm and some gymnosperm species as a sugar solution composed of water and different proportions of carbohydrates such as sucrose, fructose and glucose, small quantities of proteins (nectarins), amino acids, and volatile organic compounds (Heil 2011; Abrahamczyk et al. 2017). Distinct plants produce floral nectar with different quantities of some secondary compounds other than those cited by Heil (2011), such as flavonoids, alkaloids, terpenoids, and others (Palmer-Young et al. 2018).

The kind of sugar determines the main group of pollinators: sucrose-rich nectar is preferred by hummingbirds, butterflies, moths and long-tonged bees; hexose-rich nectar is preferred by short-tonged bees and flies; and sucrose-free nectar is preferred by birds and ants (Blüthgen and Fiedler 2004; Abrahamczyk et al. 2017). Variation in nectar volume, composition, and concentration are easily detected by pollinators and influence their behavior (Pyke 2016), indicating floral nectar is under pollinator-mediated selection (Parachnowitsch et al. 2019).

6.2.3 Floral Oils

Floral oils are fatty acids rich in calories produced in epithelial or trichomal specialized glands called elaiophores, and their composition varies among species and genera from the 11 families able to produce them (Vogel 1974; Possobom and Machado 2017). Malpighiaceae floral oils present fatty acids associated with compounds such as diacylglycerol, compounds from the acetoxy group, acetyl group and diacetoxyicosanoic acid methyl ester group (Barônio et al. 2017), while Orchidaceae floral oils present acylglycerols and relatively simple linear hydrocarbons such as diacylglycerol compounds (Reis et al. 2006). Floral oil composition is more similar within genera than between genera (Barônio et al. 2017). Floral oils are used by specialized oil-collecting bees for larval food provisioning (Michener 1974, 2007; Vogel 2009), for nest waterproof after lining brood cells (Vogel 2009) and nest construction (Alves-dos-Santos et al. 2002).

6.2.4 Floral Resins

Floral resins are composed by mucilage, starch and sugars and small amounts of lipoidal substances (Krahl et al. 2019). Floral resin from *Dalechampia*, *Clusia* (Armbruster 1984) and Orchidaceae (Krahl et al. 2019) flowers are collected by pollinating bees. The resin is used for bee nest construction as it is water repellent and has anti-pathogen chemical properties (Armbruster 1984; Roubik 1989; Oliveira et al. 1996). Sometime resins are neither a floral reward, nor are associated with pollinator attraction (Gottsberger and Amaral Jr 1984; Gottsberger 1986). Resins can act as a flower tool to facilitate pollen adhesion to the pollinator body. The spathe of *Philodendron adamantinum* produces and releases resin droplets just before pollen grains are liberated from the anthers (Gonçalves-Souza et al. 2018). The scarab beetles visiting *P. adamantinum* are smeared by the resins contacting the spathe inner side, when the inflorescence closes and the beetles are forced to leave. The resin acts as an effective adhesive for pollen grains, increasing pollination rates (Pereira et al. 2014).

6.2.5 Fragrances

Volatile organic compounds are the main attraction cues for crepuscular and nocturnal pollinators such as bats, moths and beetles. However, even diurnal flowers rely on scent to signal their locality to pollinators, such as the African milkweed *Pachycarpus grandiflorus* that attracts wasp pollinators primarily by scent rather than visual cues (Shuttleworth and Johnson 2009). An experiment with flowers from species of the genus *Thalictrum* (Ranunculaceae) showed that the volatile organic compounds of insect pollinated flowers are able to elicit larger antennal responses from *Bombus impatiens* than volatiles produced by wind-pollinated flowers, indicating that floral scent is a key characteristic to pollinator attraction (Wang et al. 2019).

6.2.6 Floral Tissue

Pollinators can use the flower as a place to develop their offspring, as seen in *Yucca*-yucca moth (Powell and Mackie 1966; Huth and Pellmyr 2000), *Glochidion*-glochidion moth (Kato et al. 2003) and *Ficus*-fig wasp associations (Anstett et al. 1997, Jansen-González et al. 2012, and see Bronstein: ► Chap. 11, and Pereira and Kejellberg: ► Chap. 12). *Yucca*, *Glochidion* and *Ficus* pollen is collected by adult female moths or wasps that carry it to another flower and deposit the grains over the stigma, but do not eat pollen grains. The pollinated flower is the oviposition place, where the moth or wasp larva grows while eating the developing seeds and part of the ovary tissue.

6.3 Pollinating Vectors

Pollen can be transported to the stigma in many distinct ways, depending on corolla size and shape, and the distinct pollen presentation in the anthers and characteristics such as the size, shape and ornamentation of the pollen grain. There are two kinds of vectors capable of pollinating: abiotic, such as wind and water, and biotic such as animals (◘ Table 6.1). Here, we will present both types of pollination, but will focus on biotic interactions, the issue of this book.

Abiotic pollination is observed in both aquatic and terrestrial plants. About 18% of all extant angiosperm families are pollinated by wind and about 3% are pollinated by water (Ackerman 2000 and references therein). Probably the local decrease of pollinator populations and/or unfavorable abiotic conditions caused entomophilous angiosperms to evolve and adapt to wind pollination as a derived condition (Culley et al. 2002), but the opposite situation also happened (Wang et al. 2019).

The vast majority of plant species need biotic pollen vectors (about 80% following Ackerman 2000). These numbers are even higher if we consider species-rich communities from warm and wet places such as tropical forests, where animal pol-

□ Table 6.1 Pollination types and main characteristics

Pollination type	Pollination vector	Floral reward	Floral type	Reference example
Hydrophily	Water	No reward	Small flowers, petals and sepals are absent, no scent	Cox (1988)
Anemophily	Wind	No reward	Small flowers, petals and sepals are absent, no scent	Friedman and Barrett (2008)
Ambophily	Wind + animals	Pollen, nectar	Small flowers, open corolla, there is scent	Tamura and Kudo (2000)
Zoophily (Entomophily)	Animals (insects)			
	Hymenoptera			
Melittophily	Bees (diurnal)	Pollen, nectar, scent, oil	Diurnal, petals and sepals ornamented, scent present. Symmetry, size and shape variable	Torezan-Silingardi and Del-Claro (1998)
Melittophily	Bees (crepuscular/ nocturnal)	Pollen	Radial symmetry, open corolla, exposed fertile structures and scent present	Wcislo et al. (2004), Warrant (2008)
—	Wasps	Nectar	Diurnal anthesis, usually with radial symmetry, scent present and ornamented corolla	Santos et al. (2010)
Myrmecophily	Ants	Pollen, nectar	Diurnal, small flowers, radial symmetry, open corolla, and exposed fertile structures	Del-Claro et al. (2019)
Cantharophily	Coleoptera	Pollen and floral tissue	Diurnal/crepuscular/ nocturnal, color is variable, scent present	Gottsberger (1989)
	Lepidoptera			
Psychophily	Butterflies	Nectar	Diurnal anthesis, tubular corolla, vibrant floral coloration, sweet scent	Goulson and Cory (1993)
Phalaenophily	Small moths	Nectar	Nocturnal/crepuscular anthesis, tubular corolla, discrete floral coloration	Freitas and Sazima et al. (2009)

▣ Table 6.1 (continued)

Sphingophily	Big moths	Nectar	Evening/crepuscular anthesis, tubular corolla, discrete floral coloration, some species have very strong pleasant scents	Amorim et al. (2014)
	Diptera			
Myophily	Flies	Nectar, pollen	Diurnal anthesis, fermentation or decomposition scent	Goldblatt and Manning (2000)
Sapromyophily	Carrion flies	No reward	Diurnal anthesis, scent of meat decomposition	Wiśniewska et al. (2019)
	Thysanop-tera	Nectar, pollen	Diurnal anthesis, small, white or yellow	Williams et al. (2001)
Zoophily (vertebrates)				
Chiropteroph-ily	Bats	Nectar (pollen)	Nocturnal flower, white, yellow or green, pollination between distant plants, some species have strong unpleasant scents	Fleming et al. (2009)
Ornitophily	Birds	Nectar	Diurnal flower, tubular flowers of vibrant colors, often red	Cronk and Ojeda (2008)
—	Non flying mammals	Nectar	Diurnal and nocturnal flowers, radial symmetry, usually near the ground, scent present	Carthew and Goldingay (1997)
—	Lizards	Nectar	Diurnal, rare pollination, flower tissue may be consumed	Olesen and Valido (2003), Sazima et al. (2005)

lination is observed in 94% of plant species (Ollerton et al. 2011; Rech et al. 2016). The reduced winds in dense forest vegetation can explain the prevalence of animal pollination, at least partially. Animals are attracted to flowers due to their fragrances (long distance cues), colors and acoustic guides (medium distance), visual guides and size/shape/morphological/tactical cues (short distance) (▣ Table 6.1 and references therein, Helversen and Helversen 1999, Raguso 2004, Streinzer et al. 2009). If the flower shape is modified, as in a corolla damaged by florivory (herbivore damage on floral parts) or a flower manipulative experiment, the visitation rate and consequent fruit formation will decrease, especially in self-incompatible species (Ferreira and Torezan-Silingardi 2013).

The image of a flower plays a key role in its identification and offers honest cues of quality to floral visitors. Flower image is mainly the result of corolla size and shape and the presence of pigments such as anthocyanins and carotenoids. However, microscopic conical cells in the adaxial surface of petals, tepals or nearby bracts and other floral parts can modify the color pattern of a pigment and render it more intense or more discreet (Moyroud and Glover 2017; Kraaij and van der Kooi 2020). The conical design permits the color intensity to be increased as more light reaches the cell vacuole and intensifies the pigment reflection, reducing surface gloss and creating a velvety appearance which permits better use of floral pigments and light scattering structures inside floral tissue (Stavenga et al. 2020). The result is a flashy and intense color. The fruit-set dependent on pollinator attraction is reduced in mutant plants that have lost the conical cells and produce flat cells instead (Dyer et al. 2007).

Each animal clade has its own ability to perceive colors, a recent review based on *Drosophila* is given by Schnaitmann et al. (2020). We humans are trichromatic (Cronin et al. 2014). Insect pollinators such as bees are trichromatic too (Briscoe and Chittka 2001), butterflies may even have 15 photoreceptors and some of them have the chromatic abilities (Chen et al. 2016). Vertebrate pollinators like mammals may be di- or trichromatic (Arrese et al. 2002) and birds are tetrachromatic (Hart and Hunt 2007). Fly vision is not completed understood, but color vision is possible due to a system with four (Lunau 2014) or five photoreceptors (Schnaitmann et al. 2020).

If pollinator diversity is restricted to just one or few groups, plants that depend on biotic pollination for reproduction and maintenance will be locally conditioned by the existing pollinators. For example, Macquarie Island is isolated in the Southern Ocean, 1200 km from its nearest neighbor, New Zealand. The island has neither bees nor birds visiting flowers, just dipteran pollinators live on the island, consequently flies have become the big ecological limiting factor for the local flora which is composed of 36 indigenous angiosperm species presenting basically white-cream-green flowers (Shrestha et al. 2016).

The convergent evolution of floral features driven primarily by distinct adaptations to the functional group of the most effective pollinator was predicted by Stebbins (1970). It was confirmed in a review with 417 plant species by Rosas-Guerrero et al. (2014), which also pointed out that secondary pollinators are not excluded by pollination syndromes. Pollination syndromes use characteristics such as floral color, shape and size, rewards and many others features to organize the huge amount of distinct floral types under a functional and ecological perspective, pointing to possible and probable pollinators (Dellinger 2020). Here we present the main pollination syndromes:

6.3.1 Water Pollination: Hydrophily

Hydrophilous pollination is known for at least 31 genera from 11 plant families, from marine and freshwater environments (Cox 1993), including dioecious, monoecious and hermaphroditic species. For example, *Vallisneria* and *Lagarosiphon* species have submerged plants that liberate the male flowers, which float to the surface of water and move with the breeze. Another submerged plant produces a long branch with the female flower at the end, floating in the surface of the water. A delicate depression in the water surface is produced by the female flower, permitting the capture of the male flower and pollen transfer to the stigma. Some genera, including *Amphibolis*, *Halodule* and *Ruppia* have a distinct pollinating process where just the pollen grains are liberated to reach the water surface, where they encounter the very long stigma, but the male and female flowers are both underwater. Other plants such as *Thalassia* and *Syringodium* can transport pollen in mucilaginous strands underwater between submerged flowers, and some species from *Phyllospadix* and *Zostera* have pollen grains that move on the surface and underwater to reach the stigma. Cox (1993) and Ackerman (2000) and the references therein present many interesting details about hydrophilous pollination.

6.3.2 Wind Pollination: Anemophily

Anemophily is reported in 60 angiosperm families, common in species with either spatial (dioecy) or temporal (dichogamy) separation of stamens and pistils (Ackerman 2000) and with dense inflorescences wherein small flowers present reduced and inconspicuous perianths (Stebbins 1970). The long and delicate filament permits the exposed anther to move easily with the wind, releasing pollen grains through large openings usually in dry conditions, common in Poaceae species (Kozub et al. 2017) and also in other families. Anemophilous pollen grains are usually spherical and small, unornamented and without pollenkitt, so they do not clump easily (Crane 1986). The pollen-ovule ratios are impressively high, greater than10^6:1 (Faegri and van der Pijl 1979) or than10^{15}:1 (Payne 1981), but just a small number of grains will reach the stigma. The brush or feather-like stigma and its extensive surface increase pollen capture, especially in windy and dry periods (Heslop-Harrison and Shivanna 1977). Anemophilous species are usually observed in higher latitudes and elevations mainly in temperate areas (Regal 1982). Wind pollination is common in self-compatible plants, in gregarious population living in open habitats, with low humidity and low precipitation (Culley et al. 2002).

6.3.3 Ambophily: Wind and Insect Pollination

The phenomenon in which species are pollinated by both wind and insects is called ambophily (Ackerman 2000). Ambophily can occur in two situations. The huge amount of pollen grains of anemophilous species can be an important food resource easily accessed by generalist insects during periods of pollen limitation for nearby entomophilous species (◻ Fig. 6.1). Or, plants usually pollinated by insects can be wind pollinated if pollinator limitation is severe and pollen features permit wind transport (Culley et al. 2002). Examples of ambophilic pollination are spread over Arecaceae (Silberbauer-Gottsberger 1990; Rios et al. 2003), Euphorbiaceae (Quirino and Machado 2014), Salicaceae (Tamura and Kudo 2000), Ranunculaceae (Wang et al. 2019) and many other plant families.

◘ Fig. 6.1 Diptera eating pollen from the longitudinal dehiscent anther of Poaceae flower. Note the white feather-like stigma and thin style near the anther. (Photo credit: Eduardo Calixto Soares)

6.3.4 Insect Pollinators

Pollination by insects is much more frequent and diversified than vertebrate pollination, both in nature and in cultivated areas. Wilson (1992) pointed out that we have about 751,000 distinct insect species, the equivalent of 72.2% of all animal species. However, Stork (2018) recently suggested that this number could surpass five million species considering that Earth contains 20 million species or more, including cryptic diversity. Thus, it is easy to presume that we could find more insect pollinators than vertebrate pollinators. But there are more reasons to justify that idea.

The immense richness of insect species needs to be considered together with other characteristics that make this group of pollinators important: (a) the reduced size usually seen in shorter life cycle species makes the development of genetic mutations more common. Mutations that increase fitness are more likely to be maintained in subsequent generations, but mutations that decrease fitness will be negatively selected for and may disappear or remain with low numbers in the population; (b) many insect species have different food niche between juvenile and adults; (c) insect dispersion is favored by winged adults, facilitating the reach of distant areas; (d) insects developed a sophisticated neuro-motor and sensory system, with special olfactory receptors able to perceive very small quantities of volatiles organic compounds far from the plants producing them. Odor is more relevant than visual cues for insect pollination in many extant gymnosperms (Rydin and Bolinder 2015 and references therein); (f) insects have distinct types of visual receptors that allow them to perceive more colors than humans do. All these traits are described in detail by Bernays (1992), Gullan and Cranston (1994), Triplehorn and Johnson (2020), Schowalter (2000) and many others.

These morphological and physiological adaptations allow insects to have many distinct interactions with plants. Of the 26 insect orders just five intensely interact with flowers, however these five orders had so many benefits from flowers that they constitute 84.5% of insect species: Coleoptera, Lepidoptera, Hymenoptera, Diptera and Thysanoptera. Most plants receive distinct floral visitors and many of them may be apt to pollinate in a mixed pollination system (Cesário and Gaglianone 2013; Shuttleworth and Johnson 2010). For instance, *Ipomoea pes-caprae*

(Convolvulaceae) is bee pollinated and snail pollinated (malacophily) (Raju et al. 2014). The most relevant types of pollination and their main characteristics are pointed out in ▢ Table 6.1.

Here we will present some insect-flower associations using the most relevant orders.

6.3.4.1 Diptera Pollination: Myiophily and Sapromyiophily

Myiophily, besides more generalized pollination by dipterans also includes sapromyiophily or pollination by attraction of carrion and dung-feeding flies. Diptera are among the most common insects that visit flowers. Among the 110 Diptera families, representatives of more than 70 families contain anthophilous species (Kearns 2001; Larson et al. 2001). Species range from opportunistic nectar and pollen consumers (the generalists), to specialists, which may be attracted to sapromyiophilous flowers that mimic carrion or dung or others that imitate fungal fragrances or even fungal forms, or others that have proboscides much longer than their body and are thus able to explore and pollinate flowers with long-tubed corollas.

Good examples of generalist foraging by Diptera can be shown for several Syrphidae, Tachinidae, Calliphoridae, Muscidae, among others. Hoverflies (Syrphidae), for example, are fond of flowers with easily accessible nectar and which are white, reddish or blue (Knuth 1898). Certain species, such as the well-known syrphid fly in Europe, *Eristalomyia tenax*, have a strong preference for yellow flowers (Kugler 1955; Lunau et al. 2005). Many Syrphidae and other flies also eat pollen grains in addition to nectar (Holloway 1976; Inouye et al. 2015) and thus forage on a great number of flowering plants, usually together with other pollinator groups.

In South Africa the specialized long-tongued flies of Nemestrinidae and Tabanidae represent a distinct pollination niche, servicing colorful and nearly scentless, long-tubed flowers. These specialist flies explore the nectar of tubular flowers with mouthparts exceeding their body lengths. One nemestrinid fly with a proboscis 57 mm long, four times its body length, pollinates the orchid *Disa draconis* (Johnson and Steiner 1997). The long floral tube makes nectar unavailable to most other insects, including other flies, bees and wasps (Manning and Goldblatt 1997; Goldblatt and Manning 2000; Potgieter and Edwards 2005). Another specialized group of flies with long tongues are the bee flies (Bombyliidae), which can gain greater nectar rewards in the long-tubed flowers than other flies in generalized flowers (Chittka and Thomson 2005).

Another specialization of flies (e.g., *Sarcophaga* spp., Sarcophagidae, and other groups) is to deposit their eggs and rear their larvae on feces of carnivorous and omnivorous mammals or cadavers. Certain flowers imitate such substrates in color and stench emission and thus evolved a deceptive fly pollination system (Bänzinger and Pape 2004). Most remarkable are the gigantic flowers of the Asian *Rafflesia* (Rafflesiaceae) species. Thus, females of calliphorid flies were found on the vivid dark colored *R. cantleyi* flowers, which emit a pungent floral scent. The analyses indicated that the species "biochemically mimics carrion and that relative ratio of oligosulfides in the floral scent play a key role in sex-biased pollinator specialization, attracting only female *Chrysomya chani* flies to the flowers" (Wee et al. 2018).

A prominent group with myiophilous and even deceptive (sapromyiophilous) systems are the Aristolochiaceae. The basic mechanism of pollination in *Aristolochia* has been known since Sprengel (1793). The typical characters associated with fly pollination in this family were summarized by Huber (1993): flower gigantism in some species; caudate perianth lobes often bearing osmophores (Vogel 1962); flower parts imitating fruiting bodies of mushrooms (including their lamellae), and the flowers indeed being pollinated by fungus gnats (Mycetophilidae) (Vogel 1978); in some species there are floral nectaries or nectarioles, which play a role in attracting certain flies (Daumann 1959; Vogel 1998a; Murugan et al. 2006), and in some cases the nectar functions as food to guarantee survival of the imprisoned pollinators, in which case the nectaries belong to the trichomatous type located inside the utricle (Vogel 1998b; Erbar 2014); dark purple, brown or black coloration, often contrasting with yellow or green backgrounds; a musky, fruit-, fungus-, urine- or carrion-like scent in several species; and the perianth tube converted into a trap, which commonly retains visitors by a smooth, oily inner surface or by stiff "trap hairs". These hairs allow the visitors to enter the basal part of the perianth tube, the utricle, harboring the stigmas and anthers, but inhibit their exit until hairs wilt after pollination. Oelschlägel et al. (2015) described an extraordinary kleptomyiophilous strategy for *A. rotunda*. The main pollinators are female chloropid flies. The flies are food thieves that feed on secretions of true bugs (Miridae) while these are eaten by arthropod predators. Freshly killed mirids and *A. rotunda* flowers release the same scent compounds that chloropid flies use to find their food sources.

Many more angiosperm species and whole groups, beyond the few mentioned above, depend on flies as generalist or specialist pollinators, such as Apocynaceae (Asclepiadoideae), Araceae or Orchidaceae and others. Therefore, Raguso (2020), when he reports about dipteran diversity and its consequences for floral ecology and evolution, certainly is correct in adverting: "Don't forget the flies". This also has a practical or economic component for global crops. Facing tremendous environmental changes, bees as pollinators of crops are in strong decline and non-bee insects, such as flies, beetles, moths, butterflies, wasps and ants, but also birds and bats, may be substitutes for crop pollination services. Rader et al. (2016) focused and synthesized 39 field studies from five continents and found that non-bees performed 25–50% of the total number of flower visits. Non-bees were less-effective pollinators than bees, but they made more visits, and thus pollination services rendered by non-bees were similar to those provided by bees. On the other hand, there are some tropical crops that can be pollinated "only" by flies. One of these crops important for the world market is *Theobroma cacao* (Malvaceae), which is exclusively fly pollinated by Ceratopogonidae and Cecidomyiidae. The hidden stigma and anthers can be reached only by flies, and *Trigona* bees are at the maximum pollen thieves but not pollinators (Westerkamp and Gottsberger 2000).

There might be a bias with regards to the anthophilous fly distribution across elevation gradients. Since Müller (1881) the idea prevails that "The predominance of flies on flowers at high elevations appears to be due to a decrease in representation of other orders of insects." (Kearns 1992, and literature mentioned therein). However, today, high altitude montane and alpine regions nearly all over the world are overgrazed by a too high number of livestock (cow, horse, sheep, goat, etc.). Such overgrazed and destroyed mountainous and alpine environments full of animal feces evidently do not have the sufficient natural plant species composition to attract hymenopterous, lepidopterous and coleopterous pollinators, but seem to be much more attractive for certain fly groups. In the Ötztaler Alps, in the Austrian province Tirol, at an altitude of approximately 2550 m, well above the timber line, we observed an area of several hectares being fenced and protected against grazing livestock for making hay. The fenced meadow was extremely rich of flowering alpine plant species and also of all kinds of hymenopterous, lepidopterous and coleopterous pollinators. Apparently, it is not the high altitude which prevents the presence and activity of non-dipterous insects but the lack of sufficient resources in overgrazed and largely destroyed environments (Gottsberger and Silberbauer-Gottsberger, pers. obs.). The old story that an increasing altitude promotes the presence of flies in detriment to other pollinators is probably a legend.

6.3.4.2 Coleoptera Pollination: Cantharophily

Pollination by beetles is a very old association between plants and insects and cantharophily was also already established long before the advent of the angiosperms. Beetles were associated with the extinct gymnosperm group Bennettitales, which existed from the Triassic to the Cretaceous. A comparison of Bennettitales with cycads and angiosperms, three plant groups which co-existed for a certain time, hints that Bennettitales, just as cycads and many early angiosperms, were indeed also insect pollinated, with their pollinators probably being beetles and in addition also flies and thrips (Delevoryas 1968; Crepet 1972; Crowson 1981). In the Cycadales, one of the oldest gymnosperm groups, and which have still about 350 extant species worldwide (Terry et al. 2012), beetles are the major group of pollinating insects and fossils indicate that beetles were already their pollinators at least 100 mya ago (Peris et al. 2017c; Alekseev and Bukejs 2017; Cai et al. 2018).

The angiosperms or flowering plants first appeared in the fossil record about 130 mya in the Early Cretaceous. Although we know that many of the early-diverging angiosperms went extinct, there still exists quite a number of representatives of these early groups. These are the ANITA grade group with *Amborella* (Amborellales), Nymphaeales and Austrobaileyales, the early monocots, the Chloranthales, probably together with the Ceratophyllales and the magnoliids. Although the early-divergent angiosperms started as pollination generalists, jointly pollinated by Diptera, Coleoptera, Thysanoptera and other insect groups, several of these angiosperms specialized either exclusively on beetles, flies, bees or thrips. Among the 13,000 to 14,000 extant basal angiosperms, there are approximately 3,600 cantharophilous species, thus exclusive beetle pollination in basal angiosperms is the most common form of specialist pollination, followed by fly pollination with an estimated 2,400 species, bee pollination with around 550 cases, thrips pollination with 160, and cockroach pollination with only a few species (Gottsberger 2016).

Some characteristics of beetle-pollinated basal angiosperms are protogyny, with the female pistillate organs of the flower being functional before the male staminate organs. These protogynous flowers produce thick, tissue-rich petals or tepals. Petals or tepals also curve over the flower center to form a dark interior, called the pollination chamber, which is the most important structure to maintain the beetles inside the flower during the first part of anthesis until pollen is shed (Gottsberger 2016). The closed, dark interior of the flowers also shields the beetles from daylight and the chamber may be warm and scented (especially during the evening and night), promoting the activities of the beetles, such as feeding, digesting and mating (Seymour and Schultze-Motel 1997; Seymour 2010), and at the same time protecting the beetles against predatory birds and lizards.

To illustrate the relationship of angiosperm flowers and beetles and the co-adaptations among them, a few examples of magnoliid representatives of the order Magnoliales are given. The largest family with about 2,300 species is the family Annonaceae, whose species are predominantly pollinated by beetles. Only some genera and species are pollinated exclusively by thrips (Thysanoptera), flies, cockroaches and even by bees. There are two major groups of beetles pollinating flowers of Annonaceae. Small beetles (Curculionidae, Nitidulidae, Staphylinidae, Chrysomelidae), with a body length of up to 7 mm, are pollinators of the majority of Annonaceae species, and large scarab beetles (Dynastinae, Rutelinae, Cetoniinae, Trichiinae), with a body length of 14–21 mm, pollinate a smaller number of species.

How do the flowers of the small beetle pollinated Annonaceae function? For example, in *Guatteria* species, the greenish open petals of the pendulous flowers expose their reproductive organs days and weeks before the flowers become anthetic and are first completely unattractive for their pollinators. When the flowers finally enter their reproductive stage (Gottsberger 1999), the hard and greenish petals become soft, change to yellow and sometimes even to brownish, and the three inner petals fold over the flower center to form a relatively small dark pollination chamber, which starts to emit a heavy acetone- and fruit-like scent (with e.g., alcohols, esters, benzenoids; Jürgens et al. 2000). These changed anthetic flowers attract small fruit-inhabiting Nitidulidae beetles by providing floral odors similar to overripe fruits. The beetles remain inside the pollination chamber until the next day when the flower enters its staminate stage and sheds its pollen. In this stage the flower drops its petals, the pollen-covered beetles are forced to leave and are then eventually attracted by another strong-smelling flower in the pistillate stage, which they pollinate when penetrating into the semi-closed pollination chamber. Although each species of small-flowered Annonaceae has its particularities, they all show the common characteristics of cantharophilous species. Some Annonaceae with small flowers have diurnally active flowers, and others have nocturnally active flowers. Especially, some of the nocturnally active species of *Anaxagorea*, *Polyalthia*, *Xylopia* and others have flowers that can raise their temperature several centigrade above the ambient air (thermogenesis). Depletion of the high concentration of starch and in some cases also lipids in the petals during the period of thermogenesis indicate that these are the source of fuel for heat production.

The species of *Annona*, *Cymbopetalum*, *Duguetia*, *Malmea* and others, with large flowers, pollinated by large nocturnally active cyclocephaline scarab beetles (Melolonthidae, Scarabaeidae) have adequate adaptations to lure and shelter their specific pollinators. Its large-sized, pendent flowers (ca. 4 cm diam.) have thick (up to about 5 mm), fleshy petals, which form a voluminous pollination chamber. There is sufficient space inside the flowers for the large beetles to serve as effective pollinators. They are nocturnally active and the flowers they visit also have a nocturnal flowering rhythm. Although the large beetles are quite voracious and start to gnaw at the petals after arrival, the flowers are not destroyed because of the enormous thick petals, which even have food bodies containing large amounts of starch and lipids. This tissue also causes the warming of the flower and even their extremely strong smell. *Annona coriacea*, a common tree in the Brazilian cerrado vegetation is an instructive example of an Annonaceae that is pollinated by large scarab beetles (Gottsberger 1989). The flowers are bisexual, strongly protogynous and show a clear temporal separation of pistillate and staminate stages. The flowering rhythm of this species lasts about 24 h, during which the temperature of the flowers rises during two nights in succession. During the first evening, the flowers enter the pistillate stage and the petals produce heat, peaking between 20:30 and 21:30 h with about 34 °C (14 °C above ambient temperature). Thereafter, the flowers cool down for the rest of the night. In the afternoon and evening of the second night, in the staminate stage, the flowers start to rise in temperature for a second time, with values similar to the first evening. This time, the highest temperature is attained approximately half an hour or even 1 h earlier than in the first night. The strong scent of the heated flowers, namely the nitrogen and sulphur-containing heterocyclic compound 4-Methyl-5-vinylthiazole (Maia et al. 2012) attracts its specific scarab beetle, *Cyclocephala atricapilla* (◘ Fig. 6.2). Once the beetles have entered the pollination chamber, they start feeding on the three inner petals, where two regions with special nutritional cells are located. The flowers warm up a second time on the second evening in the staminate stage, and because of new scent volatilizations, attract more beetles. The newcomers enter the floral chamber, joining the beetles already there, and they all become covered with liberated sticky pollen. Pollen is now also an additional food for the beetles. The beetles are suddenly released from the flower by the abscission of the petals about half an hour before other flowers in the pistillate stage attain their temperature peak. The released beetles, when flying to first-evening flowers, are the effective pollinators of this species.

The flowers, which offer their associated beetles a dark, odoriferous, often warm pollination chamber and sufficient food, can keep them as long as necessary to become efficiently pollinated. Some members of Nymphaeaceae, several Araceae, some Winteraceae and Calycanthaceae, Hydnoraceae, some Myristicaceae and Magnoliaceae, Degeneriaceae, Eupomatiaceae, many Arecaceae, and probably about 90% of extant Annonaceae diversified and function on the basis of these efficient and successful devices of beetle pollination.

Although there exists basic knowledge how beetle pollination works in different groups, this knowledge is deduced from only a very limited number of case history studies, mainly from the American tropics. Thus, future research will have to concentrate also on studies which show the situation in other continents (e.g. Ratnayake

◘ Fig. 6.2 Beetle pollination. *Cyclocephala literata* (Scarabaeidae) pollinating *Magnolia ovata* (Magnoliaceae) after entering the semi-closed flower in pistillate stage **a**. *Erioscelis emarginata* being attracted to a strongly scented inflorescence of *Philodendron selloum* (Araceae) in the pistillate stage **b** and the pollinating beetles crowded in the kettle of the inflorescence, contacting the receptive female flowers **c**. *Eulasia pareyssei* (Glaphyridae) pollinating *Cistus parviflorus* (Cistaceae) **d**. (Photo credits: Gerhard Gottsberger)

et al. 2006, 2007; Saunders 2012). How many plants function on the basis of thermogenesis is one question? Very important will be also scent analyses, as well as biotests with synthetic scent compounds to see how attractive they are. Is the whole scent bouquet of a flower attractive or only one or some specific strong compounds, as was found already in several American cantharophilous species attracting dynastid scarab beetles? (e.g. Jürgens et al. 2000; Maia et al. 2012; Goodrich 2012). The robust flowers and their beetles are well suited for manipulation. Thus, quite a lot of experiments would be essential to obtain a better insight of this interesting but still not fully understood pollination mode.

6.3.4.3 Thrips Pollination

The Fringe Wings or Thysanoptera are tiny insects of a few mm length. One habit involves pollen feeding and thus some species occur numerous in flowers and can be significant pollinators (Grimaldi and Engel 2005). Thrips, in addition to beetles were pollinators of fossil cycads (Peñalver et al. 2012) and are still co-occurring pollinators in several extant cycad species (Terry et al. 2012). One of the earlier observations of thrips as pollinators in angiosperms was done by Hagerup E. and Hagerup O. (1953) on *Erica tetralix* (Ericaceae) in Copenhagen and elsewhere. The small tubulous flowers are pollinated by the one mm long *Taeniothrips ericae* and *Frankliniella intonsa,* which can transfer pollen from flower to flower and lay their eggs in the tissue of the corolla. The males of *Taeniothrips* are rare and wingless and thus the winged females of this species fly actively from flower to flower in order to find males, becoming effective pollinators in the process. 26 years later, Fægri and van der Pijl (1979), in the third revised edition of their classic textbook still wrote that…"there is every reason to presume that both this group of insects and other small and insignificant, unobtrusive animals may prove to be of much greater importance in pollination than hitherto suspected." Nowadays one has to give consent to this prediction, because we now know that pollination by Thysanoptera is a common and quite well understood phenomenon.

In the magnoliids, three New Caledonian species of *Zygogynum* (Winteraceae. Canellales), with scented, yellow, protogynous flowers, function all in a similar manner; they are found to be pollinated by a single species of thrips, *Taeniothrips novocaledonensis.* The insects chew on the stigmas and eat the pollen (Thien 1980). Likewise, in the pantropical Monimiaceae, the two Brazilian dioecious species *Mollinedia floribunda* and *M. widgrenii,* female thrips (in *M. floribunda Liothrips seticollis* and *Heterothrips* sp.) puncture the still closed pistillate and staminate buds with their ovipositor and deposit their eggs in the interior of the receptacle. When the flowers open, their interior contains thrips eggs, larvae and adults. Movements of adult thrips from staminate to pistillate flowers cause pollination (Gottsberger 1977, 2016, see also Mound and Marullo 1996; Williams et al. 2001). Also, the self-compatible *Ocotea porosa* (Lauraceae, Laurales) appears to be exclusively pollinated by the thrips *Frankliniella gardenia* (Danieli-Silva and Varassin 2013). In the Annonaceae (Magnoliales) there are at least two well documented cases of thrips pollination. The Amazonian *Bocageopsis multiflora* (Annonaceae) has quite small, semiclosed petals, forming a kind of pollination chamber. The whitish flowers just open with small slits and emit a sweetish odor with a rancid component, recalling the odor of the fruits of *Genipa americana* (Rubiaceae). Only winged adults and unwinged nymphs of thrips occurred inside the flowers, which after the staminate stage were covered by pollen grains. The insects were forced to leave the flowers when stamens and petals were shed (Webber and Gottsberger

1995). Another annonaceous species, *Popowia pisocarpa* in lowland dipterocarp forest in Sarawak, with similar flower construction as *Bocageopsis*, having a tiny pollination chamber, also is exclusively pollinated by thrips species (Momose et al. 1998). Further thrips pollinated species in basal angiosperms are two *Chloranthus* species, *C. serratus* and *C. fortunei* in China (Luo and Li 1999).

From what can be seen in thrips pollination in basal angiosperms is, that since thrips are minute elongate insects, they like to hide, to nourish themselves and to breed in small flowers, which are closed (e.g., *Mollinedia*, Monimiaceae) or semi-closed, and thus provide a shelter, and which are light colored and produce a faint, sweet fragrance. Pollen is the main food while nectar can be present or absent. In relation to thrips as visitors or pollinators of higher angiosperms or thrips as crop pests, the literature is quite vast.

One interesting phenomenon, however, involving the flowering and pollination of the huge Dipterocarpaceae forests in Southeast Asia has still to be tackled. Dipterocarp forests in Malaysia come into heavy mass flowering at irregular inter-vals of 2–10 years. The large number of flowers produced by *Shorea* species allow certain thrips species to feed on flowers and multiply rapidly. Anthesis of *Shorea* species begins in the evening and corollas together with the thrips are shed in the morning. In the following evening, the thrips in the fallen flowers are covered with pollen, fly up to the newly opened flowers and pollinate them (Appanah and Chan 1981; Appanah 1993). Subsequent investigations have indicated an even more sophisticated situation. During mass flowering of *Shorea acuminata*, the big-eyed bug, *Geocoris* sp., a major thrips predator, was an inadvertent pollinator that importantly contributed to cross-pollination. During flowering the bugs travelled among the *Shorea* trees, attracted by the abundant thrips. The outcross pollen on thrips was about 30% that on the bugs. This indicates that thrips and big-eyed bugs are essential pollinators of *S. acuminate* (Kondo et al. 2016a). A similar situation was found in *Shorea curtisii* (Kondo et al. 2016b). On the other hand, in *Vatica yeechongii*, another Malysian dipterocarp tree, Suhaida et al. (2018) found that members of Apidae were the effective pollinators. Although thrips were the most abundant flower visitors of this species, they were poor pollen-carriers and thus probably not very effective as pollinators. Future studies certainly will reveal more about the diversity of the pollination biology of these tall and interesting tropical trees.

6.3.4.4 Bee Pollination: Melittophily

Bees depend on floral resources for their larval development and adult survival and reproduction (Nicolson 2011; Vaudo et al. 2015). In many plant communities, bee pollination is prevalent, with about 63% (Ishara and Maimoni-Rodella 2011), 65.2% (Quirino and Machado 2014) or 75% melittophilous species (Silberbauer-Gottsberger and Gottsberger 1988). Lots of pollination interactions can present more than just one pollinator type, as in *Ipomoea pes-caprae* (Convolvulaceae) that receives melittophilous and malacophilous (snail pollination) interactions (Raju et al. 2014). Pollen collection is performed by bees gleaning the longitudinal anther openings (Torezan-Silingardi and Oliveira 2004, ◘ Fig. 6.3) or buzzing the pori-cidal anthers (Silva et al. 2004), but buzzing is also used in quirky pollen collection

▣ **Fig. 6.3** Bee pollination. Centridini bee collecting oil from the elaiophores of *Byrsonima interme-dia* (Malpighiaceae), note the oil deposited on the scopa at the hind leg **a**. *Trigona spinipes* (Meli-ponini) collecting pollen in the anthers of *Campomanesia pubescens* (Myrtaceae) **b**. *Apis mellifera* (Apini) collecting nectar from *Lantana camara* (Verbenaceae) **c**. Meliponini bee collecting nectar from *Coleus* sp. (Lamiaceae) **d**. (Photo credits: Kleber Del-Claro)

of non-poricidal anthers (Torezan-Silingardi and Del-Claro 1998). Bee plumose hairs can be simple or multiple branched and are suitable for pollen, oil, resin and fragrance collection (Vogel 1963, 1966, ▣ Fig. 6.3). Hairs also decrease the amount of energy used to maintain bee temperature (Southwick 1985).

Small bees may collect nectar from small, open flowers, most times radially symmetric, usually called generalist flowers as nectar collection is not restrictive and the flower is visited by other generalist insects as wasps, butterflies, flies and beetles (▣ Fig. 6.3). Medium and big bees may visit and collect nectar during pollination in bigger flowers with profound tubular corollas or spurs, often zygomorphic. These differences are due to the tongue length (Balfour et al. 2013). Long spurs or long corolla tube flowers have one or a few specialized long-tongued bees, such as Euglossine, but long-tongued bees are not specialized on just one or a few plant species, they visit many distinct corolla lengths (Borrell 2005).

Species from the Melittidade, Ctenoplectridae, Anthophoridae and Apidae families are able to collect oil from the flower (Buchmann 1987). An example of coevolution of oil-bee and oil-flowers is observed in about 92% of all American species of Malpighiaceae, whose flowers produce disc-shaped epithelial oil glands exposed in the abaxial surface of the sepals and can be easily accessed by bees from tribes Centridini, Tetrapediini and Tapinotaspidini, their main pollinators (Vogel 1974; Michener 1974, 2007; Neff and Simpson 2017). The big or medium sized bees arrive in front of the flower, use their mandibles to cling to the banner petal and attach to the flower while inserting the fore and mid pair of legs between the petals to scratch the sepal oil glands (Vogel 1990, ◻ Fig. 6.3). During oil collection the anthers and stigmas touch the bee thorax, dusting the ventral portion of the thorax and abdomen with pollen and leading to the pollination of the same and the next visited flower (Vogel 1990). The floral conservatism of Malpighiaceae species permits specific insect-flower relationships to develop in the family (Anderson 1979; Vogel 1990). The floral oil and pollen are collected by bees in sequential flowering periods from the distinct Malpighiaceae species, permitting the maintenance of oil-bee community in the area (Barônio and Torezan-Silingardi 2017).

Another very good example of coevolution between oil collecting bees and their flowers is the case of the African female bees *Rediviva emdeorum* and *R. longimanus* that collect oils from inside the two long spurs of the flower of *Diascia longicornis* (Scrophulariaceae), the flower morphological adaptations restrict oil collection to few *Rediviva* species (Vogel and Michener 1985). Specialized bees have their forelegs elongated as much as the entire body in similar length to the visited floral spur, and also have distinct types of hairs elaborately arranged to improve the oil collection (Kuhlmann and Hollens 2014).

The volatile chemicals or odoriferous compounds of many Neotropical orchid species as *Coryanthes*, *Gongora* and *Stanhopea* are the sole reward offered to male Euglossini bee, which actively collect these compounds from the flower to use as precursors of sexual pheromones to attract females (Proctor et al. 1996). Male Euglossini store the compound in specialized and highly vascularized hind tibial structures. The new volatile formed after the union of the floral fragrance and the bee metabolites will take part in Euglossinae courtship as a faithful indicator of male quality. The high importance of the compound is enough to make *Eulaema nigrita* males fight and rob it one from the other, even from a detached hind leg (Carvalho-Filho 2010).

The deceptive pollination is a very curious process where the insect is attracted to the flower for a simulated reward that does not exist, but pollen transfer happens effectively (Darwin 1904; Kullenberg 1950). Here, male insects (bees, wasps, ants, sawflies, tachinid flies, fungus gnats and beetles) are the pollinators, but they neither receive pollen nor nectar. They were not even looking for pollen or nectar or any other flower resource, many of them were looking for a sexual partner! Many sexually deceptive orchids present a differentiated flower with a specialized labellum which simulates visual and tactical cues of a receptive insect female, but the main attractant is the mimic sexual pheromone (Streinzer et al. 2009). This system is observed in species from the Mediterranean (Dressler 1981; Jersáková et al. 2006), Australia (Peakall et al. 2010), South America (Ciotek et al. 2006; Martel et al. 2019) and South Africa (Steiner et al. 1994). Male bees patrol a restricted area near the nesting place looking for sexual partners when they observe the *Ophrys* flowers, in which the chemical/visual/tactical mimetic cues are so convincing that the male bees try to copulate with the flowers leading to pollinia transfer (Peakall and Schiestl 2004). After being pollinated, the floral scent is modified and the new fragrance is similar in composition to the Dufour's gland secretion of copulated female bees, which decreases male copulating attempts (Schiestl and Ayasse 2001).

Another interesting and extremely rare deceptive pollinating interaction is observed with *Bombus* queens visiting and pollinating the nectarless *Cypripedium tibeticum* flowers (Li et al. 2006). These orchids require animal pollination for fruit and seed production. The floral labellum forms an inner chamber due to its sharp concavity, which mimics a good nesting site for the queens. *Bombus* queens are bigger than working bees and present the proper size and shape to contact the pollinia when leaving the chamber. *Bombus* workers were less frequent in the flowers and inapt to pollinate.

Bees and wasps are predominantly diurnal, but some species prefer to visit flowers during low light incidence periods called dim-light periods (Warrant 2008; Cordeiro et al. 2017). Hypothesis were proposed to justify the nocturnal foraging, such as avoiding competition for floral resources with the higher diversified of diurnal bees when exploiting crepuscular or nocturnal flowers, or escape from natural enemies as predators and parasites are more frequent during the day (Wcislo et al. 2004; Warrant 2008). The dim-light foraging is observed in Central American bees as the crepuscular *Rhinetula* (Halictidae) and *Ptiloglossa* (Colletidae) and the nocturnal bees *Megalopta* (Halictidae) (Wolda and Roubik 1986) and in the Indian carpenter bee *Xylocopa tranquebarica* (Apidae) (Warrant 2008). They prefer to forage during two short periods each day, just before dawn and just after sunset, but they are able to visit flowers also during the day with high intensity light (Wcislo et al. 2004). The night vision is possible due to special morphological and physiological properties of the photoreceptors. For instance, the eyes of the nocturnal *Megalopta genalis* are almost 30 times more sensitive to light than the eyes of a honeybee worker (Warrant et al. 2004).

6.3.4.5 Wasp Pollination

Wasps are usually seen as predators due to their protein requirements, but they also need water and carbohydrates, which are easily found in nectar from nectaries inside or outside the flower. Some angiosperms apt to receive wasp visits can have up to seven species of social wasp collecting floral resources in the same blooming period, but their low frequency in the flowers suggests they act as opportunistic visitors and eventual pollinators (Santos et al. 2010). Social wasp-plant interactions are higher in more heterogeneous habitats, indicating a bigger tolerance to extinctions of interactions (Clemente et al. 2013), which is expected to be true in many other pollinators.

Generalist flowers with radial symmetry and with exposed pollen and nectar usually can be visited and pollinated by wasps, bees, flies and beetles, as observed in Anacardiaceae, Hyacinthaceae and many other families (Cesário and Gaglianone 2013, Shuttleworth and Johnson 2010, ◘ Fig. 6.4). Four African *Eucomis* species

◘ Fig. 6.4 Wasp and ant pollination. Wasp collecting nectar from *Hortia brasiliana* (Rutaceae) **a**. Ant *Camponotus crassus* collecting nectar from *Paepalanthus lundii* (Eriocaulaceae) **b**. (Photo credits: Kleber Del-Claro)

(Hyacinthaceae) with similar morphology, shape and size, offer exposed pollen and nectar to diurnal floral visitors. Color was not the main character to attract pollinators and many scent compounds were the same. But the presence of two sulphur compounds (dimethyl disulphide and dimethyl trisulphide) indicates fly preference and wasp avoidance, acting as the main cue to fly or wasp visits on the four plant species (Shuttleworth and Johnson 2010). This study exemplifies how genome modification can lead to an odor change with shift in the pollination system.

Wasp specialized pollination happens even in a visually/morphologically generalist flower. The African milkweed *Pachycarpus grandiflorus* (Apocynaceae) has open cryptic colored flowers and expose sugar-rich nectar, but is pollinated by just four species of spider-hunting wasps (*Hemipepsis*, Pompilidae) which are the only flower visitors able to effectively carry the pollinaria (Shuttleworth and Johnson 2009). Coleopteran, dipteran and other wasp visitors were just nectar thieves. Experiments in the field and also in the laboratory using a Y-maze indicate floral scent is more important than visual cues to the wasps. Honeybees are present in the field area, but do not visit *P. grandiflorus* flowers. Authors note nectar has an unpleasant bitter taste to humans but is palatable to wasps and unpalatable to honeybees. As *P. grandiflorus* is self-incompatible and consequently needs cross-pollination to set fruits, wasp pollination is of paramount importance. Shuttleworth and Johnson (2009) point out that plants can achieve specialized pollinating vectors even without morphological restrictions.

Specialized pollination is also seen in the South American asclepiad flowers *Morrenia odorata* with wasp pollinators and *M. brachystephana* with wasp and bee pollinators (Wiemer et al. 2012). Both plants present an intricate system of guide rails and chambers carefully positioned to facilitate the effective transport of their pollinaria. Fig wasp pollination is very well presented and discussed in this book by Bronstein (► Chap. 11) and Pereira and Kjellberg (► Chap. 12).

The antagonistic co-evolutionary relationship between deceptive orchids and 'loving wasp males' is very interesting. It is supposed that the deceptive orchid ancestors had a mutation that turned the wax metabolites similar to the sex pheromones of a determinate pollinator wasp species (Schiestl et al. 1999), and this process may have happened repeatedly in orchids (Bohman et al. 2016a, 2016b). The *Ophrys* orchid surface wax simulates female cuticular metabolities to attract the male scoliid wasp (Paulus and Gack 1990). Australian orchids of the genus *Drakaea* and *Caladenia* produce semiochemicals in their labellum to simulate the sex pheromone of the female Thynnini wasp, composed of alkylpyrazines and hydroxymethylpyrazine (Bohman et al. 2016a; Phillips et al. 2017).

Males are lured by the flower labellum semiochemicals that convincingly simulate the morphology and odor of the wingless female, attracting specific pollinators able to transfer the pollinia during the pseudocopulation process (Paulus and Gack 1990; Bohman et al. 2016b). One unique compound (chiloglottone) is responsible for the chemical signaling of the Australian sexually deceptive orchid *Chiloglottis trapeziformis* to attract the male wasps (Schiestl et al. 2003), while the European *Ophrys sphegodes* has a mixture of 14 alkanes and alkenes that produce a synergistic effect forming the signaling compound (Schiestl et al. 1999).

The deceptive processes are so effective that they result in changes in behavior. The male liberates a smaller quantity of sperm per ejaculation in areas with the deceit orchid than in areas without the orchid (Martin et al. 2020). The authors propose that (a) males are not able to distinguish between real females and orchids, so they act as if many of the females were considered as 'low-quality females' leading males to save sperm for a new 'better opportunity'; or (b) males consider both real females and orchids as high-quality sex partners and sperm is partitioned and offered to all of them.

6.3.4.6 Ant Pollination: Myrmecophily

Ants are extremely abundant and diversified insects, omnipresent in terrestrial environments. Usually ants are in the flowering plant as predators able to inhibit visiting pollinators and decrease fruit and seed formation (Assunção et al. 2014; Sousa-Lopes et al. 2020). But this is not the case at all times.

In the overall context of ant–plant mutualism, ant pollination is regarded as a rare interaction, with few studies showing the role of ants as effective pollinators (Hickman 1974; de Vega et al. 2014; Domingos-Melo et al. 2017; Del-Claro et al. 2019). The main reasons why ants are considered bad agents of pollination (especially cross-pollination) are: (1) their small size, many time smaller than most floral reproductive structures; (2) ant self cleaning behavior removes pollen grains before they can reach the stigma; (3) limited displacement as foragers-ants cannot fly and thus only visit resources near the nest; and (4) the interference in pollen viability and germination by antibiotics secreted by the metapleural glands and spread over the surface of the insect's body (Hickman 1974; Del-Claro and Torezan-Silingardi 2020).

However, in dry environments, small herbs or shrubs that produce inflorescences near the ground level, with small flowers produced in blooming and unisexual male and female flowers or bisexual flowers in a same plant, offering nectar as main reward, can be pollinated by ants lacking metapleural glands (Del-Claro and Torezan-Silingardi 2020). Ant importance as pollinators in dry ecosystems is recently attracting enormous attention of biologists (de Vega et al. 2014; Del-Claro et al. 2019; Del-Claro and Torezan-Silingardi 2020; Delnevo et al. 2020). Indeed, ants can be considered a new field for pollination studies.

6.3.4.7 Lepidoptera Pollination: Psychophily, Phalenophily and Sphingophily

The order Lepidoptera is composed by more than 160,000 species described, with butterflies (9%) and moths (91%) spread worldwide (Capinera 2008). Butterfly pollination is called psychophily, while moth pollination is differenciated as phalenophily (small moths) and sphingophily (big moths). Their mouthparts are adapted do collect fluids such as the nectar inside long and narrow corolla tubes or spurs (Ehrlich and Raven 1964, Capinera 2008, ◘ Fig. 6.5). The tongue and proboscides lengths determine from how deep in the floral tube the nectar can be collected. During the feeding processes pollen grains or pollinaria are easily deposited on their bodies, usually on their faces and proboscises (Johnson and Liltved 1997; Balducci et al. 2019). The deposition place depends on the morphology of the flower and the lepidopteran behavior. For instance, many Campanulaceae species present pendant bell-shaped flowers with exerted styles that facilitate pollen grain deposition onto the ventral side of the latter thorax and former abdomen of their moth pollinators, characterizing the unusual sternotribic pollination (Funamoto 2019). Another uncommon pollination is observed in *Gloriosa* genus (Colchicaceae) with its distinct floral morphology, such as the African *G. superba* flowers that are adapted to deposit their pollen grains on the scales from the ventral wing surface of their pollinators, mainly the Pieridae butterflies (Daniels et al. 2020).

 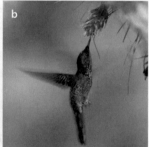 

◘ **Fig. 6.5** Nectar feeding pollinators. Butterfly *Leptophobia* sp. (Pieridae) on Asteraceae flower, note the long proboscis inside the narrow floral tube **a**. Hummingbird *Thalurania glaucopis* on cactus flowers **b**. *Coereba flaveola* on *Agave americana* (Agavaceae) flowers **c**. (Photo credits: A, B: José Sabino; C: Kleber Del-Claro)

Yucca-moths, glochidion-moths and fig-wasps present species-specific pollination mutualism with their respective plant species, and represent the singular active or intentional pollination cases currently known. The yucca-moths (*Tegeticula*, Proxidae) (Pellmyr 2003; Pellmyr and Segraves 2003; Cole et al. 2017) and the glochidion-moths (*Epicephala*, Gracillariidae) (Kato et al. 2003) do not look for nectar, as yucca (Asparagaceae, Agavoideae) and glochidion (Phyllanthaceae) flowers do not produce nectar. Female moths have exceptionally adapted mouth parts as tentacle-like appendages arising from the maxillary palps. These tentacles are used to collect pollen from the anthers and transport the pollen grains pressed like a ball up to another flower and actively pressing the grains onto the stigma. The female oviposit in the style or near the developing ovules, so the larva will be able to feed on the new formed seeds. To avoid fruit abortion, not all the seeds will be consumed by the growing larvae.

Floral traits are used differently by butterflies and moths to locate the flower. The floral scent is an important feature (Andersson et al. 2002). Visual cues such as colors are more important to butterflies which visit flowers during the day; however a weak, fresh and sweet scent may exist, produced by a group of metabolites as benzenoids and terpenoids (Dobson 2006; Kinoshita et al. 2017). Moths which visit crepuscular or nocturnal flowers depend more on the strong, heavy and sweet scents, caused usually by the acyclic terpene alcohols or benzenoid esters and monoterpenes components (Dobson 2006, Kinoshita et al. 2017). But there are moths that visit flowers also during the day and use the odor and the floral colors as important cues to find the flowers (Sakamoto et al. 2012). Few examples are given, as for instance the African *Bonatea cassidea*, which is pollinated by Hesperiidae and Pieridae butterflies (Balducci et al. 2019) and which has weakly scented flowers with diurnal anthesis and landing platforms formed by the petals. Or the African orchid *Bonatea speciosa* (Johnson and Liltved 1997), pollinated by hawkmoth. The hovering hawkmoths attracted by the strong scented evening/night flowers collect nectar at dusk and also during the night.

Some plants, such as the *Clerodendrum trichotomum* (Lamiaceae) produce protandrous flowers that can last from 2 to 4 days long, which are intensely visited by lepidopterans and bees (Sakamoto et al. 2012). The stigma is able to receive pollen during the day and the night, but not all lepidopteran have the same role in pollination. The main visitors are the several species of the butterfly genus *Papilio*, the bees *Xylocopa appendiculata* and the larcenist hawkmoths *Macroglossum pyrrhosticta*, which is the most frequent floral visitor. Butterflies and bees visit the flowers during the day, and the hawmoths visit during the night. Authors performed controlled pollination treatments and insect exclusion treatments and observed fruit formation after distinc visitation behaviors. They considered butterflies and bees as effective pollinators, as they moved between flowers from neighboring plants and the *C. trichotomum* needs crosspollination to achieve better fruit and seed formation. The high frequency of hawkmoth visits may increase self-pollination with geitonogamous pollen transfer (from distinct flowers produced in a single plant), what makes the hawmoth an inefficient pollinator to that plant species.

6.3.4.8 Hemipteran Pollination

Hemipteran species are usually known as plant antagonists acting as herbivores that suck plant tissue, or as mutualists predating small insects as the Thysanoptera, but they are more than that. Some Hemipteran species can pollinate. Hemipteran and thrips pollination was described in *Macaranga* species (Euphorbiaceae) (Fiala et al. 2011). The dioecious tree *Macaranga tanarius* produces buds and flowers with an enclosed chamber formed by floral bracts, which are the breeding sites of hemipteran species that pierce the nectaries from the adaxial surface of flower bracts to feed on nectar, and pollinate (Ishida et al. 2009).

6.3.5 Vertebrate Pollinators

Vertebrates such as bats, birds, terrestrial mammals and even lizards are also pollinators. A meta-analysis study based on 126 experiments on the vertebrate pollination evidenced their responsibility for about 63% of fruit or seed production of 91 plant species all over the world (Ratto et al. 2018). The higher degree of tropical plant specialization renders vertebrate pollinator much more relevant in tropics than in temperate areas (Trøjelsgaard and Olesen 2013).

6.3.5.1 Bats

Bats that feed on nectar and pollen present morphological adaptations such as the skull with elongated rostral portion, modified teeth, long tongue, throat muscles and stomach linings (Koopman 1981). Their tongues have many long papillae to increase nectar collection velocity. Bat-pollinated species can fail in about 83% of their fruit or seed set if bats are absent, due to the specific adaptations for quiropterophily in floral morphology and phenology that exclude many other animals, like the crepuscular/nocturnal anthesis period (□ Table 6.1) and the capacity of the fur to hold and shed easily more pollen than feathers. Consequently, plant dependence on bats as pollinators is higher than the dependence on birds.

The tropical pitayas (*Stenocereus queretaroensis*) from Mexico are pollinated by nectar-feeding bats *Leptonycteris* (Tremlett et al. 2020). If bats are experimentally excluded pollination can be done by birds and insects, but at the cost of a 35% decrease in pitaya yield and fruits 13% less sweet, obtaining a worse market value. The fruit-bats (Pteropodidae) from Asia and Australia and the leaf-nosed bats (Phyllostomidae) from the Neotropics are the most important mammal pollinators (Fleming and Muchhala 2008), already observed in about 528 plant species (Kunz et al. 2011). The Glossophaginae bats are important pollinators on Neotropical areas from Central and South Americas (Koopman 1981). *Passiflora* species (Passifloraceae) are usually pollinated by bees, and sometimes by hummingbirds that visit the radial and colorful corollas during the day looking for nectar and perform cross-pollination in self-incompatible species (Yamamoto et al. 2012). But the passionflower species *P. setacea* and *P. mucronata* have white flowers that open at dusk, with anthers and stigmas bilaterally arranged in the flowers. These two species also need cross-pollination to produce quality fruits with seeds and pulp adequate for commercialization. *Passiflora setacea* flowers offer pollen and nectar to their main pollinator, the bat *Glossophaga soricina* that visits the flowers mainly from 20 to 22 h and receives pollen on the fur of the snout, neck and back (Teixeira et al. 2019). *Passiflora mucronata* is pollinated by the bats *Glossophaga soricina* and *Carollia perspicillata* (Sazima and Sazima 1978). Other nocturnal visitors do not pollinate, including wasps and sphingid and noctuid moths.

Species can present some modifications on floral structures and physiology with time that permit an evolutionary switch in the pollen vector. The *Mimosa* genus is mostly pollinated by insects, but *M. lewisii* shows characteristics pointing to a switch from insect to bat pollen vectors (Vogel et al. 2005). The shrub is found in open areas and presents big racemes over the foliage, developing many capitula of white brush-type small flowers without scent. After nocturnal anthesis, pollen and nectar are produced and the long filaments are seen immersed in the copious nectar. Just one species of Glossophagine bat pollinates while feeding on the nectar, while hummingbirds visit the flowers during the day but are not able to pollinate as flowers are not functional anymore, and insects are absent.

6.3.5.2 Birds

There are more than 920 bird species responsible for pollinating flowers, including hummingbirds (Trochilidae), honeyeaters (Meliphagidae), sunbirds (Nectariniidae), lories (Loriinae) and honeycreepers (Coerebidae) (Stiles 1981, ◨ Fig. 6.5). Whelan et al. (2008) reviewed many studies on bird pollination and noted that the rate of pollination by birds is very low when communities are considered, with the higher values found by Brown and Hopkins (1995) that observed about 20% of bird-pollinated tree species in New Guinea. But when species are considered values are higher. For instance, a hurricane in the Bahamas was responsible for the death of the two bird species that pollinated *Pavonia bahamensis*, leading to a fruit-set decrease of 74%, indicating the great importance of bird pollination to that plant reproduction (Rathcke 2000). Bird pollination is even more important if we consider their large area coverage. As birds are able to travel greater distances than insects, another benefit of bird pollination is the higher paternal diversity observed in the seeds produced compared to insect pollination (Krauss et al. 2017).

Morphological and phenological features of bird-flowers, such as nectar production, diurnal anthesis and a long and narrow corolla, allow many of those flowers to be pollinated by animals other than birds (Hargreaves et al. 2004), then, if bird pollinators are absent some insect species could provide at least a little pollen transfer (Whelan et al. 2008). However, the degree of floral specialization is variable. Many hummingbird flowers are able to reflect mainly long-wavelength light that is not perceived by bees and other insects (Altshuler 2003). Consequently, specific floral colors can decrease insect visits and save nectar to bird pollinators (◨ Table 6.1).

Hummingbirds are the most specialized nectarivores, followed by sunbirds. Bill morphology indicates the manner of nectar extraction: by piercing, mashing or probing. Flower nectar feeding birds have some common features such as a long tongue able to extend beyond the bill tip, a tongue tip grooved and fringed, and the birds may roll the tongue into a tube to easily take up nectar. Pollen feeding birds may present a papillate tongue tip. These important pollinators look for floral resources in large and dense patches of flowers, with abundant nectar. Hummingbirds present spatial and plant preference, as observed in Costa Rica (Stiles 1981). Hermit hummingbirds' higher abundance was found in wet lowlands and adjacent foothills with abundant monocotyledonous herbs as *Heliconia*, while nonhermit hummingbirds preferred the lower middle elevations plenty of dicotyledonous and bromeliads. Ecological and evolutionary processes drive plant-hummingbird networks. Dalsgaard et al. (2011) observed that plant-hummingbird pollination network is more specialized in tropical areas than in temperate ones, evidencing a greater number of species interactions between morphologically matching partners (Sonne et al. 2020).

6.3.5.3 Ground-Dwelling Mammals

Primates, marsupials and rodents are the non-flying or ground-dwelling mammals already observed to pollinate 85 plant species (Carthew and Goldingay 1997). It is not so easy to confirm ground-dwelling mammals as pollinators, due to their sensitivity to human presence and the nocturnal activity of many species. Usually, their activities as pollinators are inferred after field-trapped animals have their fur and fecal pollen loads analyzed, but the motion-activated remote cameras can help a lot in these studies (Zoeller et al. 2016). Usually, flowers are not destroyed by mammals, which visit them to look for insects or floral resources such as sugar-rich nectar and pollen, and transport pollen grains in their fur. As many insectivore mammals have large home ranges, they benefit the plant enabling a great genetic flow between distant plants.

The capuchin monkeys from Costa Rica act as florivores (flower herbivores) mainly in the dry season, when fruits and invertebrate abundance are low (Hogan et al. 2016). The flowers visited present pollen and nectar, with high proportion of water soluble carbohydrates, proteins and also fats on floral tissues. From the nine species whose flowers were consumed by the capuchin monkeys, three of them were effectively pollinated by geitonogamous and xenogamous processes: *Luehea speciosa*, *Callistemon viminalis* and *Manilkara chicl*.

Flowers of *Protea humiflora* (Proteaceae) from South Africa are visited by three species of little ground-dwelling mammals (Fleming and Nicolson 2002). *Acomys subspinosus* and *Aethomys namaquensis* are nocturnal species that feed on pollen and possibly on nectar, and *Elephantulus edwardii* that is a diurnal insectivore visiting flowers looking for preys. The three mammal species carry pollen grains on their fur, and were considered responsible for about 56% of *P. humiflora* seed set. Seed production was significantly higher after open visitation than in caged flowers, where the small mammals were excluded.

6.3.5.4 Lizards

At least 37 lizard species are flower visitors and many of them are able to pollinate, especially in islands (Olesen and Valido 2003). Lizards find nectar, pollen and arthropods in the flower. When animal preys are rare, flower resources will be more consumed increasing pollination possibilities. For example, lizards, doves and passerine birds from Fernando de Noronha Island, Brazil, can use the floral nectar produced by plants that bloom during the dry season for their energetic requirements and water intake (Sazima et al. 2005, 2009). To access the nectar accumulated within a chamber the lizard skin and dove feathers touch the anthers and became covered with pollen grains that can be used to pollinate distinct flowers from the same plant (geitonogamy) or from other trees (xenogamy). Passerine birds also collect the nectar, but they are too small to touch the reproductive organs, so pollination is rare.

6.3.5.5 Mixed Pollination

Many species produce flowers that can benefit from distinct visitors transferring pollen, including vertebrates and insects simultaneously. For example, *Langsdorfia* sp. (Balanophoraceae) has features that permit birds, mammals and insects to visit the flowers looking for pollen and nectar (Thorogood and Santos 2020). *Langsdorffia hypogaea* produces sweetly scented, conspicuous red inflorescences near the soil. In Brazil, the flowers attract ants, beetles, Hemiptera, Dermaptera, Blattodea, Araneae (Freitas et al. 2017) and the corvid bird *Cyanocorax cyanopogon* (Santos et al. 2017). In Madagascar *Langsdorffia malagasica* flowers attract primates (Irwin et al. 2007). Further studies are needed to define the main pollinators, but most animal species appear to be able to pollinate at least partially. In another example, the red tubular nectar flowers of *Seemannia sylvatica* (Gesneriaceae) are pollinated mainly by hummingbirds and secondarily by butterflies (Camargo et al. 2011).

6.4 Pollination: Benefits and Threats

6.4.1 Beneficial Consequences of Pollination

The important benefit of pollination is the gene exchange between different flowers, especially between two distinct plants from the same species. Seeds originating from cross-pollination will present more gene variability than those from self-pollination. Variability is very advantageous in a modifiable environment and in the presence of pests and diseases (e.g., Holeski: ► Chap. 4). Seed production by agamospermy and clonal vegetative reproduction produces no genetic variation in the next generation. If the species grows pretty well in an uncommon situation of a very stable environment, with steady soil characteristics, temperature, rains, species presence and species interactions along time, then it would be advantageous to maintain this genetic pool which is so highly adapted to this specific immutable ambient. But this is a highly unusual situation.

The benefits from animal pollination in both natural areas and agricultural systems are shown in many studies (Bowers 1975, Gemmill-Herren and Ochieng 2008; Hudewenz et al. 2013; Nunes-Silva et al. 2013; Deprá et al. 2014; Lindström et al. 2015). Even self-compatible plants such as the oilseed rape produces significantly more seeds and heavier seeds with insect pollination or manual pollen supplementation than in autonomous self-pollination or wind pollination (Hudewenz et al. 2013). When honeybee hives are added near the oilseed rape crop fields, yields are

11% higher than in fields without artificial hives (Lindström et al. 2015). Eggplants produce heavier fruits after bee pollination (Nunes-Silva et al. 2013) and increase the seed numbers caused by higher quantities of pollen grains arriving on the stigma, independently if pollen was delivered by an animal or by manual pollination (Gemmill-Herren and Ochieng 2008). Some populations of tomatoes are incapable of fruiting if animal pollinators are absent (Bowers 1975). Even tomato varieties able to automatically self-pollinate can benefit a lot from animal pollination services, achieving higher fruit-set and number of seeds in open field pollination (Deprá et al. 2014).

6.4.2 Ecological Services and Crop Pollinators

Pollination failure may be caused by lack (or even excess) of pollen grains deposited by animals on the stigma, bad quality of the pollen load caused by a high number of unviable grains or heterospecific pollen grains (from two or more species mixed together) on the stigma, late deposition when stigma is no longer receptive, and self-incompatibility reactions preventing fertilization or inducing rapid seed abortion. Aspects that can decrease pollination quality are very well presented by Wilcock and Neiland (2002) and references therein.

Biotic pollination services are fundamental to sustain natural areas and also for many of our crops. Insect pollination is essential for food production globally, for example, apples, strawberries, coffee, cocoa, avocado, guarana, melon, passion fruit, pear, peach, sunflower, tomato, watermelon, alfalfa, red clover, figs, pineapple, guava, kiwifruit, vanilla, durian, cherimoya, papaya and many others fruit crops rely heavily on this important ecosystem service (Westerkamp and Gottsberger 2000; Novais et al. 2016). A lot of these crops are considered economically significant commodities on the global market (Prescott-Allen and Prescott-Allen 1990). In a study performed in 200 countries from four continents, Klein et al. (2007) found that 87 species of important crops are animal pollinated, while 28 species are not. For instance, although soy flower is able to self-pollinate before anthesis in a cleistogamous process, insect pollination significantly increases seed production (57%) and final harvest income (Chiari et al. 2005).

Information about the role of insect pollination needs to be disseminated to farmers all over the world, but especially to small farms with family agriculture in developing countries. For example, Tanzanian watermelon yield is pollen limited by insufficient floral visits and the lack of efficient pollinators (Sawea et al. 2020). Human manual pollination increased the development of mature watermelons (42%) with heavier fruits. Although manual pollination is possible, the costs are incredibly high (Westerkamp and Gottsberger 2000) and not nearly as effective as natural pollination near preserved areas. Moreover, natural pollination by animal vectors doesn't cost anything.

Aside from that, New Zealand bee populations decreased due to *Varroa destructor* infestation (Brown et al. 2018). For the kiwi being a dioecious plant (male and female unisexual flowers on separate plants), cross pollination is obligatory (Simão 1998). To support the country's most profitable horticulture product (Zespri 2016) avoiding the decrease of kiwi yields and getting better results than those provided by the handheld sprayer for manual pollination, a robotic pollinator machine was developed for kiwi plantations in New Zealand (Williams et al. 2019). Future work is needed to enhance this system and reduce its costs.

All over the world, complex landscapes such as natural preserved ecosystems present a larger overall abundance and richness of wild insects than homogeneous areas. This is the case of Sweden's oilseed rape crop, which produces heavier seeds (18%) and achieves a better market value (20%) in farms near complex landscapes (Bommarco et al. 2012). Oilseed rape crops are increased by good management practices that lead to higher pollinator abundance and functional divergence (Lindström et al. 2015). Also, Brazilian coffee crops near natural areas and with low management intensity are able to maintain higher pollinator biodiversity and increase yields up to 30%, improving a fundamental trade-off between food production and flora/fauna conservation (Hipólito et al. 2018).

A recent meta-analysis indicated that pollinator functional diversity and abundance can increase crop pollination and consequently its yield, constituting the complementarity hypothesis: pollination function should be maintained by non-overlapping trait distributions (Woodcock et al. 2019). A crop can be considered an artificially constructed community, where species richness is positively correlated with yield. Many studies can confirm the conclusions of that meta-analysis.

For instance, in apple fields in the United Kingdom, the presence of pollinating insects as bees and hoverflies is crucial to the fructification. Garratt et al. (2014) show insects increase the quantity, size and shape of the fruits, leading to higher prices on the market and a potential to improve UK output by up to £5.7 million per annum. Their study indicates how continued pollinator decline is as a serious danger to apple crops and the apple industry.

The same was observed in 191 plant species related to food production in Brazil (Wolowski et al. 2019). For 75% of plant species animal pollination is important, performed by nine distinct groups: bees (66.3%), beetles (9.2%), butterflies (5.2%), moths (5.2%), birds (4.4%), wasps (4.4%), flies (2.8%), bats (2%) and even hemipterans (0.4%). These studies show how useful animal pollination is (◘ Fig. 6.6).

The 'Evaluation Report on Pollinators, Pollination and Food Production' estimated that the global economic valuation of pollination ecosystem service ranges between US $ 235 billion and US $ 577 billion (IPBES 2016) and grows as times goes on. The financial implications of pollinators decline are already severe and will be worse in the future if we do not recover pollinator populations.

6.4.3 The Impact of Pollinator Declines: Biodiversity, Human Health and Economic Aspects

It is certain that pollinator decline has enormous potential to lead us to a scenario of expensive food and vegetal resources (Wilcock and Neiland 2002). Besides that, we need to understand that the current and intense destruction of angiosperm communities, as seen through the past years in Brazilian Amazônia/Cerrado/

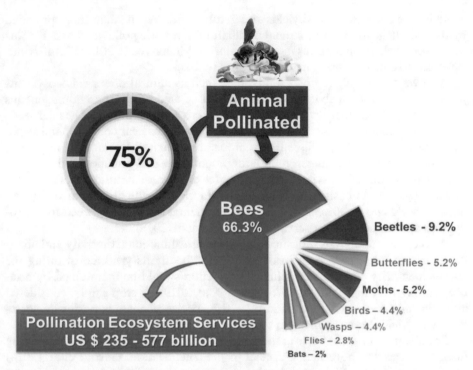

Fig. 6.6 Approximately 75% of plant species related to food production are animal pollinated, and it is estimated that the global economic valuation of pollination ecosystem service ranges between US $ 235 billion and US $ 577 billion. (Based on Wolowski et al. 2019 and IPBES 2016)

Pantanal ecosystems and in many other parts of the world, can reverse insect evolutionary diversification and success obtained in about 245 million years (Labandeira and Sepkoski Jr 1993).

The worldwide pollinator crisis refers to wild and domesticated pollinators and their population declines. Disorders in bee colonies are increasing due to viruses and other pathogens (Faurot-Daniels et al. 2020) such as the ectoparasitic mite *Varroa destructor* and the gut parasite *Nosema ceranae* (Brown et al. 2018).

The worldwide pollinator crisis presents direct consequences for the vegetal populations that rely upon them. Plants are the basis of our life, not just our food source. Plants give us wood for furniture, ethanol for car engines, many substances used in the pharmaceutical and cosmetic industries and rubber products such as rubber gloves, tires, etc. As already said, pollinator decline will have many serious consequences on natural life and on crop production. In Brazil, a country with great biodiversity richness and one of the world's agricultural leaders, 68% of 53 major food crops depend on animal pollination, occupy 59% of the total cultivated area and are responsible for 68% of all crop monetary value (Novais et al. 2016).

6.4.4 Fragmented Areas and Animal Pollination

Fragmentation of natural areas modifies the landscape and the local plant communities, creating difficulties for pollinators moving from one flowering plant to another, as long distances between fragments may create unfavorable conditions such as dry and windy open spaces without floral resources (Lázaro et al. 2020). Fragmentation also modifies nesting and oviposition sites (Williams et al. 2010; Johst et al. 2006). Natural areas near big farms of wind pollinated crops may be isolated from floral nectar and other floral resources needed by many pollinators. Meanwhile, biotically pollinated crops can produce and offer lots of homogeneous floral resources, but in a very restricted period and often with chemical defenses. When such crops are not in bloom, pollinators will need to look for food in natural areas nearby, such as ruderal areas, field margins and ecological reserves (Neumüller et al. 2020).

Ruderal weeds usually grow near crops on farms and road sides, and many are annual or biannual species able to flourish for weeks or even months, in alternating periods. The floral resources offered by ruderal species or by native wildflower plantings near crop areas can support pollinators' food requirements when crops are not in bloom (Williams et al. 2015). When two or more species coexist and share pollination services, sequential flowering is a key point to support pollinator richness and abundance, benefiting both plant fruit set and animal fitness, whether they are from the same plant family (Vilcla et al. 2018) or not (Ogilvie and Thomson 2016).

One of the important causes of fragmentation and decreases of wild pollinator communities is agricultural intensification (Klein et al. 2007). The removal of natural vegetation to open new farm areas destroys insect nests and the juveniles within. New roads to newly established planting areas can facilitate the arrival of exotic species. Many farms use pesticides like pyrethroids, neonicotinoids, azaconazole, tebufenpyrad and other chemicals that seriously affect not only herbivores but the vast majority of animals, including humans (Rigotto et al. 2014). Pesticides kill insects, including pollinators, predators and parasitoids which would otherwise benefit the harvest yield. The consequences of pesticides are well studied in bee nests, where larvae do not grow properly and adult bees loose spatial orientation and, consequently àre not able to return to the nest and die outside during the night (Goulson et al. 2008). These nest disorders reduce bee health and numbers, with the possibility of loss of the nest. In central Europe about 25% of domesticated honeybee colonies were lost from 1985 and 2005 (Potts et al. 2010). A worse situation was seen in the USA, with a loss of 59% of all honeybee colonies from 1947 and 2005 (Natural Research Council 2006).

Cities can participate in pollinator maintenance with better management of urban wildflower meadows. When roadside vegetation and urban green spaces, such as private or public gardens, have exotic woody species removed and replaced by native wildflower meadows, arthropod populations increase and maintenance costs are reduced. The native plants grown in these meadows permit colonization by native animals and support a varied insect community by offering food and nesting places (Mody et al. 2020). Available floral resources from distinct native plant families can support many dietary preferences of a high diversity of pollinators, from generalists to specialists (Kelly and Elle 2020).

6.4.5 Invasive Species

Invasive species, whether animal or vegetable, can be observed in many places all over the world, usually either by accidental or intentional human transport to a new area where the invasive species did not exist before. The invasion begins with introduction to a new area, colonization by the first individuals (animals or propagules), successful reproduction of the first generations (establishment) and more dispersal events (Mekki 2007). Some invasive species are able to promote quick and intense perturbations in the new community, and others are not. The results of the interaction of the invasive species in the new community are conditioned by the invasive species' characteristics and preferences, and by the biotic and abiotic conditions of the new place that receives the invader.

6.4.5.1 Animal Invasive Species X Pollinators

Invasive species able to interact as predators or competitors of local pollinators are considered novel stressors in the environment. They may have intense and direct harmful effects on the maintenance of pollinator communities, and also negative indirect effects on the pollination and fruiting of many plants.

Several distinct lineages of the European honeybee were brought to Brazil in the first half of the nineteenth century to produce honey and wax for candles: *Apis mellifera ligustica* from Italy, *Apis mellifera mellifera* from Germany and *Apis mellifera carnica* from Austria. Although their colonies were easy to maintain, as these European bees were very calm, the tiny honey production was insufficient. The aggressive African *Apis mellifera scutellata* was introduced by Brazilian government in the 1950s to increase honey production, but they accidentally escaped the experimental bee hives, mated with the European honeybees and spread across the Americas (Torezan-Silingardi 2008). The new hybrid, usually called Africanized *A. mellifera* caused many fatal accidents in the first years, but as time went on the bees became less aggressive and honey production increased a lot. However, *A. mellifera* has displaced many native pollinators. One of the causes of its high competitiveness and exclusion of native pollinators is that *A. mellifera* has a high degree of floral constancy (Chittka et al. 1999 and references therein).

For instance, in a Brazilian Cerrado natural reserve flowers of *Campomanesia pubescens* (Myrtaceae) were intensely visited by *A. mellifera* to collect pollen, but as these bees were too small to contact the stigma they were very inefficient as pollinators. The main pollinator of *C. pubescens* was the native bumblebee *Eulaema nigrita*, responsible for about 43% of fruit formation after buzz pollination. Bumblebees usually avoid shrubs intensely visited by *A. mellifera* due to its aggressive behavior and high density, despite the bigger size of *E. nigrita* (Torezan-Silingardi and Del-Claro 1998).

A recent animal invasion able to severely disturb beehives is the Asian Giant Hornet or *Vespa mandarinia* (Vespidae). Some months ago, in the end of 2019 the first records of the Asian Giant Hornet were reported in Canada and the United States (USDA 2020). In its original Asian Northwest areas this insect can pollinate important crops, but their aggressive behavior is a serious danger to bee and wasp nests as they are able to attack and decapitate the adults in the hive, and after that take the bee larvae and some muscular parts of the adults back to their own nests to feed their young (Matsuura 1984). North American and Canadian beekeepers and farmers that depend on pollination by *A. mellifera* will be at serious risk if the eradication of *V. mandarinia* is not successful (Alaniz et al. 2020).

6.4.5.2 Plant Invasive Species X Pollinators

Species interactions and local biodiversity may be severely impacted by invasive plants (Traveset and Richardson 2006) from the moment that invasive and native species co-occur, flourish simultaneously or at least partially, and share mainly generalist pollinators. Invasive plants compete with local plants (native or crop species) for water, nutrients, light, space and also pollinators (Tscheulin and Petanidou 2013 and references therein).

To illustrate the effects of invasive plants on plant-pollinator associations we can mention the Solanaceae family, in which many species are able to invade and adapt very well to conditions in distant places. In a Korean island, Hong et al. (2014) recently discovered *Solanum elaeagnifolium,* the tenth invasive *Solanum* species in a country that originally had just two native *Solanum* species. *Solanum elaeagnifolium* is considered a noxious weed native from the US and Mexico, but nowadays can be found in countries from South America, Europe, Africa, Asia and Australia. This species is a perennial herb able to grow and develop even in relative drought thanks to its creeping horizontal and deep vertical roots, it produces sexual and asexual propagules and is toxic to cattle, sheep and horses (Mekki 2007). The purple/blue petals of *Solanum elaeagnifolium* contrast with its yellow poricidal anthers and make the flowers very attractive to visitors, especially buzz pollinating bees from the tribes Anthophorini, Centridini, Exomalopsini and Xilocopini (Teppner 2005).

Considering that *S. elaeagnifolium* is spread all over the continents and can attract the same bees that visit other Solanaceae species, it is expected that many crop or native species will have a reduced fruit set due to competition for pollinators. Consequently, many plant species are at potential risk of having their harvest yields decreased, as their pollinators may prefer the new invasive species more than native/crop ones. *Solanum elaeagnifolium* disturbs invaded crops so much that a biological control program was developed against it (Hoffman et al. 1998) and also against other invasive *Solanum* species (Minghetti et al. 2019). *Solanum elaeagnifolium* was found to be responsible for decreasing the cotton production in Texas by about 75% (Brandon 2005) and wheat yield in Australia by about 50% (Benalla and Frankston 1998).

6.4.6 Climate Changes and Global Warming

Global temperature increases are severely affecting many countries by modifying temperatures and humidity. Recently many scientists all over the world are advising about the near future with warmer temperatures and its high impact consequences for humanity, considering urban and natural areas modifications (Del-Claro and Dirzo: ► Chap. 13). Natural area conservation is important not just for the climate, but for the preservation of biodiversity and all the still unknown uses for the biological resources present in native species, including plant and animal species and their interactions locally and beyond.

In a 10 years study from 2005 to 2014, Vilela et al. (2018) observed in a Cerrado area (Brazilian Savanna) the sequence and intensity of four Malpighiaceae species flowering in relation to an increase in the mean annual temperature by one degree Celsius. As climatic factors changed, the sequential flowering and the intensity of flowering also changed. Only one species (*Banisteriopsis malifolia*) increased flowering intensity, while other three (*B. campestris*, *B. laevifolia* and *Peixotoa tomentosa*) flowering decreased. Besides that, initially the four flowering period peaks were in sequence, after 5 years two species (*B. laevifolia* and *P. tomentosa*) were overlapping. The modified climatic conditions across 10 years had direct and indirect consequences on the phenology of the studied plants, and also on associated pollinators. Decreased flowering periods mean less pollen and oil offered to bees, which use these resources to prepare nests for generations. Coincident flowering periods indicate competition for pollinators and the possibility of heterospecific pollen loads on the stigma. So, this study points out that the intensification of global warming led to a smaller blooming and a reduction of oil and pollen offered to the bee community.

Moreover, insects depend on the environmental temperature to control their physiological processes. The ongoing global warming forces species to adapt to new and higher thermal limits, which may be achieved through genetic adaptation and phenotypic plasticity (González-Tokman et al. 2020). Consequently, global warming is a relevant factor for the decline of pollinators and associated plants.

Thomas et al. (2004) estimated the proportion of animal and plant species at risk of future extinction as a result of climate change until 2050. They also considered the effects of unprecedented CO_2 levels, habitat loss and fragmentation. They concluded that increasing temperatures are the greatest threat in most world regions. Species that flourish in the same place and time and offer the same floral resources are expected to have the same group of pollinating species. In this case, the interactions may vary from competitive to facilitative (Thomson et al. 2019). The presence of a more rewarding flower species can attract more pollinators and may benefit the visitation rates of a less rewarding flower species, as it may receive more visits and increase its fruit-set. But every situation has its costs and gains. Simultaneous flowering promotes a detrimental increase in heterospecific pollination, especially in congeneric or similar species, disturbing the pollen tube growth and fertilization.

Conclusion

Animal-plant interactions are extremely old, and among them, pollination is one of the main factors responsible for the diversification of animals and plants. Pollination is usually mutualistic in character and its results are conditioned by numerous factors, which vary from flower shape and size to pollinator morphology and behavior. Distinct animal groups are related in different way to pollination phenomena in terrestrial environments. These relationships, therefore, structure the trophic chains and networks of ecological interactions. Pollination is a vital process, not only for human survival and economic systems, but also for life on the planet as a whole. In a world undergoing deep and rapid transformation, with increasing degradation, environmental fragmentation and global warming, investigation into certain issues in pollination must be deepened or initiated, such as: How to preserve natural populations of pollinators and plants in a world with increasing rates of deforestation? What is our real capacity to raise awareness among the general population and politicians about the importance of ecosystem services and the preservation of natural environments close to crops and cities? Why do we not fight for preservation of natural vegetation, to make it possible to study the still unknown species of plants and animals and their pollination interactions, especially in tropical environments where deforestation is increasing? Pollination studies must take prerogative in decision-making by governments and sources of scientific funding, as pollination is one of the main mutualisms that maintain biodiversity.

Key Points

1. Pollination is the main process of plant reproduction, especially important in permitting gene exchange in modifying environmental conditions.
2. Mutations and genetic changes along time in plants and insects permitted the fabulous diversification of the angiosperms.
3. Animals are the main pollen vectors, especially insects and particularly bees.
4. The mutualistic relationships of plants and their pollinators are fundamental to our well-being, our health and our economy.

❓ Questions

1. Is there time to discover new pollination interactions?
2. What can be done to reconnect fragmented areas and benefit plants and pollinators?
3. What can we do to avoid the increasing deforestation and the excessive use of pesticides and not intensify the pollinator crisis, so that better crops and yields may be produced?
4. How can we make our governments aware of pollination's relevance to health and economy?

Acknowledgements   We are grateful to Brigitte Gottsberger for identifying the Glaphyridae beetle, Eduardo Andrade Botelho de Almeida for identifying the bees, André Vitor Lucci Freitas for identifying the Pieridae butterfly, Liliane Martins de Oliveira for identifying the bird *C. flaveola*. We are also grateful to Ana Paula Moraes, Felipe W. Amorim and Kleber Del-Claro for helpful comments on the text.

References

Abrahamczyk S, Kessler M, Hanley D, Karger DN, Müller MPJ, Knauer AC, Keller F, Schwerdtfeger M, Humphreys AM (2017) Pollinator adaptation and the evolution of floral nectar sugar composition. J Evol Biol 30:112–127. https://doi.org/10.1111/jeb.12991

Abrahamson WG (1989) Plant-animal interactions: an overview. In: Abrahamson WG (ed) Plant-animal interactions. Mc-Graw-Hill Publishing, New York, pp 1–22

Ackerman JD (2000) Abiotic pollen and pollination: ecological, functional, and evolutionary perspectives. Plant Syst Evol 222:167–185

Alaniz AJ, Carvajal MA, Vergara PM (2020) Giants are coming? Predicting the potential spread and impacts of the giant Asian hornet (*Vespa mandarinia*, Hymenoptera:Vespidae) in the USA. Pest Manag Sci. https://doi.org/10.1002/ps.6063

Alekseev VI, Bukejs A (2017) First fossil representatives of Pharaxonthinae Crowson (Coleoptera: Erotylidae): indirect evidence for cycad existence in Baltic amber forest. Zootaxa 4337:413–422

Altshuler DL (2003) Flower color, hummingbird pollination, and habitat irradiance in four Neotropical forests. Biotropica 35:344–355

Alves-dos-SantosI MGAR, Rozen JG (2002) Biology and immature stages of the bee tribe Tetrapediini (Hymenoptera: Apidae). Am Mus Novit 3377:1–45. https://doi.org/10.1206/0003--0082(2002)3772.0.CO;2

Amorim FW, Wyatt GE, Sazima M (2014) Low abundance of long-tongued pollinators leads to pollen limitation in four specialized hawkmoth-pollinated plants in the Atlantic Rain forest, Brazil. Naturwissenschaften 101:893–905

Anderson WR (1979) Floral conservatism in Neotropical Malpighiaceae. Biotropica 11:219–223

Andersson S, Nilsson LAA, Groth I, Bergstroem G (2002) Floral scents in butterfly-pollinated plants: possible convergence in chemical composition. Bot J Linn Soc 140:129–153

Anstett MC, Hossaert-McKey M, Kjellberg F (1997) Figs and fig pollinators: evolutionary conflicts in a coevolved mutualism. Trees 12:94–99

Appanah S (1993) Mass flowering of dipterocarp forests in the aseasonal tropics. J Biosci 18:457–474

Appanah S, Chan HT (1981) Thrips: the pollinators of some Dipterocarps. Malaysian For 44:234–252

Armbruster WS (1984) The role of resin in Angiosperm pollination: ecological and chemical considerations. Am J Bot 71:1149–1160

Arrese CA, Hart NS, Thomas N, Beazley LD, Shand J (2002) Trichromacy in Australian Marsupials. Curr Biol 12:657–660

Assunção MA, Torezan-Silingardi HM, Del-Claro K (2014) Do ant visitors to extrafloral nectaries of plants repel pollinators and cause an indirect cost of mutualism? Flora 209:244–249. https://doi.org/10.1016/j.flora.2014.03.003

Balducci MG, Niet TVD, Johnson SD (2019) Butterfly pollination of *Bonatea cassidea* (Orchidaceae): solving a puzzle from the Darwin era. S Afr J Bot 123:308–316

Balfour NJ, Garbuzov M, Ratnieks FLW (2013) Longer tongues and swifter handling: why do more bumble bees (Bombus spp.) than honey bees (Apis mellifera) forage on lavender (Lavandula spp.)? Ecol Entomol 38:323–329. https://doi.org/10.1111/een.12019

Bänzinger H, Pape T (2004) Flowers, faeces and cadavers: natural feeding and laying habits of flesh flies in Thailand (Diptera: Sarcophagidae, *Sarcophaga* spp.). J Nat Hist 38:1677–1694

6

Barônio GJ, Torezan-Silingardi HM (2017) Temporal niche overlap and distinct bee ability to collect floral resources on three species of Brazilian Malpighiaceae. Apidologie 48:168–180. https://doi.org/10.1007/s13592-016-0462-6

Barônio GJ, Haleem MA, Marsaioli AJ, Torezan-Silingardi HM (2017) Characterization of Malpighiaceae flower-visitor interactions in a Brazilian savannah: how do floral resources and visitor abundance change over time. Flora 234:126–134

Bateman RM, Hilton J, Rudall PJ (2006) Morphological and molecular phylogenetic context of the angiosperms: contrasting the 'top-down' and 'bottom-up' approaches used to infer the likely characteristics of the first flowers. J Exp Bot 57:3471–3503

Benalla DS, Frankston IF (1998) Silverleaf nightshade. In: Landcare Notes LC0227. Department of Natural Resources & Environment, State of Victoria (AU). http://www.dpi.vic.gov.au/

Bernays E (1992) Insect-plant interactions, vol IV. CRC Press Taylor & Francis Group

Blüthgen N, Fiedler K (2004) Preferences for sugars and amino acids and their conditionality in a diverse nectar-feeding ant community. J Anim Ecol 73:155–166

Bohman B, Karton A, Dixon RCM, Barrow RA, Peakall R (2016a) Parapheromones for Thynnine Wasps. J Chem Ecol 42:17–23. https://doi.org/10.1007/s10886-015-0660-0

Bohman B, Flematti GR, Barrow RA, Pichersky E, Peakall R (2016b) Pollination by sexual deception — it takes chemistry to work. Curr Opin Plant Biol 32:37–46. https://doi.org/10.1016/j.pbi.2016.06.004

Bommarco R, Marini L, Vaissière BE (2012) Insect pollination enhances seed yield, quality, and market value in oilseed rape. Oecologia 169:1025–1032. https://doi.org/10.1007/s00442-012-2271-6

Borrell BJ (2005) Long tongues and loose niches: evolution of Euglossine bees and their nectar flowers. Biotropica 37:664–669

Bowers KAW (1975) The pollination ecology of *Solanum rostratum* (Solanaceae). Am J Bot 62:633–638

Brandon LJ (2005) Weed management in roundup ready flex cotton. MSc Thesis in crop science, Graduate Faculty of Texas Tech University (US)

Briscoe AD, Chittka L (2001) The evolution of color vision in insects. Annu Rev Entomol 46:471–510

Brown ED, Hopkins MJG (1995) A test of pollinator specificity and morphological convergence between nectarivorous birds and rainforest tree flowers in New Guinea. Oecologia 103:89–100

Brown P, Newstrom-Lloyd LE, Foster BJ, Badger PH, McLean JA (2018) Winter 2016 honey bee colony losses in New Zealand. J Apic Res 57:278–291. https://doi.org/10.1080/00218839.2018.1430980

Buchmann SL (1987) The ecology of oil flowers and their bees. Annu Rev Ecol Syst 18:343–369

Cai CY, Escalona HE, Li L-Q, Yin Z-W, Huang D-Y, Engel MS (2018) Beetle pollination of cycads in the Mesozoic. Curr Biol 28:2806–2812

Camargo E, Rodrigues LC, Araujo AC (2011) Pollination biology and reproduction of *Seemannia sylvatica* (Kunth) Hanstein (Gesneriaceae) in the Serra da Bodoquena National Park, Mato Grosso do Sul. Biota Neotrop 11:125–130

Campos MGR, Bogdanov S, Almeida-Muradian LB, Szczesna T, Mancebo Y, Frigerio C, Ferreira F (2008) Pollen composition and standardisation of analytical methods. J Apic Res 47:154–161. https://doi.org/10.1080/00218839.2008.11101443

Cane JH (2016) Adult pollen diet essential for egg maturation by a solitary *Osmia* bee. J Insect Physiol 95:105–109

Capinera JL (2008) Encyclopedia of entomology, 2nd edn. Springer, 4346pp

Carthew SM, Goldingay RL (1997) Non-flying mammals as pollinators. TREE 12:104–108

Carvalho-Filho FS (2010) Scent-robbing and fighting among male orchid bees, Eulaema (Apeulaema) nigrita Lepeletier, 1841 (Hymenoptera: Apidae: Euglossini). Biota Neotrop 10:405–408. http://www.biotaneotropica.org

Cesário LF, Gaglianone MC (2013) Pollinators of *Schinus terebinthifolius* Raddi (Anacardiaceae) in vegetational formations of Restinga in Northern Rio de Janeiro state. Biosci J 29:458–467

Chen PJ, Awata H, Matsushita A, Yang EC, Arikawa K (2016) Extreme spectral richness in the eye of the common bluebottle butterfy, *Graphium sarpedon*. Front Ecol Evol 4:18

Chiari WS, Toledo VAA, Ruvolo-Takasusuki MCC et al (2005) Pollination of soybean (*Glycine max* L. Merril) by honeybees (*Apis mellifera* L.). Braz Arch Biol Technol 48:31–36

Chittka L, Thomson JD, Waser NM (1999) Flower constancy, insect psychology, and plant evolution. Naturwissenschaften 86:361–377

Chittka L, Thomson JD (eds) (2005) Cognitive ecology of pollination. Cambridge University Press, Cambridge

Ciotek L, Giorgis P, Benitez-Vieyra S, Cocucci AA (2006) First confirmed case of pseudocopulation in terrestrial orchids of South America: pollination of *Geoblasta pennicillata* (Orchidaceae) by *Campsomeris bistrimacula* (Hymenoptera, Scoliidae). Flora 201:365–369

Clemente MA, Lange D, Dáttilo W, Del-Claro K, Prezoto F (2013) Social wasp-flower visiting guild interactions in less structurally complex habitats are more susceptible to local extinction. Sociobiology 60:337–344. https://doi.org/10.13102/sociobiology.v60i3.337-344

Cole WS Jr, James AS, Smith CI (2017) First recorded observations of pollination and oviposition behavior in *Tegeticula antithetica* (Lepidoptera: Prodoxidae) suggest a functional basis for coevolution with Joshua Tree (Yucca) Hosts. Ann Entomol Soc Am 110:390–397. https://doi.org/10.1093/aesa/sax037

Cordeiro GD, Pinheiro M, Doetterl S, Alves-dosSantos I (2017) Pollination of *Campomanesia phaea* (Myrtaceae) by night-active bees: a new nocturnal pollination system mediated by floral scent. Plant Biol 19:132–139. https://doi.org/10.1111/plb.12520

Cox PA (1988) Hydrophilous pollination. Annu Rev Ecol Syst 19:261–279

Cox PA (1993) Water-pollinated plants. Sci Am 269:68–74

Crane PR (1986) Form and function in wind dispersed pollen. In: Blackmore S, Ferguson IK (eds) Pollen and spores: form and function. Academic, London, pp 179–202

Crepet WL (1972) Investigations of North American cycadeoids: pollination mechanisms in *Cycadeoidea*. Am J Bot 59:1048–1056

Crepet WL (1984) Advanced (constant) insect pollination mechanisms: patterns of evolution and implications vis-à-vis angiosperm diversity. Ann Mo Bot Gard 71:607–630

Crepet WL, Feldman GD (1991) The earliest remains of grasses in the fossil record. Am J Bot 78:1010–1014

Crepet WL, Friis EM, Nixon KC, Lack AJ, Jarzembowski EA (1991) Fossil evidence for the evolution of biotic pollination [and discussion]. Phil Trans R Soc Lond B 333:187–195. https://doi.org/10.1098/rstb.1991.0067

Crepet WL, Nixon KC (1996) The fossil history of stamens. In: D'Arcy WG, Keating RC (eds) The anther: form, function and phylogeny. Cambridge University Press, Cambridge, UK, pp 25–57

Crepet WL (2008) The fossil record of angiosperms: requiem or renaissance? Ann Mo Bot Gard 95:3–33

Cronin TW, Johnsen S, Marshall J, Warrant EJ (2014) Visual Ecology. Princeton University Press.

Cronk Q, Ojeda I (2008) Bird-pollinated flowers in an evolutionary and molecular context. J Exp Bot 59:715–727

Crowson RA (1981) The biology of the Coleoptera. Academic, London

Culley TM, Weller SG, Sakai AK (2002) The evolution of wind pollination in angiosperms. Trends Ecol Evol 17:361–369

Dalsgaard B, Magard E, Fjeldsa J, Martín González AM, Rahbek C et al (2011) Specialization in plant-hummingbird networks is associated with species richness, contemporary precipitation and quaternary climate-change velocity. PLoS One 6:e25891. https://doi.org/10.1371/journal.pone.0025891

Danieli-Silva A, Varassin IG (2013) Breeding system and thrips (Thysanoptera) pollination in the endangered tree *Ocotea porosa* (Lauraceae): implications for conservation. Plant Spec Biol 28:31–40

Daniels RJ, Johnson SD, Peter CI (2020) Flower orientation in *Gloriosa superba* (Colchicaceae) promotes cross-pollination via butterfly wings. Ann Bot 125:1137–1149. https://doi.org/10.1093/aob/mcaa048

Darwin C (1904) Fertilization of orchids, London

Daumann E (1959) Zur Kenntnis der Blütennektarien von *Aristolochia*. Preslia 31:359–372

Day S, Beyer R, Mercer A, Ogden S (1990) The nutrient composition of honeybee-collected pollen in Otago, New Zealand. J Apicult Res 29:138–146

Del-Claro K, Rodriguez-Morales D, Calixto ES, Martins AS, Torezan-Silingardi HM (2019) Ant pollination of *Paepalanthus lundii* (Eriocaulaceae) in Brazilian savanna. Ann Bot 123:1159–1165. https://doi.org/10.1093/aob/mcz021

Del-Claro K, Torezan-Silingardi HM (2020) In search of unusual interactions. A commentary on: 'Pollen adaptation to ant pollination: a case study from the Proteaceae'. Ann Bot 126:1–2

Delnevo N, Etten EJV, Clemente N, Fogu L, Pavarani E, Byrne M, Stock WD (2020) Pollen adaptation to ant pollination: a case study from the Proteaceae. Ann Bot 126:377–386. https://doi.org/10.1093/aob/mcaa058

Delevoryas T (1968) Investigations of North American cycadeoids: structure, ontogeny, and phylogenetic considerations of cones of *Cycadeoidea*. Palaeontographica 21B:122–133

Dellinger AS (2020) Pollination syndromes in the 21st century: where do we stand and where may we go? New Phytol. https://doi.org/10.1111/nph.16793

Deprá MS, Delaqua GCG, Freitas L, Gaglianone MC (2014) Pollination deficit in open-field tomato crops (*Solanum lycopersicum* l., Solanaceae) in Rio de Janeiro state, Southeast Brazil. J Poll Ecol 12:1–8

de Vega C, Herrera CM, Dötterl S (2014) Floral volatiles play a key role in specialized ant pollination. Perspect Plant Ecol Evol Syst 16:32–42

Dobson HEM (2006) Relationship between floral frangrance composition and type of pollinator. In: Dudareva N, Pichersky E (eds) Biology of floral scent. Taylor & Francis, Boca Raton, London, New York, pp 147–198

Domingos-Melo A, Nadia TL, Machado IC (2017) Complex flowers and rare pollinators: does ant pollination in Ditassa show a stable system in Asclepiadoideae (Apocynaceae)? Arthropod Plant Interact 11:339–349

Dormann C (2011) How to be a specialist? Quantifying specialisation in pollination networks. Net Biol 1:1–20

Doyle JA (2012) Molecular and fossil evidence on the origin of angiosperms. Molecular and fossil evidence on the origin of angiosperms. Annu Rev Earth Planet Sci 40:301–326. https://doi.org/10.1146/annurev-earth-042711-105313

Dressler R (1981) Orchids – natural history and classification, 1st edn. Harvard University Press, Cambridge, MA

Dyer AG, Whitney HM, Arnold SEJ, Glover BJ, Chittka L (2007) Mutations perturbing petal cell shape and anthocyanin synthesis influence bumblebee perception of *Antirrhinum majus* flower colour. Arthropod Plant Interact 1:45–55

Ehrlich P, Raven P (1964) Butterflies and plants: a study in coevolution. Evolution 18:586–608

Erbar C (2014) Nectar secretion and nectaries in basal angiosperms, magnoliids and non-core eudicots and a comparison with core eudicots. Plant Diver Evol 131:63–143

Faegri K, van der Pijl (1979) The principles of pollination ecology. 3rd revised ed. Pergamon Press, Oxford, New York

Faurot-Daniels C, Glenny W, Daughenbaugh KF, McMenamin AJ, Burkle LA, Flenniken ML (2020) Longitudinal monitoring of honey bee colonies reveals dynamic nature of virus abundance and indicates a negative impact of Lake Sinai virus 2 on colony health. PLoS One 15:e0237544. https://doi.org/10.1371/journal.pone.0237544

Ferreira CA, Torezan-Silingardi HM (2013) Implications of the floral herbivory on Malpighiacea plant fitness: visual aspect of the flower affects the attractiveness to pollinators. Sociobiology 60:323–328. https://doi.org/10.13102/sociobiology.v60i3.323-328

Fiala B, Meyer U, Hashim R, Maschwitz U (2011) Pollination systems in pioneer trees of the genus *Macaranga* (Euphorbiaceae) in Malaysian rainforests. Biol J Linn Soc 103:935–953

Fleming PA, Nicolson SW (2002) How important is the relationship between *Protea humiflora* (Proteaceae) and its non-flying mammal pollinators? Oecologia 132:361–368. https://doi.org/10.1007/s00442-002-0921-9

Fleming TH, Muchhala N (2008) Nectar-feeding bird and bat niches in two worlds: pantropical comparisons of vertebrate pollination systems. Journal of Biogeography (J. Biogeogr.) (2008) 35, 764–780.

Fleming TH, Geiselman C, Kress WJ (2009) The evolution of bat pollination: a phylogenetic perspective. Ann Bot 104:1017–1043

Freitas L, Sazima M (2009) Floral biology and mechanisms of spontaneous self-pollination in five neotropical species of Gentianaceae. Bot J Linn Soc 160:357–368. https://doi.org/10.1111/j.1095-8339.2009.00989.x

Freitas LS, Moreira LM, de Avila RS, Felestrino ÉB, Demarco D, de Sousa HC, Ribeiro SP (2017) Reproductive phenology and floral visitors of a *Langsdorffia hypogaea* (Balanophoraceae) population in Brazil. Flora 233:51–57. https://doi.org/10.1016/j.flora.2017.02.023

Friedman J, Barrett SCH (2008) A phylogenetic analysis of the evolution of wind pollination in the Angiosperms. Int J Plant Sci 169:49–58

Friedman WE (2009) The meaning of Darwin's "abominable mystery". Am J Bot 96:5–21

Fröhlich MW (2003) An evolutionary scenario for the origin of flowers. Nat Rev Genet 4:559–566

Funamoto D (2019) Precise sternotribic pollination by settling moths in *Adenophora maximowicziana* (Campanulaceae). Int J Plant Sci 180:200–208

Garratt MPD, Breeze TD, Jenner M, Polcec C, Biesmeijer JC, Potts SG (2014) Avoiding a bad apple: insect pollination enhances fruit quality and economic value. Agric Ecosyst Environ 184:34–40

Gemmill-Herren B, Ochieng AO (2008) Role of native bees and natural habitats in eggplant (*Solanum melongena*) pollination in Kenya. Agric Ecosyst Environ 127:31–36

Goldblatt P, Manning JC (2000) The long-proboscid fly pollination system in Southern Africa. Ann Mo Bot Gard 87:146–170

Gonçalves-Souza P, Schlindwein C, Paiva EAS (2018) Floral resins of *Philodendron adamantinum* (Araceae): secretion, release and synchrony with pollinators. Acta Botanica Brasilica 32:392–401. https://doi.org/10.1590/0102-33062018abb0115

González-Tokman D, Córdoba-Aguilar A, Dáttilo W, Lira-Noriega A, Sánchez-Guillén RA, Villalobos F (2020) Insect responses to heat: physiological mechanisms, evolution and ecological implications in a warming world. Biol Rev. https://doi.org/10.1111/brv.12588

Goodrich KR (2012) Floral scent in Annonaceae. Bot J Linn Soc 169:262–279

Gottsberger G (1977) Some aspects of beetle pollination in the evolution of flowering plants. Plant Syst Evol Suppl 1:211–226

Gottsberger G, Amaral A Jr (1984) Pollination strategies in Brazilian Philodendron species. Ber deutsch Bot Ges Bd 97:391–410

Gottsberger G (1986) Some pollination strategies in Neotropical savannas and forests. Plant Syst Evol 152:29–45

Gottsberger G (1989) Beetle pollination and flowering rhythm of *Annona* spp. (Annonaceae) in Brazil. Plant Syst Evol 167:165–187

Gottsberger G (1999) Pollination and evolution in Neotropical Annonaceae. Plant Spec Biol 14:143–152

Gottsberger G (2016) Generalist and specialist pollination in basal angiosperms (ANITA grade, basal monocots, magnoliids, Chloranthaceae and Ceratophyllaceae): what we know now. Plant Diver Evol 131:263–362

Goulson D, Cory JS (1993) Flower constancy and learning in foraging preferences of the green-veined white butterfly *Pieris napi*. Ecol Entomol 18:315–320

Goulson D, Lye GC, Darvill B (2008) Decline and conservation of bumble bees. Annu Rev Entomol 53:191–208

Grimaldi D (1999) The co-radiations of pollinating insects and angiosperms in the cretaceous. Ann Mo Bot Gard 86:373–406

Grimaldi D, Engel MS (2005) Evolution of the insects. Cambridge University Press, Cambridge, New York

Gullan PJ, Cranston PS (1994) The insects: an outline of entomology. Chapman & Hall. 491p

Hagerup E, Hagerup O (1953) Thrips pollination of *Erica tetralix*. New Phytol 52:1–7

Hargreaves AL, Johnson SD, Nol E (2004) Do floral syndromes predict specialization in plant pollination systems? An experimental test in an "ornithophilous" African *Protea*. Oecologia 140:295–301. https://doi.org/10.1007/s00442-004-1495-5

Hart NS, Hunt DM (2007) Avian visual pigments: characteristics, spectral tuning, and evolution. Am Nat 169:S7–S26. https://www.jstor.org/stable/10.1086/510141

6

Heil M (2011) Nectar: generation, regulation and ecological functions. Trends Plant Sci 16:191–200. https://doi.org/10.1016/j.tplants.2011.01.003

Helversen DV, Helversen OV (1999) Acoustic guide in bat-pollinated flower. Nature 398:795–796

Heslop-Harrison Y, Shivanna KR (1977) The receptive surface of the Angiosperm stigma. Ann Bot 41:1233–1258

Hickman JC (1974) Pollination by ants: a low-energy system. Science 184:1290–1292

Hipólito J, Boscolo D, Viana BF (2018) Landscape and crop management strategies to conserve pollination services and increase yields in tropical coffee farms. Agric Ecosyst Environ 256:218–225

Hoffman JH, Moran VC, Impson FAC (1998) Promising results from the firrst biological control programme against a solanaceous weed (*Solanum elaeagnifolium*). Agric Ecosyst Environ 70:145–150

Hogan JD, Melin AD, Mosdossy KN, Fedigan LM (2016) Seasonal importance of flowers to Costa Rican capuchins (*Cebus capucinus imitator*): implications for plant and primate. Am J Phys Anthropol 161:591–602. https://doi.org/10.1002/ajpa.23059

Holloway BA (1976) Pollen-feeding in hover-flies (Diptera: Syrphidae). N Z J Zool 3:339–350

Hong JR, Joo MJ, Hong MH, Jo SJ, Kim KJ (2014) *Solanum elaeagnifolium* Cav. (Solanaceae), an unrecorded naturalized species of Korean flora. Korean J Plant Taxon 44:18–21. https://doi.org/10.11110/kjpt.2014.44.1.18

Hu S, Dilcher DL, Jarzen DM, Taylor DW (2008) Early steps of angiosperm–pollinator coevolution. PNAS 105:240–245. https://doi.org/10.1073/pnas.0707989105

Huber H (1993) Aristolochiaceae. In: Kubitzki K, Rohwer JG, Bittrich V (eds) The families and genera of vascular plants II:129–137. Springer, Berlin, Heidelberg, New York

Hudewenz A, Pufal G, Bögeholz AL, Klein AM (2013) Cross-pollination benefits differ among oilseed rape varieties. J Agric Sci. https://doi.org/10.1017/S0021859613000440

Huth CJ, Pellmyr O (2000) Pollen-mediated selective abortion in Yuccas and its consequences for the plan-pollinator mutualism. Ecology 81:1100–1107

Inouye DW, Larson BMH, SSymank A, Kevan PG (2015) Flies and flowers III: ecology of foraging and pollination. J Poll Ecol 16:115–133

IPBES (2016) The assessment report of the intergovernmental science-policy platform on biodiversity and ecosystem services on pollinators, pollination and food production. In: Potts SG, Imperatriz-Fonseca VL, Ngo HT (eds) Secretariat of the intergovernmental science-policy platform on biodiversity and ecosystem services, Bonn. 552p

Irwin MT, Raharison FJ, Rakotoarimanana H, Razanadrakoto E, Ranaivoson E, Rakotofanala J, Randrianarimanana C (2007) Diademed sifakas (*Propithecus diadema*) use olfaction to forage for the inflorescences of subterranean parasitic plants (Balanophoraceae: *Langsdorffia* sp., and Cytinaceae: *Cytinus* sp.). Am J Primatol 69:471–476. https://doi.org/10.1002/ajp.20353

Ishara KL, Maimoni-Rodella RCS (2011) Pollination and dispersal systems in a Cerrado remnant (Brazilian Savanna) in Southeastern Brazil. Braz Arch Biol Technol 54:629–642

Ishida C, Kono M, Sakai S (2009) A new pollination system: brood-site pollination by flower bugs in *Macaranga* (Euphorbiaceae). Ann Bot 103:39–44. https://doi.org/10.1093/aob/mcn212

Jansen-González S, Teixeira SP, Pereira RAS (2012) Mutualism from the inside: coordinated development of plant and insect in an active pollinating fig wasp. Arthropod Plant Interact 6:601–609

Jersáková J, Johnson SD, Kindlmann P (2006) Mechanisms and evolution of deceptive pollination in orchids. Biol Rev 81:219–235. https://doi.org/10.1017/S1464793105006986

Johnson SD, Steiner KE (1997) Long-tongued fly pollination and evolution of floral spur length in the *Disa draconis* complex (Orchidaceae). Evolution 51:45–53

Johnson JD, Liltved WR (1997) Hawkmoth pollination of *Bonatea speciosa* (Orchidaceae) in a South African coastal forest. Nord J Bot 16:5–10

Johst K, Drechsler M, Thomas J, Settele J (2006) Influence of moving on the persistence of two endangered large blue butterfly species. J Appl Ecol 43:333–342

Jürgens A, Webber AC, Gottsberger G (2000) Floral scent compounds of Amazonian Annonaceae species pollinated by small beetles and thrips. Phytochemistry 55:551–558

Kato M, Inouye T (1994) Origin of insect pollination. Science 368:195

Kato M, Takimura A, Kawakita A (2003) An obligate pollination mutualism and reciprocal diversi-fication in the tree genus *Glochidion* (Euphorbiaceae). PNAS 100:5264–5267. https://doi.org/10.1073/pnas.0837153100

Kearns CA (1992) Anthophilous fly distribution across an elevation gradient. Am Midl Nat 127:172–182

Kearns CA (2001) North American dipteran pollinators: assessing their value and conservation sta-tus. Conserv Ecol 5. (Online) URL: http://www.consecol.org/vol5/iss1/art5

Kelly TT, Elle E (2020) Investigating bee dietary preferences along a gradient of floral resources: how does resource use align with resource availability? Insect Sci 00:1–11. https://doi.org/10.1111/1744-7917.12785

Kevan PG, Chaloner WG, Savile DBO (1975) Interrelationships of early terrestrial arthropods and plants. Palaeontology 18:391–417

Kinoshita M, Stewart FJ, Omura H (2017) Multisensory integration in Lepidoptera: insights into flower-visitor interactions. BioEssays 39:1600086. https://doi.org/10.1002/bies.201600086

Klavins SD, Kellogg DW, Krings M, Taylor EL, Taylor TN (2005) Coprolites in a middle triassic cycad pollen cone: evidence for insect pollination in early cycads? Evol Ecol Res 7:479–488

Klein AM, Vaissière BE, Cane JH, Steffan-Dewenter I, Cunningham SA, Kremen C, Tscharntke T (2007) Importance of pollinators in changing landscapes for world crops. Proc R Soc B 274:303–313. https://doi.org/10.1098/rspb.2006.3721

Knuth P (1898) Handbuch der Blütenbiologie. Band 1: Einleitung und Litteratur. Wilhelm Engelmann, Leipzig

Kondo T, Nishimura S, Tani N, Ng KKS, Lee SL, Muhammad N, Okuda T, Tsumura Y, Isagi Y (2016a) Complex pollination of a tropical Asian rainforest canopy tree by flower-feeding thrips and thrips-feeding predators. Am J Bot 103:1912–1920

Kondo T, Otani T, Lee SL, Tani N (2016b) Pollination system of *Shorea curtisii*, a dominant species in hill dipterocarp forests. J Trop For Sci 28:318–323

Koopman KF (1981) The distributional patterns of New World nectar-feeding bats. Ann Mo Bot Gard 68:352–369

Kozub C, Barboza K, Galdeano F, Quarin CL, Cavagnaro JB, Cavagnaro PF (2017) Reproductive biology of the native forage grass *Trichloris crinita* (Poaceae, Chloridoideae). Plant Biol 19:444–453

Krahl AH, Holanda ASS, Krahl DRP, Martucci MEP, Gobbo-Neto L, Webber AC, Pansarin ER (2019) Study of the reproductive biology of an Amazonian *Heterotaxis* (Orchidaceae) demon-strates the collection of resin-like material by stingless bees. Plant Syst Evol 305:281–291. https://doi.org/10.1007/s00606-019-01571-9

Krauss SL, Phillips RD, Karron JD, Johnson SD, Roberts DG, Hopper SD (2017) Novel conse-quences of bird pollination for plant mating. Trends Plant Sci 22:395–410. https://doi.org/10.1016/j.tplants.2017.03.005

Kugler H (1955) Zum Problem der Dipterenblumen. Österreichische Botanische Zeitschrift 102:529–541

Kuhlmann M, Hollens H (2014) Morphology of oil-collecting pilosity of female *Rediviva* bees (Hymenoptera: Apoidea: Melittidae) reflects host plant use. J Nat Hist 49:561–657

Kullenberg B (1950) Investigations on the pollination of *Ophrys* species. Oikos 2:1–19

Kunz TH, Torrez EB, Bauer D, Lobova T, Fleming TH (2011) Ecosystem services provided by bats. Ann N Y Acad Sci 1223:1–38. https://doi.org/10.1111/j.1749-6632.2011.06004.x

Labandeira CC, Sepkoski JJ Jr (1993) Insect diversity in the fossil record. Science 261:310–315

Labandeira CC (1998) Early history of arthropod and vascular plant associations. Annu Rev Earth Planet Sci 26:329–377

Labandeira CC, Yang Q, Santiago-Blay JA, Hotton CL, Monteiro A, Wang YJ, Goreva Y, Shih C, Siljeström S, Rose TR et al (2016) The evolutionary convergence of mid-Mesozoic lacewings and Cenozoic butterflies. Proc Biol Sci 283:20152893

Larson BMH, Kevan PG, Inouye DW (2001) Flies and flowers: taxonomic diversity of anthophiles and pollinators. Can Entomol 133:439–465

Lázaro A, Fuster F, Alomar D, Totland Ø (2020) Disentangling direct and indirect effects of habitat fragmentation on wild plants' pollinator visits and seed production. Ecol Appl 30(5):e02099

Li P, Luo YB, Bernhardt P, Yang XQ, Kou Y (2006) Deceptive pollination of the Lady's Slipper *Cypripedium tibeticum* (Orchidaceae). Plant Syst Evol 262:53–63

Li HT, Yi TS, Gao LM, Ma PF, Zhang T, Yang JB, Gitzendanner MA et al (2019) Origin of angiosperms and the puzzle of the Jurassic gap. Nat Plants 5:461–470. https://doi.org/10.1038/s41477-019-0421-0

Lindström SAM, Herbertsson L, Rundlöf M, Smith HG, Bommarco R (2015) Large-scale pollination experiment demonstrates the importance of insect pollination in winter oilseed rape. Oecologia. https://doi.org/10.1007/s00442-015-3517-x

Lloyd DG, Wells MS (1992) Reproductive biology of a primitive angiosperm, *Pseudowintera colorata* (Winteraceae), and the evolution of pollination systems in the Anthophyta Pl. Syst Evol 181:77–95

Lomáscolo SB, Giannini N, Chacoff NP, Castro-Urgal R, Vázquez DP (2019) Inferring coevolution in a plant–pollinator network. Oikos 128:775–789. https://doi.org/10.1111/oik.05960

Lunau K, Hofmann N, Valentin S (2005) Response of the hoverfly species *Eristalis tenax* towards floral dot guides with colour transition from red to yellow (Diptera : Syrphidae). Entomol Gen 27:249–256

Lunau K (2014) Visual ecology of flies with particular reference to colour vision and colour preferences. J Comp Physiol A. https://doi.org/10.1007/s00359-014-0895-1

Luo Y-B, Li Z-Y (1999) Pollination ecology of *Chloranthus serratus* (Thunb.) Roem. et Schult. and *Ch. fortunei* (A. Gray) Solms-Laub. (Chloranthaceae). Ann Bot 83:489–499

Kraaij M, van der Kooi C (2020) Surprising absence of association between flower surface microstructure and pollination system. Plant Biol 22:177–183. https://doi.org/10.1111/plb.13071

Maia ACD, Dötterl S, Kaiser R, Silberbauer-Gottsberger I, Teichert H, Gibernau M, Navarro DMAF, Schlindwein C, Gottsberger G (2012) The key role of 4-methyl-5-vinylthiazole in the attraction of scarab beetle pollinators: a unique olfactory floral signal shared by Annonaceae and Araceae. J Chem Ecol 38:1072–1080

Manning JC, Goldblatt P (1997) The *Moegistorhynchus longirostris* (Diptera: Nemestrinidae) pollination guild: long-tubed flowers and a specialized long-proboscid fly pollination system in southern Africa. Plant Syst Evol 206:51–69

Marquínez X, Cepeda J, Lara K, Sarmiento R (2010) Arañas asociadas a la floración de *Drimys granadensis* (Winteraceae). Revista Colombiana de Entomología 36:172–175

Martel C, Francke W, Ayasse M (2019) The chemical and visual bases of the pollination of the Neotropical sexually deceptive orchid *Telipogon peruvianus*. New Phytol 223:1989–2001. https://doi.org/10.1111/nph.15902

Martin ALB, O'Hanlon JC, Gaskett AC (2020) Orchid sexual deceit affects pollinator sperm transfer. Funct Ecol 00:1–9. https://doi.org/10.1111/1365-2435.13551

Matsuura M (1984) Comparative biology of the five Japanese species of the genus Vespa. Bull Fac Agr Mie Univ 69:1–132

McCall AC, Irwin RE (2006) Florivory: the intersection of pollination and herbivory. Ecol Lett 9:1351–1365. https://doi.org/10.1111/j.1461-0248.2006.00975.x

Mekki M (2007) Biology, distribution and impacts of silverleaf nightshade (*Solanum elaeagnifolium* Cav.). Bull OEPP/EPPO 37:114–118

Michener CD (1974) The social behavior of the bees: a comparative study. Harvard University Press, 404 p

Michener CD (2007) The bees of the world, 2nd edn. Johns Hopkins University Press, Baltimore

Minghetti E, Olivera L, Montemayor SI (2019) Ecological niche modelling of *Gargaphia decoris* (Heteroptera), a biological control agent of the invasive tree *Solanum mauritianum* (Solanales: Solanaceae). Pest Manag Sci. https://doi.org/10.1002/ps.5637

Mody K, Lerch D, Müller A-K, Simons NK, Blüthgen N, Harnisch M (2020) Flower power in the city: replacing roadside shrubs by wildflower meadows increases insect numbers and reduces maintenance costs. PLoS One 15(6):e0234327. https://doi.org/10.1371/journal.pone.0234327

Momose K, Nagamitsu T, Inoue T (1998) Thrips cross-pollination of *Popowia pisocarpa* (Annonaceae) in a lowland dipterocarp forest in Sarawak. Biotropica 30:444–448

Moseley FY (1898) What is a flower? Asa Gray Bull 6:9–11

Mound LA, Marullo R (1996) The thrips of Central and South America: an introduction. Mem Entomol Int 6:1–488

Moyroud E, Glover BJ (2017) The physics of pollinator attraction. New Phytol 216:350–354. https://doi.org/10.1111/nph.14312

Müller H (1881) Alpenblumen, ihre Befruchtung durch Insekten und ihre Anpassungen an dieselben. Wilhelm Engelmann, Leipzig

Murugan R, Shivanna KR, Rao RR (2006) Pollination biology of *Aristolochia tagala*, a rare species of medicinal importance. Curr Sci 91:795–798

Nabozhenko MV (2019) The fossil record of Darkling Beetles (Insecta: Coleoptera: Tenebrionidae). Geosciences 9:514. https://doi.org/10.3390/geosciences9120514

Natural Research Council (2006) Status of pollinators in North America. National Academic Press, Washinton

Neff JL, Simpson BB (1981) Oil-collecting structures in the Anthophoridae (Hymenoptera): morphology, function and use in systematics. Journal of Kansas Entomological Society 54:95–123.

Neff JL, Simpson BB (2017) Vogel's great legacy: the oil flower and oil-collecting bee syndrome. Flora 232:104–116

Neumüller U, Burger H, Krausch S, Bluthgen N, Ayasse M (2020) Interactions of local habitat type, landscape composition and flower availability moderate wild bee communities. Landsc Ecol 35:2209–2224

Nicolson SW (2011) Bee food: the chemistry and nutritional value of nectar, pollen and mixtures of the two. Afr Zool 46:197–204

Niklas KJ (1997) The evolutionary biology of plants. University of Chicago Press, Chicago

Novais SMA, Nunes CA, Santos NB et al (2016) Effects of a possible pollinator crisis on food crop production in Brazil. PLoS One 11:e0167292. https://doi.org/10.1371/journal.pone.0167292

Nunes CEP, Peñaflor MFGV, Bento JMS, Salvador MJ, Sazima M (2016) The dilemma of being a fragrant flower: the major floral volatile attracts pollinators and florivores in the euglossine-pollinated orchid *Dichaea pendula*. Oecologia 182:933–946. https://doi.org/10.1007/s00442-016--3703-5

Nunes-Silva P, Hrncir M, Silva CI, Roldão YS, Imperatriz-Fonseca VL (2013) Stingless bees, *Melipona fasciculata*, as efficient pollinators of eggplant (*Solanum melongena*) in greenhouses. Apidologie. https://doi.org/10.1007/s13592-013-0204-y

Nyman T, Onstein RE, Silvestro D, Wutke S, Taeger A, Wahlberg N et al (2019) The early wasp plucks the flower: Disparate extant diversity of sawfly superfamilies (Hymenoptera: 'Symphyta') may reflect asynchronous switching to angiosperm hosts. Biol J Linn Soc 14:128. https://doi.org/10.1093/biolinnean/blz071

Oelschlägel B, Nuss M, von Tschirnhaus M, Pätzold C, Neinhuis C, Dötterl S, Wanke S (2015) The betrayed thief – the extraordinary strategy of *Aristolochia rotunda* to deceive its pollinators. New Phytol 206:342–351

Ogilvie JE, Thomson JD (2016) Site fidelity by bees drives pollination facilitation in sequentially blooming plant species. Ecology 97:1442–1451

Oliveira CMA, Porto AM, Bittrich V, Vencato I, Marsaioli AJ (1996) Floral resins of *Clusia* spp.: chemical composition and biological function. Tetrahedron Lett 37:6427–6430

Ollerton J, Killick A, Lamborn E, Watts S, Whiston M (2007) Multiple meanings and modes: on the many ways to be a generalist flower. Taxon 56:717–728

Ollerton J, Coulthard E (2009) Evolution of animal pollination. Science 326:808–809

Ollerton J, Winfree R, Tarrant S (2011) How many flowering plants are pollinated by animals? Oikos 120:321326. https://doi.org/10.1111/j.1600-0706.2010.18644.x

Ollerton J (2017) Pollinator diversity: distribution, ecological function, and conservation. Annu Rev Ecol Evol Syst 48:353–376

Olesen JM, Valido A (2003) Lizards as pollinators and seed dispersers: an island phenomenon. Trends Ecol Evol 18:177–181

Palmer-Young EC, Farrell IW, Adler LS, Milano NJ, Egan PA, Junker RR, Irwin RE, Stevenson PC (2018) Chemistry of floral rewards: intra- and interspecific variability of nectar and pollen secondary metabolites across taxa. Ecol Monograph 0:1–20. e01335. https://doi.org/10.1002/ecm.1335

Parachnowitsch AL, Manson JS, Sletvold N (2019) Evolutionary ecology of nectar. Ann Bot 123:247–261. https://doi.org/10.1093/aob/mcy132

Paulus HF, Gack C (1990) Pollinators as prepollinating isolation factors: evolution and speciation in *Ophrys*. Israel J Bot 39:43–79

Payne WW (1981) Structure and function in angiosperm pollen wall evolution. Rev Palaeobot Palynol 35:39–59

Peakall R, Schiestl FP (2004) A mark-recapture study of male *Colletes cunicularius* bees: implications for pollination by sexual deception. Behav Ecol Sociobiol 56:579–584

Peakall R, Ebert D, Poldy J, Barrow RA, Francke W, Bower CC, Schiestl FP (2010) Pollinator specificity, floral odour chemistry and the phylogeny of Australian sexually deceptive Chiloglottis orchids: implications for pollinator-driven speciation. New Phytol 188:437–450. https://doi.org/10.1111/j.1469-8137.2010.03308.x

Pellmyr O (2003) Yuccas, yucca moths and coevolution: a review. Ann Missouri Bot Gard 90:35–55

Pellmyr O, Segraves KA (2003) Pollinator divergence within an obligate mutualism: two yucca moth species (Lepidoptera; Prodoxidae: Tegeticula) on the Joshua Tree (Yucca brevifolia; Agavaceae). Ann Entomol Soc Am 96:716–722

Peñalver E, Labandeira CC, Barrón E, Delclòs X, Nel P, Nel A, Tafforeau P, Soriano C (2012) Thrips pollination of Mesozoic gymnosperms. Proc Natl Acad Sci U S A 109:8623–8628

Peñalver E, Arillo A, Perez-de la Fuente R, Riccio ML, Delclòs X, Barrón E, Grimaldi DA (2015) Long-proboscid flies as pollinators of Cretaceous gymnosperms. Curr Biol 25:1917–1923

Pereira J, Schlindwein C, Antonini Y, Maia ACD, Dötterl S, Martins C, Navarro DMAF, Oliveira R (2014) *Philodendron adamantinum* (Araceae) lures its single cyclocephaline scarab pollinator with specific dominant floral scent volatiles. Biol J Linn Soc 111:679–691

Peris D, Fuente RP, Peñalver E, Delclòs X, Barrón E, Labandeira CC (2017a) False blister beetles and the expansion of gymnosperm-insect pollination modes before angiosperm dominance. Curr Biol 27:1–8. https://doi.org/10.1016/j.cub.2017.02.009

Peris D, Labandeira CC, Peñalver E, Delclòs X, Barrón E, Fuente RP (2017b) The case of *Darwinylus marcosi* (Insecta: Coleoptera: Oedemeridae): a cretaceous shift from a gymnosperm to an angiosperm pollinator mutualism. Commun Integr Biol 10(4):e1325048. https://doi.org/10.1080/19420889.2017.1325048

Peris D, Perez-de la Fuente R, Peñalver E, Delclòs X, Barrón E, Labandeira CC (2017c) False blister beetles and the expansion of gymnosperm-insect pollination modes before angiosperm dominance. Curr Biol 27:897–904

Phillips RD, Brown GR, Dixon KW, Hayes C, Linde CC, Peakall R (2017) Evolutionary relationships among pollinators and repeated pollinator sharing in sexually deceptive orchids. J Evol Biol 30:1674–1691. https://doi.org/10.1111/jeb.13125

Poinar G Jr (2016) Beetles with orchid pollinaria in Dominican and Mexican amber. Am Entomol 62:172–177. https://doi.org/10.1093/ae/tmw055

Possobom CCF, Machado SR (2017) Elaiophores: their taxonomic distribution, morphology and functions. Acta Botanica Brasilica 31:503–524. https://doi.org/10.1590/0102-33062017abb0088

Potgieter CJ, Edwards TJ (2005) The *Stenobasipteron wiedemanni* (Diptera, Nemestrinidae) pollination guild in eastern Southern Africa. Ann Mo Bot Gard 92:254–267

Potts SG, Roberts SP, Dean R, Marris G, Brown M, Jones R, Neumann P, Settele J (2010) Declines of managed honeybees and beekeepers in Europe? J Apic Res 49:15–22

Powell JA, Mackie RA (1966) Biological interrelationships of moths and *Yucca whipplei* (Lepidoptera: Gelechiidae, Blastobasidae, Prodoxidae). Univ Calif Publ Entomol 42:1–46

Prescott-Allen R, Prescott-Allen C (1990) How many plants feed the world? Conserv Biol 4:365–374

Proctor M, Yeo P, Lack A (1996) The natural history of pollination. Harper Collins Publishers, London. 479 pp.

Pyke GH (2016) Floral nectar: pollinator attraction or manipulation? Trends Ecol Evol 31:339–341

Quirino ZGM, Machado IC (2014) Pollination syndromes in a Caatinga plant community in northeastern Brazil: seasonal availability of floral resources in different plant growth habits. Br J Biol 74:62–71. https://doi.org/10.1590/1519-6984.17212

Rader R, Bartomeus I, Garibaldi LA, Garratt MPD, Howlett BG et al (2016) Non-bee insects are important contributors to global crop pollination. Proc Natl Acad Sci U S A 113:146–151

Raguso RA (2004) Flowers as sensory billboards: progress towards an integrated understanding of floral advertisement. Curr Opin Plant Biol 7:434–440

Raguso RA (2020) Don't forget the flies: dipteran diversity and its consequences for floral ecology and evolution. Appl Entomol Zool 55:1–7

Raju AJS, Raju PS, Ramana KV (2014) Mellitophily and malacophily in *Ipomoea pes-caprae* (Convolvulaceae). Taprobanica 6:90–99

Rathcke BJ (2000) Hurricane causes resource and pollination limitation of fruit set in a bird-pollinated shrub. Ecology 81:1951–1958

Ratnayake RMCS, Gunatilleke IAUN, Wijesundara DSA, Saunders RMK (2006) Reproductive biology of two sympatric species of *Polyalthia* (Annonaceae) in Sri Lanka. I. Pollination by curculionid beetles. Int J Plant Sci 167:483–493

Ratnayake RMCS, Gunatilleke IAUN, Wijesundara DSA, Saunders RMK (2007) Pollination ecology and breeding system of *Xylopia championii* (Annonaceae): curculionid beetle pollination, promoted by floral scents and elevated floral temperatures. Int J Plant Sci 168:1255–1268

Ratto F, Simmons BI, Spake R, Zamora-Gutierrez V, MacDonald MA, Merriman JC, Tremlett CJ, Poppy GM, Peh KS-H, Dicks LV (2018) Global importance of vertebrate pollinators for plant reproductive success: a meta-analysis. Front Ecol Environ. https://doi.org/10.1002/fee.1763

Rech AR, Dalsgaard B, Sandel B, Sonne J, Svenning JC, Holmes N, Ollerton J (2016) The macroecology of animal versus wind pollination: ecological factors are more important than historical climate stability. Plant Ecol Diver 9:253–262. https://doi.org/10.1080/17550874.2016.1207722

Regal PJ (1982) Pollination by wind and animals: ecology of geographic patterns. Annu Rev Ecol Syst 13:497–524

Reis MG, Singer RB, Gonçalves R, Marsaioli AJ (2006) The chemical composition of *Phymatidium delicatulum* and *P. tillandsioides* (Orchidaceae) floral oils. Nat Prod Commun 1:757–761

Ren D, Labandeira CC, Santiago-Blay JA, Rasnitsyn A, Shih C, Bashkuev A, Logan MAV, Hotton CL, Dilcher D (2009) A probable pollination mode before Angiosperms: Eurasian, long-proboscid Scorpionflies. Science 326(5954):840–847. https://doi.org/10.1126/science.1178338

Rigotto RM, Vasconcelos DP, Rocha MM (2014) Pesticide use in Brazil and problems for public health. Cad Saude Publ 30:1360–1362

Rios LD, Fuchs EJ, Hodel DR, Cascante-Marín A (2003) Neither insects nor wind: ambophily in dioecious *Chamaedorea* palms (Arecaceae). Plant Biol. https://doi.org/10.1111/plb.12119

Rosas-Guerrero V, Aguilar R, Martén-Rodríguez S, Ashworth L, Lopezaraiza-Mikel M, Bastida JM, Quesada M (2014) A quantitative review of pollination syndromes: do floral traits predict effective pollinators? Ecol Lett 17:388–400. https://doi.org/10.1111/ele.12224

Roubik DW (1989) Ecology and natural history of tropical bees. Cambridge University Press, New York. Tropical Biology Series

Roulston TH, Cane JH, Buchmann SL (2000) What governs protein content of pollen: pollinator preferences, pollen-pistil interaction, or phylogeny? Ecol Monogr 70:617–643

Rydin C, Bolinder K (2015) Moonlight pollination in the gymnosperm Ephedra (Gnetales). Biol. Lett. 11: 20140993. http://dx.doi.org/10.1098/rsbl.2014.0993

Sakamoto RL, Ito M, Kawakubo N (2012) Contribution of pollinators to seed production as revealed by differential pollinator exclusion in *Clerodendrum trichotomum* (Lamiaceae). PLoS One 7(3):e33803. https://doi.org/10.1371/journal.pone.0033803

Santos GMM, Aguiar CML, Mello MAR (2010) Flower-visiting guild associated with the Caatinga flora: trophic interaction networks formed by social bees and social wasps with plants. Apidologie 41:466–475. https://doi.org/10.1051/apido/2009081

Santos JC, Nascimento ART, Marzinek J, Leiner N, Oliveira PE (2017) Distribution, host plants and floral biology of the root holoparasite *Langsdorffia hypogaea* in the Brazilian savannah. Flora 226:65–71. https://doi.org/10.1016/j.flora.2016.11.008

Saunquet H, von Balthazar M, Magallón S et al (2017) The ancestral flower of angiosperms and its early diversification. Nat Commun 8:16047. https://doi.org/10.1038/ncomms16047

Saunders RMK (2012) The diversity and evolution of pollination systems in Annonaceae. Bot J Linn Soc 169:222–244

Sawea T, Nielsenb A, Totlandc Ø, Macriced S, Eldegard K (2020) Inadequate pollination services limit watermelon yields in northern Tanzania. Basic Appl Ecol 44:35–45

Sazima M, Sazima I (1978) Bat pollination of the passion flower, *Passiflora mucronata*, in Southeastern Brazil. Biotropica 10:100–109

Sazima I, Sazima C, Sazima M (2005) Little dragons prefer flowers to maidens: a lizard that laps nectar and pollinates trees. Biota Neotropica 5:185–192. https://doi.org/10.1590/S1676-06032005000100018

Sazima I, Sazima C, Sazima M (2009) A catch-all leguminous tree: Erythrina velutina visited and pollinated by vertebrates at an oceanic island. Aust J Bot 57:26–30

Schiestl FP, Ayasse M (2001) Post-pollination emission of a repellent compound in a sexually deceptive orchid: a new mechanism for maximising reproductive success? Oecologia 126:531–534

Schiestl FP, Ayasse M, Paulus HF, Löfstedt C, Hansson BS, Ibarra F, Francke W (1999) Orchid pollination by sexual swindle. Nature 399:421–422

Schiestl FP, Peakall R, Mant JG, Ibarra F, Schulz C, Franke S, Francke W (2003) The chemistry of sexual deception in an orchid-wasp pollination system. Science 302:437–438. https://doi.org/10.1126/science.1087835

Schnaitmann C, Pagni M, Reiff DF (2020) Color vision in insects: insights from *Drosophila*. J Comp Physiol A. https://doi.org/10.1007/s00359-019-01397-3

Schowalter TD (2000) Insect ecology: an ecosystem approach. Academic, San Diego, 483 pp

Seymour RS (2010) Scaling of heat production by thermogenic flowers: limits to floral size and maximum rate of respiration. Plant Cell Environ 33:474–1485

Seymour RS, Schultze-Motel P (1997) Heat-producing flowers. Endeavour 21:125–129

Shrestha M, Lunau K, Dorin A, Schulze B, Bischo M, Burd M, Dyer AG (2016) Floral colours in a world without birds and bees: the plants of Macquarie Island. Plant Biol. https://doi.org/10.1111/plb.12456

Shuttleworth A, Johnson SD (2009) The importance of scent and nectar filters in a specialized wasp-pollination system. Funct Ecol 23:931–940. https://doi.org/10.1111/j.1365-2435.2009.01573.x

Shuttleworth A, Johnson SD (2010) The missing stink: sulphur compounds can mediate a shift between fly and wasp pollination systems. Proc R Soc B 277:2811–2819. https://doi.org/10.1098/rspb.2010.0491

Silberbauer-Gottsberger I, Gottsberger G (1988) A polinização de plantas do cerrado. Braz J Biol 48:651–663

Silberbauer-Gottsberger I (1990) Pollination and evolution in palms. Phyton (Horn, Austria) 30:213–233

Silva AC, Kinupp VF, Absy ML, Ferr WE (2004) Pollen morphology and study of the visitors (Hymenoptera, Apidae) of Solanum stramoniifolium Jacq. (Solanaceae) in Central Amazon. Acta Botanica Brasilica 18:653–657

Simão S (1998) Tratado de fruticultura. FEALQ, Piracicaba, 760p

Simpson BB, Neff JL (1981) Floral rewards: alternatives to pollen and nectar. Ann Mo Bot Gard 68:301–322

Smyth DR (2018) Evolution and genetic control of the floral ground plan. New Phytologist (2018) 220:70–86. https://doi.org/10.1111/nph.15282

Sonne J et al (2020) Ecological mechanisms explaining interactions within plant–hummingbird networks: morphological matching increases towards lower latitudes. Proc R Soc B 287:20192873

Southwick EE (1985) Bee hair structure and the effect of hair on metabolism at low temperature. J Apic Res 24:44–149. https://doi.org/10.1080/00218839.1985.11100664

Sousa-Lopes B, Calixto ES, Torezan-Silingardi HM, Del-Claro K (2020) Effects of ants on pollinator performance in a distylous pericarpial nectary-bearing Rubiaceae in Brazilian Cerrado. Sociobiology 67:173–185. https://doi.org/10.13102/sociobiology.v67i2.4846

Sprengel CK (1793) Das entdeckte Geheimniss der Natur im Bau und in der Befruchtung der Blumen. Vieweg, Berlin

Stavenga DG, Staalb M, van der Kooi C (2020) Conical epidermal cells cause velvety colouration and enhanced patterning in *Mandevilla* flowers. Faraday Discuss. https://doi.org/10.1039/d0fd00055h

Stebbins GL (1970) Adaptive radiation of reproductive characteristics in Angiosperms, I: Pollination mechanisms. Annu Rev Ecol Syst 1:307–326

Steiner KE, Whitehead VB, Johnson SD (1994) Floral and pollinator divergence in two sexually deceptive South African orchids. Am J Bot 81:185–194

Stiles FG (1981) Geographical aspects of bird-flower coevolution, with particular reference to Central America. Ann Mo Bot Gard 68:323–351

Stork NE (2018) How many species of insects and other terrestrial arthropods are there on earth? Annu Rev Entomol 63:31–45

Streher NS, Bergamo PJ, Ashman TL, Wolowski M, Sazima M (2020) Effect of heterospecific pollen deposition on pollen tube growth depends on the phylogenetic relatedness between donor and recipient. AoB Plants 12(4)

Streinzer M, Paulus HF, Spaethe J (2009) Floral colour signal increases short-range detectability of a sexually deceptive orchid to its bee pollinator. J Exp Biol 212:1365–1370. https://doi.org/10.1242/jeb.027482

Suhaida M, Haron NW, Chua LSL, Chung RCK (2018) Floral phenology and pollination biology of *Vatica yeechongii* (Dipterocarpaceae). J Trop For Sci 30:497–508

Tamura S, Kudo G (2000) Wind pollination and insect pollination of two temperate willow species, *Salix miyabeana* and *Salix sachalinensis*. Plant Ecol 147:185–192

Teixeira TPO, Ferreira INM, Borges JPR, Torezan-Silingardi HM, Silva-Neto CM, Franceschinelli EV (2019) Reproductive strategy and the effect of floral pillagers on fruit production of the passion flower *Passifora setacea* cultivated in Brazil. Rev Bras Bot 42:63–71. https://doi.org/10.1007/s40415-018-00512-z

Terry I, Tang W, Blake AST, Donaldson JS, Singh R, Vovides AP, Jarámillo AC (2012) An overview of cycad pollination studies. Mem N Y Bot Gard 106:352–394

Teppner H (2005) Pollinators of tomato, *Solanum lycopersicum* (Solanaceae), in Central Europe. Phyton (Horn, Austria) 45:217–235

Theissen G, Melzer R (2007) Molecular mechanisms underlying origin and diversification of the angiosperm flower. Ann Bot 100:603–619

Thien LB (1980) Patterns of pollination in the primitive angiosperms. Biotropica 12:1–13

Thomas CD, Cameron A, Green RE, Bakkenes M, Beaumont LJ et al (2004) Extinction risk from climate change. Nature 427:145–148

Thompson JN (2014) Interaction and coevolution. University of Chicago Press, Chicago

Thomson JD, Fung HF, Ogilvie JE (2019) Effects of spatial patterning of co-flowering plant species on pollination quantity and purity. Ann Bot 123:303–310. https://doi.org/10.1093/aob/mcy120

Thorogood C, Santos JC (2020) *Langsdorffia*: creatures from the deep? Plants, people. Plan Theory 2:181–185. https://doi.org/10.1002/ppp3.10102

Torezan-Silingardi HM, Del-Claro K (1998) Behavior of visitors and reproductive biology of *Campomanesia pubescens* (Myrtaceae) in cerrado vegetation. Ciência e Cultura 50:281–284

Torezan-Silingardi HM (2011) Predatory behavior of *Pachodynerus brevithorax* (Hymenoptera: Vespidae, Eumeninae) on endophytic herbivore beetles in the Brazilian Tropical Savanna. Sociobiology 57:181–189

Torezan-Silingardi HM, Oliveira PEAM (2004) Phenology and reproductive biology of *Myrcia rostrate* and *M. tomentosa* (Myrtaceae) in Central Brazil. Phyton (Horn, Austria) 44:23–43

Torezan-Silingardi HM (2008) The danger of introducing bee species, a case study on Brazilian tropical savanna. EOLSS – eolss.nt

Traveset A, Richardson DM (2006) Biological invasions as disruptors of plant reproductive mutualisms. Trends Ecol Evol 21:208–216

Tremlett CJ, Moore M, Chapman MA, Zamora-Gutierrez V, Peh KSH (2020) Pollination by bats enhances both quality and yield of a major cash crop in Mexico. J Appl Ecol 57:450–459. https://doi.org/10.1111/1365-2664.13545

Triplehorn N, Johnson CA (2020) Borror and Delong's introduction to the study of insects. Cengage Learning Group

Trøjelsgaard K, Olesen JM (2013) Macroecology of pollination networks. Glob Ecol Biogeogr 22:149–162

Tscheulin T, Petanidou T (2013) The presence of the invasive plant *Solanum elaeagnifolium* deters honeybees and increases pollen limitation in the native co-flowering species *Glaucium flavum*. Biol Invasions 15:385–393. https://doi.org/10.1007/s10530-012-0293-y

(USDA) New Pest Response Guidelines *Vespa mandarinia* Asian giant hornet (2020). https://cms.agr. wa.gov/WSDAKentico/Documents/PP/PestProgram/Vespa_mandarinia_ NPRG_10Feb2020-(002).pdf

van der Kooi C, Ollerton J (2020) The origins of flowering plants and insect pollination. Science 368(6497):1306–1308. https://doi.org/10.1126/science.aay3662

Vanderplanck M, Moerman R, Rasmont P, Lognay G, Wathelet B, Wattiez R, Michez D (2014) How does pollen chemistry impact development and feeding behaviour of polylectic bees? PLoS One 9:e86209

Vaudo AD, Tooker JF, Grozinger CM, Patch HM (2015) Bee nutrition and floral resource restoration. Curr Opin Insect Sci 10:133–141

Vaudo AD, Tooker JF, Patch HM, Biddinger DJ, Coccia M, Crone MK, Fiely M, Francis JS, Hines HM et al (2020) Pollen protein: lipid macronutrient ratios may guide broad patterns of bee species floral preferences. Insects 11:132. https://doi.org/10.3390/insects11020132

Vilela AA, Del Claro VTS, Torezan-Silingardi HM, Del-Claro K (2018) Climate changes affecting biotic interactions, phenology, and reproductive success in a savanna community over a 10-year period. Arthropod Plant Interact 12:215–227

Vitali MJ, Machado VLL (1994) Visitantes florais de *Murraya exótica* L. (Rutaceae). Semina 15(2):153–169

Vogel S (1962) Duftdrüsen im Dienste der Bestäubung.: Über Bau und Funktion der Osmophoren. Abhandlung der Mathematisch-Naturwissenschaftlichen Klasse, Akademie der Wissenschaften. Mainz 10:599–763

Vogel S (1963) Duftdrüsen im Dienste der Bestäubung: Über Bau und Funktion der Osmophoren, vol. 10. Akademie der Wissenschaften und der Literatur Mainz Abhandlungen der Mathematisch-Naturwissenschaftlichen Klasse, pp. 601–763

Vogel S (1966) Parfümsammelnde Bienen als Bestäuber von Orchidaceen und Gloxinia. Plant Syst Evol 113:302–361

Vogel, S (1974) Olblumen und olsammelnde Bienen. 267 S., 76 Abb., 8 Tab. Franz Steiner Verlag GmbH. Wiesbaden

Vogel S (1978) Pilzmückenblumen als Pilzmimeten. Flora 167:329–366

Vogel S, Michener C (1985) Long bee legs and oil-producing floral spurs, and a new *Rediviva* (Hymenoptera, Melittidae; Scrophulariaceae). J Kansas Entomol Soc 58(2):359–364

Vogel S (1990) History of the Malpighiaceae in the light of pollination ecology. Mem N Y Bot Gard 55:130–142

Vogel S (1998a) Remarkable nectaries: structure, ecology, organophyletic perspectives III. Nectar ducts Flora 193:113–131

Vogel S (1998b) Remarkable nectaries: structure, ecology, organophyletic perspectives IV. Miscellaneous cases. Flora 193:225–248

Vogel S, Lopes AV, Machado IC (2005) Bat pollination in the NE Brazilian endemic *Mimosa lewisii*: an unusual case and first report for the genus. Taxon 54:693–700

Vogel S (2009) The non-African oil-flowers and their bees: a brief survey. S Afr J Bot 75:389–390. https://doi.org/10.1016/j.sajb.2009.02.018

Wang TV, Clifford MR, Martínez-Gómez J, Johnson JC, Riffell JA, Stilio VSD (2019) Scent matters: differential contribution of scent to insect response in flowers with insect vs. wind pollination traits. Ann Bot 123:289–301. https://doi.org/10.1093/aob/mcy131

Warrant EJ, Kelber A, Gislen GB, Ribi W, Wcislo WT (2004) Nocturnal vision and landmark orientation in a Tropical Halictid bee. Curr Biol 14:1309–1318. https://doi.org/10.1016/j. cub.2004.07.057

Warrant EJ (2008) Seeing in the dark: vision and visual behaviour in nocturnal bees and wasps. J Exp Biol 211:1737–1746

Waser N, Chittka L, Price M, Williams N, Ollerton J (1996) Generalization in pollination systems, and why it matters. Ecology 77:1043–1060

Wcislo WT, Arneson L, Roesch K, Gonzalez V, Smith A, Fernández H (2004) The evolution of nocturnal behaviour in sweat bees, *Megalopta genalis* and *M. ecuadoria* (Hymenoptera: Halictidae): an escape from competitors and enemies? Biol J Linn Soc 83:377–387

Webber AC, Gottsberger G (1995) Floral biology and pollination of *Bocageopsis multiflora* and *Oxandra euneura* in Central Amazonia, with remarks on the evolution of stamens in Annonaceae. Feddes Repert 106:515–524

Wee SL, Tan SB, Jürgens A (2018) Pollinator specialization in the enigmatic *Rafflesia cantleyi*: a true carrion flower with species-specific and sex-biased blow fly pollinators. Phytochemistry 153:120–128

Wellman CH, Osterloff PL, Mohiuddin U (2003) Fragments of the earliest land plants. Lett Nat 425:282–285. https://doi.org/10.1038/nature01884

Westerkamp C, Gottsberger G (2000) Diversity pays in crop pollination. Crop Sci 40:1209–1222

Whelan CJ, Wenny DG, Marquis RJ (2008) Ecosystem services provided by birds. Ann N Y Acad Sci 1134:25–60. https://doi.org/10.1196/annals.1439.003

Wiemer AP, Sérsic AN, Marino S, Simões AO, Cocucci AA (2012) Functional morphology and wasp pollination of two South American asclepiads (Asclepiadoideae–Apocynaceae). Ann Bot 109:77–93. https://doi.org/10.1093/aob/mcr268

Wilcock C, Neiland R (2002) Pollination failure in plants: why it happens and when it matters. Trends Plant Sci 7(6):270–277

Williams GA, Adam P, Mound LA (2001) *Thrips* (Thysanoptera) pollination in Australian subtropical rainforests, with particular reference to pollination of *Wilkiea huegeliana* (Monimiaceae). J Nat Hist 35:1–21

Williams H, Nejati M, Hussein S, Penhall N, Lim JY et al (2019) Autonomous pollination of individual kiwifruit flowers: toward a robotic kiwifruit pollinator. J Field Robot. https://doi.org/10.1002/rob.21861

Williams JH, Friedman WE (2002) Identification of diploid endosperm in an early angiosperm lineage. Nature 415:522–526

Williams NM, Crone EE, Roulston TH, Minckley RL, Packer L, Potts SG (2010) Ecological and life history traits predict bee species responses to environmental disturbances. Biol Conserv 143:2280–2291

Williams NM, Ward KL, Pope N, Isaacs R, Wilson J, May EA, Ellis J et al (2015) Native wildflower plantings support wild bee abundance and diversity in agricultural landscapes across the United States. Ecol Appl 25:2119–2131

Wilson EO (1992) The diversity of life. Harvard University Press, 424pp, Massachusets

Wiśniewska N, Lipińska MM, Gołębiowski M, Kowalkowska AK (2019) Labellum structure of *Bulbophyllum echinolabium* J.J. Sm. (section Lepidorhiza Schltr., Bulbophyllinae Schltr., Orchidaceae Juss.). Protoplasma 256:1185–1203

Wolda H, Roubik DW (1986) Nocturnal bee abundance and seasonal bee activity in a Panamanian forest. Ecology 67:426–433

Wolowski M, Agostini K, Rech AR, Varassin IG, Maués M, Freitas L, Carneiro LT, Bueno RO, Consolaro H, Carvalheiro L, Saraiva AM, Silva CI, Padgurschi MCG (Org.). 1ª edição. BPBES/REBIPP (2019) Relatório temático sobre Polinização, Polinizadores e Produção de Alimentos no Brasil. Editora Cubo. 184 páginas, São Carlos, SP. https://doi.org/10.4322/978-85-60064-83-0

Woodcock BA, Garratt MPD, Powney GD, Shaw RF, Osborne JL et al (2019) Meta-analysis reveals that pollinator functional diversity and abundance enhance crop pollination and yield. Nat Commun 10:1481. https://doi.org/10.1038/s41467-019-09393-6

Yamamoto M, Silva CI, Augusto SC, Barbosa AAA, Oliveira PE (2012) The role of bee diversity in pollination and fruit set of yellow passion fruit (*Passiflora edulis* forma *flavicarpa*, Passifloraceae) crop in Central Brazil. Apidologie 43:515–526. https://doi.org/10.1007/s13592-012-0120-6

Zespri (2016) Zespri annual report. http://www.zespri.com/ZespriInvestorPublications/Annual-Report-2016-17.pdf

Zoeller KC, Steenhuisen SL, Johnson SD, Midgley JJ (2016) New evidence for mammal pollination of *Protea* species (Proteaceae) based on remote-camera analysis. Aust J Bot 64:1–7. https://doi.org/10.1071/BT15111

Frugivory and Seed Dispersal

Richard T. Corlett

Contents

© Springer Nature Switzerland AG 2021
K. Del-Claro, H. M. Torezan-Silingardi (eds.), *Plant-Animal Interactions*, https://doi.org/10.1007/978-3-030-66877-8_7

> **Learning Objectives**
> After completing the chapter, you should have an understanding of the following:
> - Why frugivory and seed dispersal are so important.
> - How fleshy-fruited plants and frugivorous animals have adapted to each other.
> - The value of seed dispersal networks for describing and understanding communities.
> - The ongoing debate about the relative importance of coevolution and ecological fitting in the origin of seed dispersal networks.
> - The vulnerability of seed dispersal to human impacts.

7.1 Introduction

A frugivore is any animal that eats fruit, although in ecology this term is usually only applied to animals that eat a lot of fruit. Fruit in this definition means seeds surrounded by edible flesh, whatever they are called in strict botanical terminology, and usually excludes dry (non-fleshy) fruits, such as those of grasses or oaks. Seed dispersal is the movement of seeds away from the parent plant. This chapter deals with seed dispersal by frugivores, a mutualistic relationship in which the animal benefits from the nutritional value of the fruit flesh and the plant benefits by having its seeds dispersed.

Fleshy fruits are one of the few plant foods eaten mostly by vertebrates. Most birds and mammals, and many reptiles, eat at least some fruits, and frugivores account for a large proportion of bird and mammal biomass in tropical forests. Seasonality in fruit production has impacts on the behavior, breeding, and migration of frugivores, while irregular, supra-annual fruit famines can have dramatic effects on frugivore communities, including reduced breeding and increased mortality (Wright et al. 1999; Wong et al. 2005; Vogel et al. 2012). On evolutionary timescales, a high dependence on fruits affects almost all aspects of animal biology from the sensory capabilities to morphology, physiology, and behavior (Corlett 2011a).

Seed dispersal is an essential step in the life cycle of the majority of seed plants, although many herbs have small, poorly dispersed, dormant seeds that appear to be 'dispersed in time' more than in space (Chen et al. 2020a). Wind and, to a lesser extent, water disperse the seeds of many plant species, and some seeds are dispersed externally on animals, but dispersal by frugivorous animals dominates in woody plants and in forests. Dispersal away from the parent plant reduces competition from conspecific seedlings and the density- and distance-dependent impact of pathogens and pests. In a forest in Amazonia, most saplings were established from seeds dispersed at least several crown widths away from parent trees (Swamy et al. 2011). Disproportionate dispersal to favorable microsites for germination and growth would increase plant fitness but, while there are now multiple examples of this 'directed dispersal' known (Wenny 2001; Bravo and Cueto 2020), it is still unclear if it is of widespread significance.

Seed dispersal is also a major contributor to gene flow—probably as important as pollen movement in tropical forests, where both pollen and seeds usually have animal vectors (Corlett 2019)—and the only mechanism by which species can colonize new areas. It is thus the only way that plants can track climate change across the landscape (Corlett and Westcott 2013). Loss of dispersal agents can, in the short term, lead to increased clumping and reduced regeneration (Harrison et al. 2013; Pérez-Méndez et al. 2016), while in the longer term it can result in population declines and potentially extinction. Note that dispersal is not the only benefit of frugivory for a plant. Seeds may also obtain other benefits from passage through a vertebrate gut, including removal of fruit pulp and scarification of the seed coat, both of which may enhance germination, and deposition in nutrient-rich fecal matter (Fricke et al. 2013; Stringer et al. 2020).

7.2   Origins and Evolution of Frugivory and Seed Dispersal

In the modern world, frugivory and seed dispersal are associated with warm wet climates, and with trees and forests (Eriksson 2016; Chen et al. 2017). They also occur elsewhere, but both fleshy fruits and frugivores are most diverse and abundant in these environments. Forests favor large seeds with the food reserves needed to establish in deep shade, and large seeds favor dispersal by animals. The convergent evolution of fleshy rewards in numerous plant lineages led to the evolution of frugivory in many lineages of animals and, where fruits were found year-round, the evolution of specialist frugivores (see also Del-Claro and Torezan-Silingardi, ▶ Chap. 1).

The first trees and forests arose in the Late Devonian, approximately 390 million years ago, and the first seeds appear soon afterwards. Seeds are found in fossil guts and feces of animals from the Late Permian, around 270 million years ago (Tiffney 2004). This is evidence for endozoochory (dispersal of seeds in animal guts), but probably accidental from the animals' viewpoint, with the seeds swallowed while consuming other plants parts. Seeds with external flesh appeared early on, however, and by the Mesozoic were widespread, in cycads, *Gingko*, and other gymnosperms. The first evidence suggestive of at least seasonal specialization on flesh-covered seeds comes from a dinosaur fossil, from around 170 million years ago (Salgado et al. 2017). Early angiosperms mostly had small seeds, but a Middle to Late Jurassic (>164 million years ago) fleshy fruit fossil from Inner Mongolia—probably from an angiosperm—was an 11 × 8 mm drupe (Chen et al. 2020b, ▣ Fig. 7.1). Such a fruit would still attract dispersal agents today. The diversity of seed and fruit sizes, and of fruit types, peaked in the Eocene fossil record, 50–55 m years ago (Eriksson 2016), but many modern groups of frugivores, including the fruit bats and passerine birds, diversified more recently.

7.3   Some Basic Concepts

Seeds are dispersed away from the parent plant to varying distances. Information on dispersal distances (and directions, if not symmetrical) can be summarized as a two-dimensional density plot describing the probability of a seed being deposited

⬛ **Fig. 7.1** A fossil fruit, *Jurafructus daohugouensis*, from the Jurassic of Inner Mongolia, China, >164 million years ago. The 11 × 8 mm fruit consists of a single seed surrounded by flesh like a modern angiosperm drupe. (Photograph courtesy of Xin Wang and also used in Chen et al. (2020b))

at each point. This probability density function is called a *dispersal kernel*. For a given plant species, the dispersal kernel will be different for each dispersal agent. The total dispersal kernel is the sum of all the kernels for all dispersal agents for an individual plant, population, species, or community (Rogers et al. 2019).

However, the movement of seeds is not, by itself, of ecological interest, except for the weight it adds to the frugivore's body and the space it occupies in its guts. Seed dispersal is interesting only if it results in the establishment of new individuals of the plant species, i.e. if dispersal is effective. *Seed dispersal effectiveness* is the product of the number of seeds dispersed (the quantity component of seed dispersal) and the probability that each dispersed seed will produce a new reproductive adult (the quality component) (Schupp et al. 2010). The two concepts—dispersal kernels and seed dispersal effectiveness—can be combined as the *effective dispersal kernel*, which multiplies the probability of a seed being deposited at each point with the probability of seedling establishment from a deposited seed, and the total effective dispersal kernel, which sums this for all dispersal agents (Rogers et al. 2019). In practice, describing even an approximation to the total effective dispersal kernel for a single plant species is a major challenge.

7.4 How to Be a Frugivore

Fruit is usually advertised by color or scent, minimally defended, rapidly digested, and does not fight or run away. It is an easy food. Most bird and mammal species, many reptiles, and at least one frog (da Silva and de Britto-Pereira 2006) eat at least some fruit. However, while a small amount of fruit in an animal's diet may contribute valuable vitamins and minerals, eating a lot of fruit and, in particular, eating mostly fruit requires special adaptations for efficient location, harvest, and digestive processing (Moermond and Denslow 1985; Wang et al. 2020a). A major problem is that fruit flesh is nutritionally dilute, with particularly low concentrations of nitrogen (Levey et al. 2000; Donati et al. 2017). Most, but not all, frugivorous vertebrates supplement a fruit diet with animal protein (Corlett 2017; Orr et al. 2016). Two groups of Old-World pigeons appear to eat only fruit, but pigeons produce protein- and fat-rich 'crop milk' as a supplementary food for their young (Kozlowski et al. 2016). The Neotropical oilbirds also feed their young on an exclusively frugivorous diet, but the nestlings grow very slowly as result (Thomas et al. 1993). The advantages and disadvantages of relying on fruit are illustrated by the global distribution of passerine birds with female-only parental care (Barve and La Sorte 2016). Most such species are frugivorous, reflecting the relative ease of finding enough fruit for a family, compared with alternative diets, but they are largely confined to areas with exceptionally long fruiting seasons, where multiple small broods are possible.

Unfortunately, most chemical analyses of fruit flesh so far have been very crude; typically, dividing components into sugars, lipid, fiber, and 'crude protein', with the latter simply a multiple of the total nitrogen content. Most flying frugivores—birds and bats—have short guts and rapid gut passage, favoring the absorption of small molecules such as simple sugars and free amino acids (Moreno et al. 2019). A major clade of birds, including the Muscicapidae, Turdidae, Sturnidae, and Mimidae, is unable to digest sucrose and avoids foods that contain it (Zungu and Downs 2016). Sucrose is the major transport sugar in plants (Lohaus and Schwerdtfeger 2014), so the dominance of the hexose sugars, glucose and fructose, in fruits eaten by birds (Ko et al. 1998) suggests strong selection against sucrose. The sucrose-intolerant clade is not important enough in most areas to account for this, suggesting a more general preference for hexoses. Lipids are preferred by other bird species (Stiles 1993) or at certain times of the year (Bairlein 2002).

All fruits are nutritionally unbalanced as a food for vertebrates, even when supplemented with non-fruit foods, and all frugivores eat many fruit species. The extent to which this is deliberate 'nutrient balancing' is unclear, but it could potentially result in an advantage for rare plant species, with frugivores eating proportionally more of rare fruits because they are needed for a balanced diet (Morán-López et al. 2018). Nutrient balancing requires that the frugivore can detect the nutritional value of available fruits. Fine discrimination is likely to involve linking pre-ingestive (before swallowing) sensory cues—from vision, smell, touch, and taste—and spatial memory to the post-ingestive (after swallowing) consequences of eating a particular fruit at a particular stage of ripeness (Corlett 2011a; John et al. 2016). Rapid gut passage in many frugivores should allow physiological feedbacks to provide a continuous update on the costs and benefits of fruit choice. Although we know that many animals are able to do this, there have been few studies of how learning and memory influence frugivory and seed dispersal, resulting in a major gap in our ability to predict the impacts of global change on these processes (John et al. 2016). This contrasts with studies of seed dispersal by scatter-hoarding birds and rodents, where excellent spatial memory is assumed, and with pollination studies, where even bees are known to rely heavily on learning and memory.

Most research has focused on sensory cues, which are easy to study in the lab and are likely to be most important for frugivores faced with a novel fruit resource, as will occur frequently in naïve juveniles and migrants, but less often in territorial adults. Cues will be most useful in these situations if they provide honest information on fruit nutritional value, rather than simply increasing conspicuousness (Albrecht et al. 2018). The currently available evidence suggests that this is true in some cases but not in others, and that, even when there is an association between a cue and a reward, it tends to be weak (Stournaras et al. 2015; Albrecht et al. 2018). Even weak associations may be useful, however, since repeated interactions will allow a frugivore to verify their reliability and adjust their future responses accordingly. Birds (Duan et al. 2014) and macaques (Skalníková et al. 2020) have apparently innate color preferences for food items—red for both but also black for birds. However, the diversity of fruits eaten by most frugivore species in the wild suggests that any innate preferences have little impact on what is actually eaten, which presumably reflects subsequent learning and memory. Conspicuous fruit crops (e.g. those of many *Ilex* species) may persist as long as there are alternatives, because frugivores have learned to avoid them, while inconspicuous crops may be stripped as soon as they ripen. Individual female orangutans in Sumatra living in the same area, presumably with the same senses, have strikingly different fruit diets, which may reflect social learning by infants from their mothers (Hardus et al. 2013). Signals of social learning are also seen in the foraging of young golden lion tamarins (Troisi et al. 2020).

7.5 How to Be a Fruit

Frugivores are mobile and can actively choose among fruit species, or between fruits and other foods, but an adult plant is fixed in the ground. If the fruit supply is limited, however, and frugivores differ in their effectiveness as seed dispersal agents, then a plant can increase recruitment by choosing among frugivores. Plants have two advantages over frugivores in this process. First, the fruits of many plants—but by no means all—can persist on the plant in a viable, edible form for days, weeks, or in some cases months, if not eaten (Tang et al. 2005; Cazetta et al. 2008). Second, most plants are longer-lived than frugivores and can survive for years without dispersing any seeds at all. Small endothermic frugivores need to eat every day but plants can wait years to be dispersed.

Fruit flesh must be protected from invertebrates and microbes, particularly in species with fruits which persist after they become ripe (Tang et al. 2005; Cazetta et al. 2008). Moreover, deterring unsuitable vertebrate frugivores, including seed predators, is likely to be an important function of fruit flesh. Fruit size alone can exclude narrow-gaped birds, which will usually disperse seeds less far, although the stronger-billed of these may steal flesh without swallowing the seeds. Mechanical protection—an indehiscent husk, rind, or other covering—can exclude most birds and weak-jawed mammals. Classic 'primate fruits', with a protective outer layer that primates remove with teeth and hands, are hard for other frugivores to handle. Indeed, the spatulate (shovel-shaped) front teeth of monkeys and apes have been viewed as an adaptation for efficiently peeling such fruits (Valenta et al. 2020). Very large size and enhanced protection characterize so-called megafaunal fruits, >4 cm in diameter, which today are largely consumed by elephants in Africa and Asia, but are assumed to have been eaten by the diverse global megafauna that survived on all continents and near-shore islands until the end of the Pleistocene (Galetti et al. 2018; Lim et al. 2020) and through most of the Holocene on Madagascar (Albert-Daviaud et al. 2020).

There is also evidence for selective chemical exclusion—directed deterrence—in a few cases, including the well-known example of capsaicin in chili peppers, which deters mammals but not birds (Naves et al. 2019). It is likely that many less obvious examples have been overlooked. A recent study found evidence for this in the relationship between understory *Piper* species in tropical forests and their principal dispersal agents, bats in the genus *Carollia* (Baldwin et al. 2020). Amides in the ripe fruits reduced gut retention times for seeds and delayed fruit removal, reducing dispersal distances.

Fruits may cheat, providing the expected sensory cues but little or no nutritional reward, but the learning abilities of vertebrate frugivores must limit the success of this strategy. Despite this, mimetic seeds, which look like fleshy fruits but have no flesh, are quite widespread in the angiosperms, particularly in the legumes (Fabaceae) (Brancalion et al. 2010). They usually have very low removal rates and persist in an attractive and viable state for months on the plant or on the ground beneath in compensation. Although it has been suggested that they are consumed mostly by naïve young birds, a study of *Rhynchosia melanocarpa* in the Brazilian cerrado found that peak removal was early in the fruiting season when young birds were absent (Pizo et al. 2020). Migratory birds provide an alternative source of naivety, but it is also likely that resident frugivores simply forget between seasons.

7.6   Internal Dispersal of Non-fleshy Fruits by Animals

Most frugivory and seed dispersal studies have been on birds, bats, and/or primates, resulting in a focus on fruit flesh as the main attractant for seed dispersal agents. A corollary of this has been an unstated assumption that the dispersal of non-fleshy, inconspicuous, 'dry' fruits and seeds by animals in their guts is somehow less legitimate, less evolved, less effective, less common, and overall, less important. However, viable seeds of dry fruits are common in the dung of fish, pheasants, ducks, elephants, rodents, ungulates, and many other animals (Corlett 2017), and dry fruits with no other obvious dispersal mechanism dominate in herb floras worldwide (e.g. Janzen 1984; Corlett 2011b). Janzen (1984) suggested that the vegetative parts of some dry-fruited plants may be the functional equivalent of fruit flesh, selected for attraction as well as photosynthesis, i.e. 'the foliage is the fruit'. Large herbivores tend to have large ranges, and some migrate, so potential dispersal distances are large. Globally, the biomass of domestic cattle now far exceeds that of wild animals (Bar-On et al. 2018) and these may now dominate the internal dispersal of dry seeds in many areas (personal observations).

For granivorous animals, the attractant is the seed itself, so any dispersal of viable seeds is a consequence of incomplete digestion. Viable seeds have been found in the feces of many species of granivorous birds (Corlett 1998, 2017; Orlowski et al. 2016; Kleyheeg et al. 2019). Most seeds are destroyed, but those that survive may be dispersed long distances because of long gut passage times and dispersal may be directed to suitable habitats (e.g. wetland to wetland). Moreover, many granivores are migratory. For example, modeling studies suggest that migratory mallard ducks (*Anas platyrhynchos*) in Europe could disperse seeds between wetlands hundreds of kilometers apart (Kleyheeg et al. 2019). Migratory grassland birds could potentially do the same for grassland plants.

7.7 Diplochory: Sequential Dispersal by Two Different Dispersal Agents

Diplochory is common in seed dispersal and may account for an even higher proportion of effective dispersal events (Vander Wall and Longland 2004). In cases where the first step is dispersal in the guts of a frugivore, the second phase typically involves removal of intact seeds from the feces (or regurgitate) of the frugivore by ants or scatter-hoarding rodents, or removal of seeds within part of the feces by dung beetles. In general, the first phase accounts for most of the final dispersal distance from the parent tree, while the second phase may influence establishment success. Simply spreading the seeds initially deposited close together in feces over a larger area will reduce competition and possibly reduce mortality from predators and pathogens. The survival, germination, and establishment of seeds is also often enhanced by burying them: in ant nests (e.g. Passos and Oliveira 2002), rodent seed caches (e.g. Longland and Dimitri 2016), or dung balls (e.g. Culot et al. 2018). Ants and dung beetles are more likely to bury small seeds (<5 mm) while rodents are more likely to scatter-hoard larger ones.

However, the impact of seed removal from the dung of the first-phase dispersal agent is not always positive. Different ant species may move seeds very different distances, scatter-hoarding rodents are also seed predators, and different functional groups (dwellers, tunnellers, and rollers) of dung beetles differ in effectiveness. All three groups of second-phase dispersal agents may bury seeds too deep for germination. The overall impact on plant fitness will thus be species and context dependent. Diplochory may involve obvious dual adaptations, as when seeds embedded in pulp attractive to frugivores also have an elaiosome to attract ants (Aronne and Wilcock 1994), but in most cases it is not clear if natural selection on fruit and seed traits has also influenced the second phase.

7.8 The Frugivores

7.8.1 Invertebrates

Endozoochory by invertebrates has usually been regarded as a curiosity, but many plant species bear tiny-seeded, fleshy fruits near ground level which are likely to be candidates for dispersal by invertebrates. Distances will be short, unless the dispersal agent can fly, but if seeds are dispersed among suitable microhabitats even short-distance movements may enhance fitness. Tiny seeds with few or no resources for establishment are likely to be exceptionally vulnerable to centimeter-scale variation in the deposition environment, so short-distance movements may be critical. At least some seeds retain viability after gut passage through Orthoptera (wetas, crickets etc.) (Suetsugu 2018a, b, 2020), particularly large individuals (King et al. 2011; Larsen and Burns 2012; ◘ Fig. 7.2), some Coleoptera (beetles) (de Vega et al. 2011), some Blattodea (cockroaches) (Uehara and Sugiura 2017), some Gastropoda (slugs and snails) (Calvino-Cancela and Rubido-Bará 2012; Türke et al. 2012), and earthworms (Clause et al. 2017). Dispersal by invertebrates may often be a supplement to dispersal by vertebrate frugivores, but in other cases invertebrates appear to be the main or only dispersal agent (Uehara and Sugiura 2017; Suetsugu 2018a, b, 2020). In some cases, dispersal may be directed to favorable sites, e.g. the beetle *Pimelia costata* may deposit seeds of the Mediterranean root parasite, *Cytinus hypocistis*, underground near to suitable roots (de Vega et al. 2011). Evidence that some plant

◘ **Fig. 7.2** An alpine scree weta, *Deinacrida connectens*, from New Zealand. These insects consume the fruits of *Gaultheria depressa* and disperse the seeds, with the largest individuals dispersing thousands of seeds per night. (Photograph by NZSnowman from Wikimedia Commons used under the Creative Commons CC BY-SA3.0 license)

species have an evolved relationship with invertebrates includes the presentation of inconspicuous fruits on or near the ground, the small sizes of the seeds (<1 mm) so invertebrates can swallow them, and a hard, lignified seed coat (testa) which protects the seed during gut passage.

7.8.2 Fish

Frugivory and seed dispersal by fish is mostly a tropical phenomenon, although the role of temperate fish species may have been underestimated (Horn et al. 2011). It has been studied largely in the Neotropics, where fish play a major role in seed dispersal in seasonally flooded forests and savannas (Correa et al. 2015). Many plants in these systems produce fleshy fruits that are dropped into the water in the flood season. Similar ecosystems exist across the tropics, in Africa, Asia, New Guinea, and Australia, and frugivorous fish are found in them all (Horn et al. 2011; Corlett 2017). Frugivory may be genuinely less common outside the Neotropics, for historical and/or biogeographical reasons, but unequal study effort makes it impossible to be sure at present. Some fruit-eating fish show apparent morphological, behavioral, and physiological adaptations to a fruit diet (Correa et al. 2015), and there is evidence of varied fruit preferences and degrees of specialization (Araujo et al. 2020), but the degree of interdependence of fruits and fish is unclear. Many fish disperse small seeds, but only large fish disperse large seeds, which can be moved over long distances (>1 km) (Costa-Pereira et al. 2018).

7.8.3 Reptiles

The consumption of fruit has been widely reported in lizards (Valido and Olesen 2019) and in turtles and tortoises (Falcón et al. 2020), and is at least seasonally an important part of the diet in some species. Most seeds survive gut passage so seed dispersal can occur. However, there is little evidence that reptiles are quantitatively important as seed dispersal agents, except on some oceanic islands, where frugivorous lizards are common (Pérez-Méndez et al. 2016) and large tortoises—now extinct on most islands where they used to occur—have diverse fruit diets and very long gut-passage times (Corlett 2017; Falcón et al. 2020).

7.8.4 Birds

Globally, most seeds in fleshy fruits are dispersed by birds. Modern birds lack teeth and, although specialist granivores destroy seeds in their bills or guts, frugivory and seed dispersal are more widespread than seed predation across the bird phylogeny. Major frugivores include the cassowaries and tinamous in the basal Palaeognathae, and members of the orders Columbiformes (pigeons and doves), Musophagiformes (turacos), Bucerotiformes (hornbills), Piciformes (barbets and toucans), and Passeriformes (many groups of passerine songbirds) in the Neoaves.

■ **Fig. 7.3**   A great hornbill, *Buceros bicornis*, carrying food to feed its chicks in India. This is one of the largest flying frugivores. (Photograph by Angadachappa from Wikimedia Commons used under the Creative Commons CC BY-SA 4.0 license)

In most other bird orders, at least one member is recorded as eating some fruits. Specialist frugivores range in size from 5-g flowerpeckers (Dicaeidae) to 70-kg cassowaries (Casuariidae), with the numbers and maximum sizes (<6 mm to >50 mm) of the seeds dispersed, and maximum dispersal distances (10s to 1000s of meters), varying roughly in proportion. The largest flying frugivores are the hornbills (<3000 g; ■ Fig. 7.3), which can swallow whole fruits >5 cm in diameter.

Not all bird characteristics relevant to seed dispersal scale with size, however, so it matters to a plant which medium-sized bird eats its fruit. In the Neotropics, many frugivorous suboscines—manakins, cotingas, tyrant flycatchers—take fruits on the wing, swallowing them whole, while tanagers and related oscines have narrow gapes, take fruits from a perch, and then crush them in the bill, squeezing out all but the smallest seeds (Moermond and Denslow 1985; Corlett and Primack 2011). Neither of these fruit acquisition and processing behaviors—taking fruits on the wing and 'mashing' out seeds—is common in other bird groups and other parts of the world, with unclear consequences for seed dispersal. As a result of such idiosyncrasies, biogeography matters since most major frugivorous clades arose too late to spread easily between continents (Corlett and Primack 2011). Only the Neotropics has toucans, guans, tanagers, and manakins, for example, while only Africa has turacos, Asia and Africa share hornbills and bulbuls, and New Guinea has birds of paradise and shares cassowaries with Australia.

Day-flying birds have tetrachromatic vision, with the same range of cone types as people, plus an additional cone most sensitive in the near ultraviolet (or violet, in most non-passerines), and vision is usually their dominant sense for locating things. As a result, 'bird fruits' are often red, although the predominance of black fruits in some local floras is hard to explain. They usually appear odorless to our senses. Lack of teeth means that fruits are typically swallowed whole, with the indigestible seeds and digestible flesh separated in the guts before the seeds are defecated or regurgitated.

7.8.5 Mammals

Mammals are second only to birds as seed dispersal agents and, as with the birds, most mammalian orders have records of fruit consumption but only a few are widely important. Among the placental mammals, these are the Proboscidea (elephants and extinct relatives), Primates (most species), Chiroptera (fruit bats), Artiodactyla (pigs, deer, and cattle), Perissodactyla (rhinoceroses and tapirs), and Carnivora (several families). Marsupials (Marsupialia) have received less attention than placentals, but many species eat at least some fruits and disperse seeds. Most frugivorous mammals are red-green colorblind dichromats with an excellent sense of smell, many forage at night, and many of the fruits they consume appear visually inconspicuous but strong-smelling to human senses.

Mammals have teeth and strong jaws, so fruit processing is much more diverse than in birds, with very varied consequences for seed dispersal. Elephants swallow most fruits whole without any processing, including the largest fleshy fruits known (>10 cm diameter), and their long gut-passage times can result in very long dispersal distances for seeds that survive (Campos-Arceiz and Blake 2011; Bunney et al. 2017). Carnivores have limited chewing ability and relatively simple digestive systems, so seeds usually pass through undamaged. Many smaller carnivores consume a lot of fruit, at least seasonally, with the families Procyonidae (raccoons, coatis, kinkajous, and their relatives; Corlett and Primack 2011), in the New World, and Viverridae (civets; Corlett 2017), in the Old World, probably most important in seed dispersal. Fruit is also seasonally important in the diets of most bears (Ursidae), many members of the dog family (Canidae), and most Mustelidae, except the otters.

Primates have hands as well as teeth, making the removal of an inedible peel easier. Most primates swallow most seeds whole, but the Neotropical Pitheciinae (sakis, bearded sakis, and uakaris) are specialized seed predators and the Old World colobine monkeys (langurs, leaf monkeys, snub-nosed monkeys etc.) destroy all but a few small seeds in the fruits they eat. Cercopithecine monkeys (macaques, mangabeys, baboons, and guenons) spit out most larger seeds, often after dispersing fruits away from fruiting trees in their unique cheek pouches (Corlett 2017; McConkey 2018). Unlike most other frugivorous mammals, Old World apes and monkeys have human-like trichromatic color vision, which has an advantage over dichromacy when searching for red or reddish fruits (Skalníková et al. 2020).

Fruit bats lack hands but New World phyllostomids use the wrists and thumbs of their wings to manipulate fruits and many Old World pteropodids use one foot for the same purpose (Vandoros and Dumont 2004). Most fruit bats carry fruits away from the parent tree to a nearby feeding roost for processing. Here larger seeds are usually dropped along with other inedible fruit parts, and only the fruit pulp and smallest seeds swallowed. Swallowed seeds may then be defecated in flight, in contrast to birds which usually defecate from a perch.

The even-toed ungulates (Artiodactyla; pigs, deer, cattle, and their relatives) can feed only on fallen fruits or those borne on low-growing plants, but they disperse many seeds in a variety of ways, and their importance has probably been underestimated (Corlett 2017; Baltzinger et al. 2019). They disperse many seeds from dry fruits, presumably with foliage by accident, but also consume many fleshy fruits and the nutrient-rich, hard-seeded, sometimes scented, indehiscent pods of some legumes (Miller 1996; Corlett 2017). These pods are functionally fleshy fruits, even though not 'fruit-shaped'. Deer, chevrotains, and cattle may regurgitate large, hard seeds hours after swallowing (Delibes et al. 2019), and all ungulates also defecate some smaller seeds.

7.9 The Fruits

Fleshy fruits have evolved independently numerous times across the angiosperm phylogeny, with the flesh derived from a wide range of different tissues, including the seed, the fruit in the strict sense, and a variety of accessory structures. Fleshy fruits are found in basal angiosperms (*Kadsura, Schisandra*), magnoliids (Annonaceae, Lauraceae, Myristicaceae), monocots (Arecaceae, Pandanaceae), and in all major clades of the eudicots. Fruit flesh does not fossilize well so we know rather little about the evolutionary history of these fruits. The family Lauraceae, for example, was abundant in the Cretaceous, but the oldest known fruit fossils—indistinguishable from modern ones—are from the Miocene (Wang et al. 2019). Given that all branches of the family have single-seeded fleshy fruits today, it seems very likely that Cretaceous family members also did. In contrast, the family Rosaceae produces a great diversity of fruit types which appear to have originated independently from ancestors with dry fruits (Xiang et al. 2017). There is some evidence that small fleshy fruits have proliferated since the Eocene (Eriksson 2016; Yao et al. 2021), perhaps in response to the spread and diversification of the small and medium-sized passerines which are now the single most important group of frugivores (Corlett 2017; Oliveros et al. 2019).

7.10 Seed Dispersal Networks

Interactions between a particular frugivore species and a particular fleshy fruit species take place in communities that have many other species of frugivores and fruits. The number of possible pairwise interactions increases rapidly with the number of species involved, but only some of these interactions occur in nature and some of these are more important than others. The interactions between fruits and frugivores in a community can be summarized in a plant-frugivore network (see also Luna and Dáttilo, ► Chap. 10). Simply showing what eats what—a binary network— is a lot less informative than a quantitative network, where the thickness of the link between two species indicates the quantitative importanceof this interaction. In most plant-frugivore networks, quantification is by the frequency of visitation by a frugivore species to a fruiting plant species. The frequency of an interaction does not necessarily indicate its importance, to either partner, but it the easiest general proxy to measure (Vázquez et al. 2005). Measuring seed removal is better, since seeds removed per visit varies a lot among frugivore species, but it is a lot more effort.

If we are interested in seed dispersal, rather than just frugivory, it makes sense to use natural history information to remove from the network those interactions where the visitor is consuming pulp or seeds, and not dispersing seeds (Simmons et al. 2018). Unfortunately, in species-rich tropical systems, we often do not know this for many frugivore-plant pairs. Pigeons and doves (Columbidae), for example, include species which apparently grind up all seeds they consume and species which pass all seeds intact, as well as many in which seed fate seems to depend on seed size and hardness (Corlett 2017). Even where information on interaction natural history is available, this usually only accounts for the fruit removal stage of the dispersal process, and frugivores also differ in where they deposit seeds after gut passage, with potentially large consequences for dispersal effectiveness. Ideally, we want to construct a dispersal effectiveness network, with links weighted by both the quantity and quality components of dispersal, but this would be a massive undertaking!

Incomplete sampling can result in a significant underestimate of the number of interactions in a network and thus overestimates in the degree of specialization (Valdovinos 2019). The problem of 'missing links' can be reduced by the use of complementary sources of information, such as both field observations of fruit removal and fecal analyses. Stable isotopes can give a complementary picture of the assimilated diet over several weeks (Bosenbecker and Bugoni 2020), but networks in species-rich systems will never be complete. Networks also underestimate the plasticity of interactions, which allows for rewiring of interactions when a species is lost. Frugivores can adapt to novel fruit sources if a preferred source is unavailable, and plants may attract new dispersal agents when a stronger competitor is absent. It is also difficult to illustrate seasonal changes, although most plant species fruit for only part of the year and many frugivores are migratory (Ramos-Robles et al. 2018). Networks are also often deliberately truncated, for example by focusing only on birds, even though bird and mammal diets overlap, and most exclude dry fruits and invertebrate frugivores.

Despite the limitations inherent in summarizing complex natural history in a single measure of interaction strength, network studies have transformed studies of frugivory and seed dispersal. They identify both the most important links in the community and the interactions which do not occur. They have shown that networks are modular, with groups of species which interact more frequently with each other than with other species in the network. Another striking feature of frugivory networks is the great variation in how many interactions each species is involved in, ranging from specialists, involved in very few interactions, to generalists which eat—or are eaten by—many species. Interestingly, generalist populations of fruits and frugivores may consist of relatively specialized individuals. The fruits on individual plants may differ significantly in the traits assessed by frugivores, and thus in the frugivore species they attract (Crestani et al. 2019), while individual frugivores may differ significantly in their diets (Pires and de Melo 2020) and their uses of space (Kerches-Rogeri et al. 2020). Networks of individuals are also modular, with groups of individuals sharing similar fruit or frugivore resources.

Frugivory networks are also more or less nested, with the interactions of the more specialist species (of both animals and of plants) a subset of the interactions of the more generalist species. As a result, specialists tend to interact mostly with generalists, rather than other specialists. This should make networks relatively resilient to the loss of specialists, since no other species depend entirely on them, while very vulnerable to the loss of generalists. Despite the vast literature on nestedness in mutualistic networks, it is still not clear if it is an emergent property of the network itself or a simple consequence of the fact that communities contain species spread across the spectrum from extreme specialists to extreme generalists (Mariani et al. 2019; Payrató-Borràs et al. 2019). In any case, the usefulness of networks does not depend on simple descriptive metrics, but on their ability to provide a single, easily understood, visualization of relationships in a community.

Local networks, in turn, form regional and global meta-networks, with compartmentalization by biogeographic barriers increasing at the larger spatial scales (Fricke and Svenning 2020). Indeed, local networks rarely have sharp boundaries, since wide-ranging and migratory frugivores providing links at different spatial scales for different plant species. The difficulties of tracking seeds over long distances limit our ability to quantify seed export beyond local networks, but molecular techniques make it theoretically possible to assess what proportion of local plants were established from non-local sources (Rogers et al. 2019).

7.11 Plant-Frugivore Coevolution and Dispersal Syndromes

A recent review states that the 'pervasive role of evolution' is manifest within ecological networks (Segar et al. 2020a), although this view was immediately challenged (Sagoff 2020; Rossberg 2020; and see Segar et al. 2020b). A major point of disagreement is whether the species in a network share enough evolutionary history to have influenced each other's adaptive evolution, or if the species composition is simply a 'snapshot' of a structure that changes continuously in space and time. On this second view, trait matches between interacting species largely reflect 'ecological fitting' between species that have recently met and did not evolve together. These opposing views represent different parts of a spectrum, from networks that are entirely structured by coevolutionary processes—although nobody has suggested that these exist in the real world—through to networks composed of fortuitous combinations of species with independent distributions and evolutionary histories, and traits that evolved for other reasons (Segar et al. 2020b). It is striking that most of the evidence for evolution within interaction networks comes from antagonistic networks, such as those of plants and their herbivores, or from pollination. What about seed dispersal networks?

At first sight, choosy frugivores interacting with choosy plants sounds like a recipe for pairwise coevolution, but two factors mitigate against this. The long lifespans of vertebrate frugivores in comparison with individual fruit crops means that a frugivore species cannot specialize on—and evolve adaptations to—only one fruit species. Also, on a much longer timescale, plants evolve more slowly than vertebrates, and many plant families are older and more widespread than vertebrate families. The family Lauraceae, for example, has spread globally in warmer environments since the Cretaceous and its uniformly single-seeded fleshy fruits must have interacted with numerous different groups of vertebrate frugivores across space and time. Together, these factors mean that the type of reciprocal specialization which drives the few well-documented examples of mutualistic co-evolution, such as figs and their pollinating fig wasps, cannot occur with fruits and frugivores.

We know that the properties of seed dispersal networks described in the previous section do not necessarily require prolonged co-evolution, since the same structural features are found in Hawai'i in communities dominated by introduced species in which all interactions are novel (Vizentin-Bugoni et al. 2019). Instead, the general properties of seed dispersal networks seem to largely reflect trait matches and mismatches between fruits and frugivores—i.e. ecological fitting—plus lack of phenological overlap which forbids some potential links (Bender et al. 2018; Valdovinos 2019). However, fleshy fruits would not have evolved without the need to attract frugivores to eat them, or specialist frugivores evolved their multiple adaptations to a fruit-dominated diet without fruits to eat. Clearly the diversities of fruits and frugivores have evolved in concert in some sense.

Diffuse coevolution between groups of taxa is more plausible than pairwise coevolution, but also much more difficult to demonstrate convincingly. It requires that a group of frugivores evolves a set of shared traits in reciprocity with the evolution of matching traits in a group of fruits. The modular structure of most plant dispersal networks shows that, at least locally, groups of frugivore and fruit species do interact more with each other than with other species in the area. The modular structure of seed dispersal networks has, in turn, been attributed to the existence of seed dispersal syndromes: suites of fruit traits that have apparently co-evolved with matching frugivore traits, including sensory capabilities, morphology, and behavior (Valenta and Nevo 2020). There is good evidence that at least some fruit traits—particularly size, color, and scent—are adaptive, and thus likely to have been influenced during evolution by directional selection by frugivores (Valenta and Nevo 2020). However, evidence for syndromes that combine multiple, coevolved, traits is still largely anecdotal. Moreover, the idea of multiple species interacting and coevolving together is surely an oversimplification, given the great variation in the numbers of interactions between pairs of species in networks. The role of other factors, including phylogenetic and developmental constraints, abiotic conditions, and selection by pathogens and fruit and seed predators, is also unclear.

A phylogenetic comparative study of fruit diversity in *Ficus* (the figs) in Papua New Guinea showed that at least six fruit traits in this genus have evolved in concert across different clades of figs into two contrasting syndromes, whose traits match those expected for the frugivorous birds and bats respectively observed to consume them (Lomáscolo et al. 2010). 'Bird figs' were smaller, softer, redder, darker, less odorous, and borne among the foliage, while 'bat figs' were larger, harder, greener, lighter, smellier, and borne away from the foliage. A recent study of the genus *Viburnum* found evidence for the correlated evolution of fruit color, reward, and morphology, with blue fruit having a high lipid content and a large, round endocarp while red fruits were juicy, with low lipid and a smaller, flatter endocarp (Sinnott-Armstrong et al. 2020). However, the selective pressures driving the evolution of these two syndromes were unclear, as fruits with both syndromes were primarily dispersed by birds. Within the genus *Viburnum*, red and blue are largely honest signals of fruit qualities, but blue fruits are relatively rare in other plant families and the bright metallic blue of lipid-rich *Viburnum* fruits is a structural color, formed from lipids, which is so far unknown from other taxa (Middleton et al. 2020). Non-fruit traits may also be associated with the evolution of seed dispersal syndromes. In Neotropical palms, the evolution of fruit color and size are correlated, resulting in two, more or less distinct, syndromes: smaller, colorful fruits that are dispersed mainly by birds and larger, dull-colored fruits dispersed mainly by mammals (do Nascimento et al. 2020). Large fruit size was also positively associated with spines on the leaves, bracts, and stems, which are interpreted as a defense against the now-extinct megafauna, which may otherwise have been tempted to eat more than just the fruits.

7.12 Long-Distance Dispersal

Most seed dispersal is within the local population and genetic neighborhood, but often a small proportion of seeds is dispersed further: long-distance dispersal (LDD). There is no standard definition of LDD, with some authors setting arbitrary thresholds for the dispersal distance or the proportion of events, while Jordano (2017) defined three types: within the geographic limits of the local population but outside the genetic neighborhood, within the genetic neighborhood but outside the local population, and 'strict-sense' LDD which is outside both. LDD is important for connecting populations that are naturally or anthropogenically fragmented, for colonizing new areas made available by environmental changes, and, through these processes, for allowing global survival in the face of local extinctions (Jordano 2017). Rapid plant migration in response to rapid natural or anthropogenic climate change probably usually depends on LDD.

Rare events are inherently difficult to study, particularly when sparsely spread across huge areas as will often be true in the LDD of seeds, but advances in biotelemetry have made it increasingly practical to follow the movements of frugivores over relevant scales while advances in molecular methods have made it easier to identify sources of individual seeds or plants (Rogers et al. 2019). Although the correlation is not perfect, these studies have identified frugivore body size as the best single predictor of effective LDD. In tropical Asia, for example, LDD appears to be dominated by the largest frugivorous birds (fruit pigeons and hornbills, ▣ Fig. 7.3), the largest fruit bats (flying foxes, ▣ Fig. 7.4), and the largest mammals (elephants, ungulates, bears) (Corlett 2017). Large Old-World fruit bats (flying foxes) may fly tens of kilometers in a night (Epstein et al. 2009; Randhawa et al.

Fig. 7.4 An Indian flying fox, *Pteropus giganteus*, in Madhya Pradesh, India. These bats are the largest flying mammals and a similar size to large hornbills. (Photograph by Charles J. Sharp from Wikimedia Commons used under the Creative Commons CC BY-SA 4.0 license)

2020; Welbergen et al. 2020), potentially with seeds retained in their guts (Shilton et al. 1999). Unfortunately, body size is also a good predictor of vulnerability to hunting and other human impacts, making LDD particularly vulnerable (see ► Sect. 7.14). As discussed earlier, granivores may also perform LDD for any surviving seeds in their guts, particularly where long gut passage times are combined with long-distance movements. Indeed, even small migrant birds may be significant if they carry occasional viable seeds in their guts (Viana et al. 2016).

In fragmented habitats, it is also important to know what land-cover types the dispersal agents are willing and able to cross. Hornbills, fruit pigeons, flying foxes, and many migratory birds will cross both cleared areas and open sea, while many other birds and most forest mammals avoid both. In Australia, flying foxes potentially link fragmented forests over vast distances (Welbergen et al. 2020).

7.13 The Potential Velocity of Plant Spread

In response to climate change, plant populations must either acclimate (plastic change), adapt (genetic change), or move so as to stay within their climate envelope (Corlett and Westcott 2013). Evidence for genetic change is limited, but plastic responses in phenology and physiology have been widely observed. Movement is more difficult for plants than most animals because plants can only move once per generation and generation times can be much longer than for animals. The fossil record suggests that movement, rather than acclimation or adaptation, was the most common response to previous periods of climate change, although species responses were idiosyncratic, and plant communities did not shift as a block. However, the fossil record usually has a low temporal resolution—centuries, at best—and most species are not represented. We really do not know how most plant species are adjusting to the rapid climate changes experienced over recent decades and expected to continue in the immediate future. Moreover, rapid movement is difficult enough for plants in pristine natural landscapes but even more challenging in the fragmented landscapes in which most wild species now live.

How fast do they need to move? A useful concept is the 'velocity of climate change', which is the speed and direction that something needs to move to keep within its current 'climate envelope', i.e. to stay within the range of climate conditions it can tolerate (Brito-Morales et al. 2018). Velocities of temperature change are expected to exceed 1 km/year over much of the Earth's land surface during the twenty-first century, but they will be lower in areas of steep topography where movements over short vertical distances are enough to compensate for expected warming. Velocities of rainfall change are projected to be similar. In reality, plant species will likely respond individualistically to changes in multiple different climate parameters, but a standard definition of climate velocity based only on widely available physical parameters has the advantage of simplicity and generality. Moreover, areas of relatively low temperature velocity over the last 20,000 years support more narrow-range species of plants and animals, suggesting that these standard definitions are a useful indicator of future threats (Sandel et al. 2011; Harrison and Noss 2017).

Can plants keep up with climate change? Converting dispersal distances into potential velocities requires knowledge of the number of dispersal events in the time period under consideration, which will depend on the time from dispersed seed to reproductive adult. Thus, if most plant species can disperse 50–1500 m at a time and can grow from seed to adult in 1–30 years, the range of plant velocities in unfragmented habitats with no barriers to movement would be 1.7–1500 m per year (Corlett and Westcott 2013). However, potential movement velocities for most forest plant species are likely to be at the lower end of this range. Projected velocities of climate change and estimated potential velocities of plant movement thus broadly overlap, but most plant species will probably only be able to fully track climate change in steep topography where climate velocities are low.

7.14 Vulnerability

Any process that depends on vertebrates is vulnerable to human impacts. Larger fish (Correa et al. 2018), reptiles (Pérez-Méndez et al. 2016), and birds and mammals (Corlett 2017), consume larger fruits and disperse more seeds longer distances. These larger species are the main targets of hunters (Benítez-López et al. 2017; Correa et al. 2018) and are also more likely to require large areas of unfragmented habitat. Defaunation (the loss of animals from a habitat) and downsizing (the disproportionate loss of large species) are thus the hallmarks of human impacts and would be predicted to have large, selective, impacts on frugivory and seed dispersal. Downsizing of disperser faunas started in the Late Pleistocene with the megafaunal extinctions and continues to the present day (Galetti et al. 2018). Average maximum body sizes of mammalian frugivore assemblages would be more than 11-fold higher without the Late Pleistocene and Holocene extinctions (Lim et al. 2020). Population declines of large vertebrates are common even in protected areas, particularly in poorer countries (Rija et al. 2020), and large areas of the tropics have now lost most or all large species.

The consequences of downsizing for seed dispersal are neatly illustrated in the Canary Islands, where large-bodied frugivorous lizards (*Gallotia*, Lacertidae, ◗ Fig. 7.5) were abundant before human settlement but have declined subsequently (Pérez-Méndez et al. 2016). Seed dispersal distances now decrease along a defaunation and downsizing gradient, while intrapopulation spatial genetic structure increases. In the Hawaiian Islands, the loss of most native frugivores and their replacement by smaller-sized, smaller-gaped, introduced species has added to the conservation risks faced by larger (>8.1 mm) seeded plants (Case and Tarwater 2020). The shorter-term consequences of downsizing in more complex communities have also been shown in several studies. In lowland dipterocarp forest in Borneo, for example, the elimination of most large frugivores over 15 years by hunters resulted in a decline in the diversity of frugivore assemblages at fruiting fig trees, and both increased clustering of saplings and a relative decline in recruitment for animal-dispersed tree species (Harrison et al. 2013). Fleshy fruits with larger

◗ **Fig. 7.5** A frugivorous lizard, *Gallotia galloti*, on La Palma in the Canary Islands. Larger conspecifics are already extinct. (Photograph by Uwezz from Wikimedia Commons used under the Creative Commons CC by 3.0 license)

seeds showed greater declines in cluster size and recruitment than those with small seeds. In the Brazilian Atlantic forest, there is evidence that a bird-dispersed palm, *Euterpe edulis*, has evolved significantly smaller fruits in less than a century in areas that have lost large-gaped avian frugivores (Galetti et al. 2013).

In the Central Amazon, fragmentation of the forest reduced and altered seed dispersal through multiple pathways: reduced bee pollination, loss of dispersal by primates and large birds, and reduction in large-seeded, rare, and mature-forest tree species (Hooper and Ashton 2020). Reductions in fragments were large: a three-fold reduction in seed production by non-pioneer species; a six-fold reduction in the density of dispersed seeds; and a 9.6-fold reduction of the functional richness of dispersed seeds in the smallest, 1 ha, fragments. Roads and trails through forests can also impact seed-dispersing birds (da Silva and Silva 2020). Urbanization is an even more extreme form of human impact, leading to low diversity frugivore communities of well-adapted generalists which consume all available fruits, and networks with high interaction evenness and low nestedness (Schneiberg et al. 2020).

Species introductions also influence seed dispersal. The proportion of the animal and plant species in local networks which are introduced is greatest on islands and in disturbed habitats (Fricke and Svenning 2020). In tropical Asia, many successful bird invaders are bulbuls (Pycnonotidae), babblers (Timaliidae and Leiothrichidae), or mynas (Sturnidae), all of which are partly frugivorous (Corlett et al. 2020). Introduced species reduce biogeographic compartmentalization and, since introduced plants and animals are more likely to interact with introduced partners, may promote invasion, resulting in 'invasional meltdown' (MacFarlane et al. 2016; Thibault et al. 2018). The combined effect of these human impacts risks leading to a future in which a few well-connected generalists, often introduced, replace many, more specialized, native species. However, seed dispersal networks in Hawai'i, dominated by introduced species and consisting entirely of novel interactions, were surprisingly similar to networks elsewhere in specialization, modularity, and nestedness (Vizentin-Bugoni et al. 2019).

The interactions between climate change and seed dispersal were discussed in the previous section. Climate change could also, potentially, reduce seed dispersal by causing a loss of synchrony between migrating frugivores and the ripening of seasonal fruits if they respond to different climatic cues (Rafferty et al. 2015). There is, as yet, no evidence for this and it is possible that either fruit timing is not that precise or birds will show behavioral flexibility.

7.15 The Way Forward in Seed-Dispersal Research

Pollination envy has long been a problem for seed dispersal studies, with ideas from pollination ecology sometimes uncritically extrapolated to 'the other plant-animal mutualism'. Despite the morphological continuity between flowers and fruits, pollination and seed dispersal are very different processes, however, carried out largely by very different animals. Pollination has a precise target—a conspecific stigma—and a degree of specificity is maintained by 'payment on delivery' of pollen, which gives the plants some control over pollinators. Moreover, pollinators are mostly smaller than or a similar size to flowers, so access to rewards can be mechanically restricted. Seed dispersal agents, in contrast, must be 'paid in advance', making them impossible to direct to particular targets, even if these were predictable, and frugivores are much bigger than fruits and cannot easily be mechanically excluded. These differences largely account for the amazing diversity of flowers and relative uniformity of fruits, at least in morphology (Herrera 2002).

Frugivory and seed dispersal studies are still dominated by more or less systematically collected records of this eating that. In contrast to pollination ecology, quantitative assessments of dispersal effectiveness have been rare and manipulative experiments almost nonexistent. There are good reasons for this: the diffuseness of fruit-frugivore interactions and their variability in space and time, the shyness of many dispersal agents which makes them hard to observe, and the very large spatial scales over which seed dispersal operates. Experiments have been widely used in studies of dispersal by scatter-hoarding rodents, however, making this the best-understood seed dispersal mutualism (e.g. Kuprewicz and García-Robledo 2019; Wang et al. 2020b), and opportunities for experimentally manipulating fleshy fruits and/or frugivores should be explored further. We need to move past simply describing and explaining observed patterns of frugivory and seed dispersal, and towards making useful predictions in a changing world: predictions that can help with real-world management decisions.

Two very different approaches look most promising at present. One is trait matching, based on the assumption that the observed patterns largely reflect trait matches and mismatches between fruits and frugivores. Trait-based approaches have had considerable success in explaining patterns of frugivory and seed dispersal in a variety of situations (Bender et al. 2018; Ramos-Robles et al. 2018; Valdovinos 2019; Case and Tarwater 2020; Sorensen et al. 2020). Most success so far has been with frugivorous birds, where size matching between fruit and gape constrains frugivory, while other potentially relevant traits of fruits and frugivores have usually been represented by crude proxies. The use of more and better-chosen traits is likely to increase predictive power, although the incorporation of memory and learning into trait-based approaches may be difficult. The second, complimentary, approach uses fossil-calibrated phylogenies to investigate the evolution of fruit traits, in relation to each other and to the physical environment (Lomáscolo et al. 2010; Onstein et al. 2019; do Nascimento et al. 2020; Sinnott-Armstrong et al. 2020). Although an understanding of trait evolution does not have the immediate application that empirical trait-matching can, it may be more robust to changes in local conditions and thus useful for predicting responses to novel futures.

Conclusions

Defaunation and downsizing of disperser communities are increasingly the 'new normal', not the exception. Although the general consequences for seed dispersal are well understood, we are not yet in a position to make detailed recommendations that can be directly applied in local conservation management. Despite an explosion of new work over the last decade, studies of frugivory and seed dispersal are still dominated by description and correlation, while experiments are rare, so our ability to make useful predictions for novel future conditions is still limited. Most studies look only at visitation and fruit removal, and other components of effective seed dispersal have received much less attention. The focus has been on birds and primates, and thus on fleshy fruits, but internal dispersal after fruit consumption encompasses considerably more than this: indehiscent pods and a variety of dry fruits, ungulates and other mammals, plus reptiles, fish, and many species of invertebrates. Fruits are typically described only in terms of external dimensions, along with human perceived color and odor, while defensive and nutritional chemistry have been neglected. Frugivores have also been viewed in a simplistic way, ignoring their ability to learn from experience. Frugivory and seed dispersal studies need to expand their focus to include all fruits and fruit-eaters, and to follow seeds from fruit maturation to the establishment of a new plant.

7

Key Points

- Frugivory and seed dispersal are very different processes from pollination: more diffuse, more variable in space and time, less coevolved, and in many ways more difficult to study.
- Observed patterns of frugivory and seed dispersal are best explained by trait matches and mismatches between fruits and frugivores, and phenology, i.e. ecological fitting rather than coevolution.
- Over evolutionary time, however, it is clear that fruit traits have evolved, more or less diffusely, in response to frugivore traits and presumably, to some extent, *vice versa*.
- Dependence on vertebrates makes frugivory and seed dispersal very sensitive to hunting and habitat degradation.

? Questions

- Which fruit and frugivore traits are most important in controlling patterns of frugivory and seed dispersal?
- How do memory and learning influence frugivore choices and seed dispersal outcomes?
- To what extent can frugivory and seed dispersal enable plant populations to track anthropogenic climate change?

References

Albert-Daviaud A, Buerki S, Onjalalaina GE et al (2020) The ghost fruits of Madagascar: identifying dysfunctional seed dispersal in Madagascar's endemic flora. Biol Conserv 242:108438

Albrecht J, Hagge J, Schabo DG et al (2018) Reward regulation in plant-frugivore networks requires only weak cues. Nat Commun 9:4838

Araujo JM, Correa SB, Anderson J, Penha J (2020) Fruit preferences by fishes in a Neotropical floodplain. Biotropica 52:1131–1141

Aronne G, Wilcock CC (1994) First evidence of myrmechochory in fleshy-fruited shrubs in the Mediterranean region. New Phytol 127:781–788

Bairlein F (2002) How to get fat: nutritional mechanisms of seasonal fat accumulation in migratory songbirds. Naturwissenschaften 89:1–10

Baldwin JW, Dechmann DKN, Thies W, Whitehead SR (2020) Defensive fruit metabolites obstruct seed dispersal by altering bat behavior and physiology at multiple temporal scales. Ecology 101(2):e02937

Baltzinger C, Karimi S, Shukla U (2019) Plants on the move: hitch-hiking with ungulates distributes diaspores across landscapes. Front Ecol Evol 7:38

Bar-On YM, Phillips R, Milo R (2018) The biomass distribution on Earth. Proc Natl Acad Sci U S A 115:6506–6511

Barve S, La Sorte FA (2016) Fruiting season length restricts global distribution of female-only parental care in frugivorous passerine birds. Plos One 11:e0154871

Bender IMA, Kissling WD, Blendinger PG et al (2018) Morphological trait matching shapes plant-frugivore networks across the Andes. Ecography 41:1910–1919

Benítez-López A, Alkemade R, Schipper AM et al (2017) The impact of hunting on tropical mammal and bird populations. Science 356:180–183

Bosenbecker C, Bugoni L (2020) Trophic niche similarities of sympatric *Turdus* thrushes determined by fecal contents, stable isotopes, and bipartite network approaches. Ecol Evol 10:9073–9084

Brancalion PHS, Novembre A, Rodrigues RR, Marcos J (2010) Dormancy as exaptation to protect mimetic seeds against deterioration before dispersal. Ann Bot 105:991–998

Bravo SP, Cueto VR (2020) Directed seed dispersal: the case of howler monkey latrines. Perspectives in Plant Ecology Evolution and Systematics 42:125509

Brito-Morales I, Molinos JG, Schoeman DS et al (2018) Climate velocity can inform conservation in a warming world. Trends Ecol Evol 33:441–457

Bunney K, Bond WJ, Henley M (2017) Seed dispersal kernel of the largest surviving megaherbivore-the African savanna elephant. Biotropica 49:395–401

Calvino-Cancela M, Rubido-Bará M (2012) Effects of seed passage through slugs on germination. Plant Ecol 213:663–673

Campos-Arceiz A, Blake S (2011) Megagardeners of the forest - the role of elephants in seed dispersal. Acta Oecol 37:542–553

Case SB, Tarwater CE (2020) Functional traits of avian frugivores have shifted following species extinction and introduction in the Hawaiian Islands. Funct Ecol 12:2467–2476

Cazetta E, Schaefer HM, Galetti M (2008) Does attraction to frugivores or defense against pathogens shape fruit pulp composition? Oecologia 155:277–286

Chen SC, Cornwell WK, Zhang HX, Moles AT (2017) Plants show more flesh in the tropics: variation in fruit type along latitudinal and climatic gradients. Ecography 40:531–538

Chen L-j, Hou Y-m, Yin P-f, Wang X (2020b) An edible fruit from the Jurassic of China. China Geol 3:8–15

Chen SC, Poschlod P, Antonelli A et al (2020a) Trade-off between seed dispersal in space and time. Ecol Lett 23:1635–1642

Clause J, Forey E, Eisenhauer N et al (2017) Seed selection by earthworms: chemical seed properties matter more than morphological traits. Plant and Soil 413:97–110

Corlett RT (1998) Frugivory and seed dispersal by vertebrates in the Oriental (Indomalayan) Region. Biol Rev 73:413–448

Corlett RT (2011a) How to be a frugivore (in a changing world). Acta Oecol 37:674–681

Corlett RT (2011b) Seed dispersal in Hong Kong, China: past, present and possible futures. Integr Zool 6:97–109

Corlett RT (2017) Frugivory and seed dispersal by vertebrates in tropical and subtropical Asia: an update. Glob Ecol Conserv 11:1–22

Corlett RT (2019) The ecology of tropical East Asia. Oxford University Press, Oxford, UK

Corlett RT, Primack RB (2011) Tropical rain forests: an ecological and biogeographical comparison, 2nd edn. Wiley-Blackwell, Oxford, UK

Corlett RT, Westcott DA (2013) Will plant movements keep up with climate change? Trends Ecol Evol 28:482–488

Corlett RT, Leven MR, Yong DL, Eaton JA, Round PA (2020) Continental analysis of invasive birds: Asia. In: Downs CT, Hart LA (eds) Invasive birds: global trends and impacts. CABI, Wallingford, UK, pp 315–340

Correa SB, Costa-Pereira R, Fleming T et al (2015) Neotropical fish-fruit interactions: eco-evolutionary dynamics and conservation. Biol Rev 90:1263–1278

Costa-Pereira R, Correa SB, Galetti M (2018) Fishing-down within populations harms seed dispersal mutualism. Biotropica 50:319–325

Crestani AC, Mello MAR, Cazetta E (2019) Interindividual variations in plant and fruit traits affect the structure of a plant-frugivore network. Acta Oecol 95:120–127

Culot L, Huynen MC, Heymann EW (2018) Primates and dung beetles: two dispersers are better than one in secondary forest. Int J Primatol 39:397–414

da Silva HR, de Britto-Pereira MC (2006) How much fruit do fruit-eating frogs eat? An investigation on the diet of Xenohyla truncata (Lissamphibia: Anura: Hylidae). J Zool 270:692–698

da Silva BG, Silva WR (2020) Impacts of park roads and trails on a community of Atlantic Forest fruit-eating birds. Trop Ecol 61:371–386

de Vega C, Arista M, Ortiz PL et al (2011) Endozoochory by beetles: a novel seed dispersal mechanism. Ann Bot 107:629–637

Delibes M, Castañeda I, Fedriani JM (2019) Spitting seeds from the cud: a review of an endozoochory exclusive to ruminants. Front Ecol Evol 7:265

do Nascimento LF, Guimãraes PR, Onstein RE et al (2020) Associated evolution of fruit size, fruit colour and spines in Neotropical palms. J Evol Biol 33:858–868

Donati G, Santini L, Eppley TM et al (2017) Low levels of fruit nitrogen as drivers for the evolution of Madagascar's primate communities. Sci Rep 7:14406

Duan Q, Goodale E, Quan RC (2014) Bird fruit preferences match the frequency of fruit colours in tropical Asia. Sci Rep 4:5627

Epstein JH, Olival KJ, Pulliam JRC et al (2009) Pteropus vampyrus, a hunted migratory species with a multinational home-range and a need for regional management. J Appl Ecol 46:991–1002

Eriksson O (2016) Evolution of angiosperm seed disperser mutualisms: the timing of origins and their consequences for coevolutionary interactions between angiosperms and frugivores. Biol Rev 91:168–186

Falcón W, Moll D, Hansen DM (2020) Frugivory and seed dispersal by chelonians: a review and synthesis. Biol Rev 95:142–166

Fricke EC, Svenning JC (2020) Accelerating homogenization of the global plant-frugivore meta-network. Nature 585:74–78

Fricke EC, Simon MJ, Reagan KM et al (2013) When condition trumps location: seed consumption by fruit-eating birds removes pathogens and predator attractants. Ecol Lett 16:1031–1036

Galetti M, Guevara R, Côrtes MC et al (2013) Functional extinction of birds drives rapid evolutionary changes in seed size. Science 340:1086–1090

Galetti M, Moleón M, Jordano P et al (2018) Ecological and evolutionary legacy of megafauna extinctions. Biol Rev 93:845–862

Hardus ME, de Vries H, Dellatore DF et al (2013) Socioecological correlates of inter-individual variation in orangutan diets at Ketambe, Sumatra. Behav Ecol Sociobiol 67:429–437

Harrison S, Noss R (2017) Endemism hotspots are linked to stable climatic refugia. Ann Bot 119:207–214

Harrison RD, Tan S, Plotkin JB et al (2013) Consequences of defaunation for a tropical tree community. Ecol Lett 16:687–694

Herrera CM (2002) Correlated evolution of fruit and leaf size in bird-dispersed plants: species-level variance in fruit traits explained a bit further? Oikos 97:426–432

Hooper ER, Ashton MS (2020) Fragmentation reduces community-wide taxonomic and functional diversity of dispersed tree seeds in the Central Amazon. Ecol Appl 30:e02093

Horn MH, Correa SB, Parolin P et al (2011) Seed dispersal by fishes in tropical and temperate fresh waters: the growing evidence. Acta Oecol 37:561–577

Janzen DH (1984) Dispersal of small seeds by big herbivores - foliage is the fruit. Am Nat 123:338–353

John EA, Soldati F, Burman OHP et al (2016) Plant ecology meets animal cognition: impacts of animal memory on seed dispersal. Plant Ecol 217:1441–1456

Jordano P (2017) What is long-distance dispersal? And a taxonomy of dispersal events. J Ecol 105:75–84

Kerches-Rogeri P, Niebuhr BB, Muylaert RL, Mello MAR (2020) Individual specialization in the use of space by frugivorous bats. J Anim Ecol 89(11):2584–2595

King P, Milicich L, Burns KC (2011) Body size determines rates of seed dispersal by giant king crickets. Popul Ecol 53:73–80

Kleyheeg E, Fiedler W, Safi K et al (2019) A comprehensive model for the quantitative estimation of seed dispersal by migratory mallards. Front Ecol Evol 7:40

Ko IWP, Corlett RT, Xu RJ (1998) Sugar composition of wild fruits in Hong Kong, China. J Trop Ecol 14:381–387

Kozlowski CP, Vickerman E, Sahrmann J et al (2016) Parent-offspring behavior of Jambu fruit doves (*Ptilinopus jambu*). Zoo Biol 35:120–127

Kuprewicz EK, García-Robledo C (2019) Deciphering seed dispersal decisions: size, not tannin content, drives seed fate and survival in a tropical forest. Ecosphere 10:e02551

Larsen H, Burns KC (2012) Seed dispersal effectiveness increases with body size in New Zealand alpine scree weta (*Deinacrida connectens*). Austral Ecol 37:800–806

Levey DJ, Bissell HA, O'Keefe SF (2000) Conversion of nitrogen to protein and amino acids in wild fruits. J Chem Ecol 26:1749–1763

Lim JY, Svenning J-C, Göldel B et al (2020) Frugivore-fruit size relationships between palms and mammals reveal past and future defaunation impacts. Nat Commun 11:4904

Lohaus G, Schwerdtfeger M (2014) Comparison of sugars, iridoid glycosides and amino acids in nectar and phloem sap of *Maurandya barclayana, Lophospermum erubescens*, and *Brassica napus*. Plos One 9:e87689

Lomáscolo SB, Levey DJ, Kimball RT et al (2010) Dispersers shape fruit diversity in *Ficus* (Moraceae). Proc Natl Acad Sci U S A 107:14668–14672

Longland WS, Dimitri LA (2016) Are western juniper seeds dispersed through diplochory? Northwest Sci 90:235–244

MacFarlane AET, Kelly D, Briskie JV (2016) Introduced blackbirds and song thrushes: useful substitutes for lost mid-sized native frugivores, or weed vectors? N Z J Ecol 40:80–87

Mariani MS, Ren ZM, Bascompte J, Tessone CJ (2019) Nestedness in complex networks: observation, emergence, and implications. Phys Rep 813:1–90

McConkey KR (2018) Seed dispersal by primates in Asian habitats: from species, to communities, to conservation. Int J Primatol 39:466–492

Middleton R, Sinnott-Armstrong M, Ogawa Y et al (2020) *Viburnum tinus* fruits use lipids to produce metallic blue structural color. Curr Biol 30:3804–3810

Miller MF (1996) Dispersal of *Acacia* seeds by ungulates and ostriches in an African savanna. J Trop Ecol 12:345–356

Moermond TC, Denslow JJ (1985) Neotropical avian frugivores: patterns of behavior, morphology, and nutrition with consequences for fruit selection. Ornithol Monogr 36:865–897

Morán-López T, Carlo TA, Amico G, Morales JM (2018) Diet complementation as a frequency-dependent mechanism conferring advantages to rare plants via dispersal. Funct Ecol 32:2310–2320

Moreno SA, Gelambi M, Biganzoli A, Molinari J (2019) Small nutrient molecules in fruit fuel efficient digestion and mutualism with plants in frugivorous bats. Sci Rep 9:19376

Naves ER, Silva LD, Sulpice R et al (2019) Capsaicinoids: pungency beyond capsicum. Trends Plant Sci 24:109–120

Oliveros CH, Field DJ, Ksepka DT et al (2019) Earth history and the passerine superradiation. Proc Natl Acad Sci U S A 116:7916–7925

Onstein RE, Kissling WD, Chatrou LW et al (2019) Which frugivory-related traits facilitated historical long-distance dispersal in the custard apple family (Annonaceae)? J Biogeogr 46:1874–1888

Orlowski G, Czarnecka J, Golawski A et al (2016) The effectiveness of endozoochory in three avian seed predators. J Ornithol 157:61–73

Orr TJ, Ortega J, Medellin RA et al (2016) Diet choice in frugivorous bats: gourmets or operational pragmatists? J Mammal 97:1578–1588

Passos L, Oliveira PS (2002) Ants affect the distribution and performance of seedlings of *Clusia criuva*, a primarily bird-dispersed rain forest tree. J Ecol 90:517–528

Payrató-Borràs C, Hernández L, Moreno Y (2019) Breaking the spell of nestedness: the entropic origin of nestedness in mutualistic systems. Phys Rev X 9:031024

Pérez-Méndez N, Jordano P, García C, Valido A (2016) The signatures of Anthropocene defaunation: cascading effects of the seed dispersal collapse. Sci Rep 6:24820

Pires LP, de Melo C (2020) Individual-resource networks reveal distinct fruit preferences of selective individuals from a generalist population of the Helmeted Manakin. Ibis 162:713–722

Pizo MA, Fontanella ABA, Canassa G et al (2020) Decoding Darwin's puzzle: avian dispersal of mimetic seeds. Ecology 101:e03005

Rafferty NE, CaraDonna PJ, Bronstein JL (2015) Phenological shifts and the fate of mutualisms. Oikos 124:14–21

Ramos-Robles M, Dattilo W, Díaz-Castelazo C, Andresen E (2018) Fruit traits and temporal abundance shape plant-frugivore interaction networks in a seasonal tropical forest. Sci Nat 105:29

Randhawa N, Bird BH, VanWormer E et al (2020) Fruit bats in flight: a look into the movements of the ecologically important *Eidolon helvum* in Tanzania. One Health Outlook 2:16

Rija AA, Critchlow R, Thomas CD, Beale CM (2020) Global extent and drivers of mammal population declines in protected areas under illegal hunting pressure. PloS One 15:e0227163

Rogers HS, Beckman NG, Hartig F et al (2019) The total dispersal kernel: a review and future directions. AoB Plants 11:plz042

Rossberg AG (2020) What are the fundamental questions regarding evolution in ecological networks? Trends Ecol Evol 35:863–865

Sagoff M (2020) Ecological networks: response to Segar *et al.* Trends Ecol Evol 35:862–863

Salgado L, Canudo JI, Garrido AC et al (2017) A new primitive Neornithischian dinosaur from the Jurassic of Patagonia with gut contents. Sci Rep 7:42778

Sandel B, Arge L, Dalsgaard B et al (2011) The influence of Late Quaternary climate-change velocity on species endemism. Science 334:660–664

Schneiberg I, Boscolo D, Devoto M et al (2020) Urbanization homogenizes the interactions of plant-frugivore bird networks. Urban Ecosyst 23:457–470

Schupp EW, Jordano P, Gómez JM (2010) Seed dispersal effectiveness revisited: a conceptual review. New Phytol 188:333–353

Segar ST, Fayle TM, Srivastava DS et al (2020a) The role of evolution in shaping ecological networks. Trends Ecol Evol 35:454–466

Segar ST, Fayle TM, Srivastava DS et al (2020b) On the perils of ignoring evolution in networks. Trends Ecol Evol 35:865–866

Shilton LA, Altringham JD, Compton SG, Whittaker RJ (1999) Old World fruit bats can be long-distance seed dispersers through extended retention of viable seeds in the gut. Proc R Soc B Biol Sci 266:219–223

Simmons BI, Sutherland WJ, Dicks LV et al (2018) Moving from frugivory to seed dispersal: incorporating the functional outcomes of interactions in plant-frugivore networks. J Anim Ecol 87:995–1007

Sinnott-Armstrong MA, Lee C, Clement WL, Donoghue MJ (2020) Fruit syndromes in *Viburnum*: correlated evolution of color, nutritional content, and morphology in bird-dispersed fleshy fruits. BMC Evol Biol 20:7

Skalníková P, Frynta D, Abramjan A et al (2020) Spontaneous color preferences in rhesus monkeys: what is the advantage of primate trichromacy? Behav Processes 174:104084

Sorensen MC, Donoso I, Neuschulz EL et al (2020) Community-wide seed dispersal distances peak at low levels of specialisation in size-structured networks. Oikos 11:1727–1738

Stiles EW (1993) The influence of pulp lipids on fruit preference by birds. Vegetatio 108:227–235

Stournaras KE, Prum RO, Schaefer HM (2015) Fruit advertisement strategies in two Neotropical plant-seed disperser markets. Evol Ecol 29:489–509

Stringer SD, Hill RA, Swanepoel L et al (2020) Assessing the role of a mammalian frugivorous species on seed germination potential depends on study design: a case study using wild samango monkeys. Acta Oecol 106:103584

Suetsugu K (2018a) Seed dispersal in the mycoheterotrophic orchid *Yoania japonica*: further evidence for endozoochory by camel crickets. Plant Biol 20:707–712

Suetsugu K (2018b) Independent recruitment of a novel seed dispersal system by camel crickets in achlorophyllous plants. New Phytol 217:828–835

Suetsugu K (2020) A novel seed dispersal mode of Apostasia nipponica could provide some clues to the early evolution of the seed dispersal system in Orchidaceae. Evol Lett 4:457–464

Swamy V, Terborgh J, Dexter KG et al (2011) Are all seeds equal? Spatially explicit comparisons of seed fall and sapling recruitment in a tropical forest. Ecol Lett 14:195–201

Tang AMC, Corlett RT, Hyde KD (2005) The persistence of ripe fleshy fruits in the presence and absence of frugivores. Oecologia 142:232–237

Thibault M, Masse F, Pujapujane A et al (2018) "Liaisons dangereuses": the invasive red-vented bulbul (*Pycnonotus cafer*), a disperser of exotic plant species in New Caledonia. Ecol Evol 8:9259–9269

Thomas DW, Bosque C, Arends A (1993) Development of thermoregulation and the energetics of nestling oilbirds (*Steatornis caripensis*). Physiol Zool 66:322–348

Tiffney BH (2004) Vertebrate dispersal of seed plants through time. Annu Rev Ecol Evol Syst 35:1–29

Troisi CA, Hoppitt WJE, Ruiz-Miranda CR, Laland KN (2020) The role of food transfers in wild golden lion tamarins (Leontopithecus rosalia): support for the informational and nutritional hypothesis. Primates 62:207–221

Türke M, Andreas K, Gossner MM et al (2012) Are gastropods, rather than ants, important dispersers of seeds of myrmecochorous forest herbs? Am Nat 179:124–131

Uehara Y, Sugiura N (2017) Cockroach-mediated seed dispersal in *Monotropastrum humile* (Ericaceae): a new mutualistic mechanism. Bot J Linn Soc 185:113–118

Valdovinos FS (2019) Mutualistic networks: moving closer to a predictive theory. Ecol Lett 22:1517–1534

Valenta K, Nevo O (2020) The dispersal syndrome hypothesis: how animals shaped fruit traits, and how they did not. Funct Ecol 34:1158–1169

Valenta K, Daegling DJ, Nevo O et al (2020) Fruit selectivity in anthropoid primates: size matters. Int J Primatol 41:525–537

Valido A, Olesen JM (2019) Frugivory and seed dispersal by lizards: a global review. Front Ecol Evol 7:49

Vander Wall SB, Longland WS (2004) Diplochory: are two seed dispersers better than one? Trends Ecol Evol 19:155–161

Vandoros JD, Dumont ER (2004) Use of the wings in manipulative and suspensory behaviors during feeding by frugivorous bats. J Exp Zool A Ecol Integr Physiol 301A:361–366

Vázquez DP, Morris WF, Jordano P (2005) Interaction frequency as a surrogate for the total effect of animal mutualists on plants. Ecol Lett 8:1088–1094

Viana DS, Santamaria L, Figuerola J (2016) Migratory birds as global dispersal vectors. Trends Ecol Evol 31:763–775

Vizentin-Bugoni J, Tarwater CE, Foster JT et al (2019) Structure, spatial dynamics, and stability of novel seed dispersal mutualistic networks in Hawai'i. Science 364:78–7+

Vogel ER, Knott CD, Crowley BE et al (2012) Bornean orangutans on the brink of protein bankruptcy. Biol Lett 8:333–336

Wang ZX, Sun FK, Wang JD et al (2019) New fossil leaves and fruits of Lauraceae from the Middle Miocene of Fujian, southeastern China differentiated using a cluster analysis. Hist Biol 31:581–599

Wang K, Tian SL, Galindo-Gonzalez J et al (2020a) Molecular adaptation and convergent evolution of frugivory in Old World and neotropical fruit bats. Mol Ecol 29(22):4366–4381

Wang ZY, Wang B, Yan C et al (2020b) Neighborhood effects on the tannin-related foraging decisions of two rodent species under semi-natural conditions. Integr Zool 15(6):569–577

Welbergen JA, Meade J, Field HE et al (2020) Extreme mobility of the world's largest flying mammals creates key challenges for management and conservation. BMC Biol 18:101

Wenny DG (2001) Advantages of seed dispersal: a re-evaluation of directed dispersal. Evol Ecol Res 3:51–74

Wong ST, Servheen C, Ambu L, Norhayati A (2005) Impacts of fruit production cycles on Malayan sun bears and bearded pigs in lowland tropical forest of Sabah, Malaysian Borneo. J Trop Ecol 21:627–639

Wright SJ, Carrasco C, Calderon O, Paton S (1999) The El Niño Southern Oscillation variable fruit production, and famine in a tropical forest. Ecology 80:1632–1647

Xiang YZ, Huang CH, Hu Y et al (2017) Evolution of Rosaceae fruit types based on nuclear phylogeny in the context of geological times and genome duplication. Mol Biol Evol 34:262–281

Yao X, Song Y, Yang JB et al (2021) Phylogeny and biogeography of the hollies (*Ilex* L., Aquifoliaceae). J Syst Evol 59:73–82

Zungu MM, Downs CT (2016) Digestive efficiencies of Cape white-eyes (*Zosterops virens*), red-winged starlings (*Onychognathus morio*) and speckled mousebirds (*Colius striatus*) fed varying concentrations of equicaloric glucose or sucrose artificial fruit diets. Comp Biochem Physiol A Mol Integr Physiol 199:28–37

7

Plant-Mediated Above-Belowground Interactions: A Phytobiome Story

*Frédérique Reverchon
and Alfonso Méndez-Bravo*

Contents

© Springer Nature Switzerland AG 2021
K. Del-Claro, H. M. Torezan-Silingardi (eds.), *Plant-Animal Interactions*, https://doi.org/10.1007/978-3-030-66877-8_8

Learning Objectives

After reading this chapter, you should have a better understanding of the following topics:

- The concept of "phytobiome" and its implication for the study of plant-mediated above- belowground interactions
- The mechanisms through which plant-associated above- and belowground organisms influence each other
- The importance of the plant induced systemic resistance (ISR) in mediating above- belowground interactions
- The role of the "soil legacy effect" in shaping the plant rhizosphere microbiome and affecting the performance of aboveground herbivores
- The potential of manipulating the phytobiome to enhance plant performance

8.1 Introduction

Plants are an important link between the aboveground and belowground compartments as their roots and aerial parts connect both habitats (Bezemer and van Dam 2005). They also interact with a wide range of organisms both above- and belowground that can influence each other, even though they do not come into direct contact (Heil 2011). Together, the plant and its interacting organisms form the so-called "phytobiome" (Leach et al. 2017). Interactions within the phytobiome may be beneficial, neutral or harmful for the host plant (Pieterse and Dicke 2007; Méndez-Bravo et al. 2018). Beneficial plant-microbe relationships, especially at the rhizosphere level, are important for plant health and productivity as they enhance nutrient acquisition, promote plant growth through the production of phytohormones, or act as a protection against phytopathogens (Philippot et al. 2013; Báez-Vallejo et al. 2020). As a result, they alter the plant fitness and physiology, thereby affecting the performance of plant-associated aboveground organisms (Pineda et al. 2017; Heinen et al. 2018a). On the other hand, harmful plant herbivores and pathogens induce defense responses that are often systemic, thus modifying the whole plant nutritional and defense levels and subsequently influencing other plant-associated organisms both above- and belowground (Bernaola et al. 2018). Furthermore, pest attacks or pathogen infections influence the quantity and composition of root exudates, which induces shifts in root-associated microbes (Bais et al. 2006). Plant-associated organisms therefore have the potential to form complex interactions that are mediated through nutrient cycling and a wide array of signaling compounds (Leach et al. 2017).

Belowground organisms can affect aboveground insects through a chain of molecular and chemical reactions in the host plant (Pineda et al. 2017) that will culminate in the modification of the plant phenotype. Root herbivores, for instance, have been reported to modify flower sizes, numbers and flowering period, thus affecting flower visiting insects and pollinators (Poveda et al. 2005). Furthermore, root herbivores also alter the performance of aboveground herbivores, although it is not yet clear whether their influence is positive or negative. The stress response induced by root herbivory may either result in an accumulation of defense secondary metabolites in the plant aerial tissues, hence repelling aboveground herbivores, or in increased leaf carbohydrate and nitrogen contents, which enhances plant palatability (Bezemer et al. 2002; Poveda et al. 2005). Soil microbes, in turn, may promote plant growth, enhance its survival under harsh environmental conditions, and induce plant immunity by triggering systemic resistance against biotic stressors, including aboveground insects and pathogens (Van Wees et al. 2008; Pineda et al. 2010). However, their effect on aboveground herbivores seems to be largely species-dependent (Bezemer et al. 2005) and to be indirectly controlled by responses of the host plant metabolome to soil microbes (Badri et al. 2013), which calls for more studies looking at unravelling the complex role of the soil microbiome in shaping plant-insect interactions.

Recently, an increasing body of literature has been aiming at deciphering the plant-mediated mechanisms through which soil organisms influence aboveground insects. Evidences seem to point at three main pathways. First, soil biota can modify plant nutritional value and tissue quality (Bezemer and van Dam 2005) which in turn will affect the performance of aboveground insects, either positively through an accumulation of carbohydrates or nitrogen, or negatively through increased content of plant secondary defense metabolites (De Deyn and van der Putten 2005; Heinen et al. 2018a). Secondly, soil microorganisms can enhance the immune system of the plant by inducing its systemic resistance, thus affecting shoot defense levels, or by modifying its emission of volatile compounds that enable the plant to attract herbivore natural enemies (Bezemer and van Dam 2005). Finally, soil microbes also alter plant performance by modifying functional traits such as plant biomass, chemical composition, flower size and stamen number (De Deyn and van der Putten 2005; Heinen et al. 2018a), thereby affecting the behavior of herbivores, parasitoids and flower visiting insects.

Plant interactions with aboveground insects can also have an effect on soil organisms and belowground processes. Aboveground herbivores have been reported to modify plant nutrient and defensive compound contents (Bezemer and van Dam 2005), alter the quantity and quality of root exudates and carbon allocation which results in changes in plant biomass and physiology, and by influencing plant litter inputs to the soil (Bardgett et al. 1998). As herbivores modify leaf nutrient and secondary metabolite concentrations, these changes in leaf litter quality are likely to produce shifts in the soil biota communities. These top-down effects appear to be dependent on the herbivore feeding strategy. For example, plant parasitic nematodes have been shown to be positively affected by leaf-chewing insects while sap-feeding insects seem to reduce their abundance (Hoysted et al. 2018).

The objective of the present chapter is to review the current information on plant-mediated above- belowground interactions, in order to provide a more holistic view of the complex biological interactions occurring within the phytobiome. Such interactions influence plant health and productivity at the local level and therefore have great significance for ecological processes at the ecosystem scale. We will mainly focus on interactions involving soil microbes, as the advent of new sequencing technologies has allowed for a more in-depth understanding of the soil microbiome effect on plant – aboveground herbivore interactions. Above- belowground interactions involving earthworms, arthropods and nematodes have been thoroughly reviewed elsewhere (Wondafrash et al. 2013; Biere and Goverse 2016; Wurst et al. 2018). In this chapter, we will address these finely-tuned plant-mediated bottom-up and top-down interactions and review the current knowledge on the mechanisms regulating the networks connecting aboveground and belowground phytobiome members (◘ Fig. 8.1).

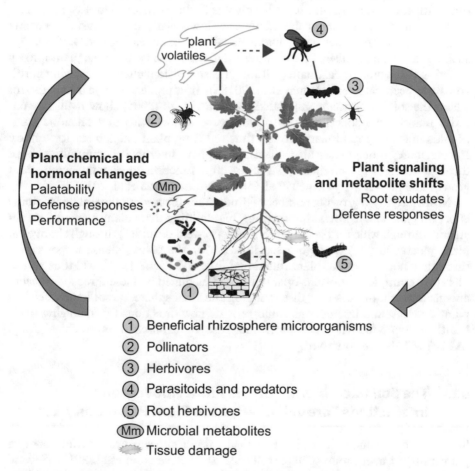

◘ **Fig. 8.1** An overview of plant-mediated above- belowground interactions. In bottom-up interactions, the soil microbiome affects plant aboveground interactions with beneficial and herbivore insects by inducing changes in plant palatability and performance. Diffusible and volatile metabolites produced by beneficial rhizosphere microbes (plant growth-promoting bacteria and fungi, and arbuscular mycorrhizal fungi) induce systemic resistance (ISR), mediated mainly by jasmonic acid and ethylene-dependent signaling pathways. The activation of ISR positively modifies pollinator visitations and attraction of foliar insects' natural enemies by increasing plant volatile emissions. Damage caused by root herbivores affects the floral visiting rates by pollinators and incidence of leaf herbivores. In top-down interactions, sap-sucking and leaf-chewing insects modify the diversity and functions of soil microorganisms by inducing plant defense responses and modifying the composition of root exudates. Intermittent arrows indicate plant-mediated impacts

8.2 Bottom-Up: Belowground Plant Interactions Influence Aboveground Plant-Associated Insects

It is becoming increasingly clear that plant-associated belowground organisms have a profound influence on plant-insect interactions at the aboveground level. The first evidences came from studies investigating the role of root herbivores, decomposers or parasites in modifying the behavior of leaf herbivores or pollinators, with contrasting results. Root herbivores and earthworms, for example, have been shown to increase aphid abundance and flower visits in wild mustard (Poveda et al. 2005). On the other hand, plant-parasitic nematodes negatively affect sap-sucking insects, most likely by triggering a defense response from the plant and by affecting tissue nutritional quality, although no consistent effect on chewing herbivores has been detected (Heinen et al. 2018a). In a pioneer one-year experiment where the soil biota was manipulated, Bezemer et al. (2005) showed that nematodes reduced the concentrations of nitrogen, amino acids and phenolic compounds in aboveground plant tissues, thus reducing plant palatability for aphids. Furthermore, root grazers also affect rhizosphere microbial communities, both through changes in community structure and function, thus influencing plant growth and productivity (Leach et al. 2017; Medina-Sauza et al. 2019).

More recently, a growing number of studies has focused on the soil microbiome effect on plant aboveground interactions, highlighting three plant-mediated mechanisms through which these bottom-up effects could occur: (1) through changes in plant palatability, (2) through a sensitization of the plant defense response, (3) through a modification of plant functional traits (■ Fig. 8.1). We will thus review literature examples of bottom-up interactions mediated by these three main mechanisms and will discuss how disentangling the rhizosphere microbiome effect on plant aboveground interactions could provide new directions for managing plant health and productivity. The particular case of arbuscular mycorrhizal fungi (AMF) will also be addressed.

8.2.1 The Soil Microbiome Affects Plant Aboveground Interactions Through Changes in Plant Palatability

Soil microbial communities play a critical role in nutrient cycling and hence in plant nutrient acquisition (Chaparro et al. 2012; Reverchon et al. 2015). Several studies have reported that, by modifying plant nutrient content and palatability, soil microbes indirectly affect aboveground insects. For example, soil fungal communities under *Senecio jacobaea* influenced amino-acid concentrations in the phloem sap, thereby affecting the performance of the aphid *Aphis jacobaea* (Kos et al. 2015a). Hol et al. (2010) also reported that plant nutritional quality was influenced by soil microbial community assemblages, and especially by the presence of rare taxa. These authors showed that a reduction in rare microbes increased the concentration of nitrogen, sugar and amino-acids in plant tissues and promoted aphid performance, whilst plants grown in soils where rare microbes were maintained had higher levels of defense compounds. These findings are important because they emphasize the relevance of rare soil microbes as an insurance value for the capacity of the soil microbiome to mitigate the impact of aboveground herbivores (Hol et al. 2010).

The soil microbiome has been shown to alter the whole leaf metabolome. Badri et al. (2013) reported that different soil microbiomes distinctively influenced the production of phenolic compounds, amino acids and carbohydrates in leaves of *Arabidopsis thaliana* and partially determined the feeding behavior of insects aboveground. They found that leaf amino-acid concentration was positively correlated to the degree of insect herbivory and recommended to further investigate the potential of bacterial taxa such as *Balneimonas*, *Skermanella* and *Nocardioides* to mitigate the impact of herbivores, as they were associated with low leaf tissue amino-acid concentrations (Badri et al. 2013). However, plant palatability and attractiveness to herbivore insects not only depend on tissue concentrations of amino acids and sugars, but also rely on their secondary metabolite content, particularly those associated with the plant defense response. Glucosinolates and their breakdown products, for instance, have been associated with a loss in leaf tissue palatability to generalist herbivores, although not to specialist herbivores (Mosleh Arany et al. 2008). By enhancing the plant defense response and thus the concentration of defense compounds in the leaves, beneficial symbionts such as nitrogen-fixing bacteria reduced plant palatability for insects as shown for the Mexican bean beetle and for pavement ants in lima bean (Thamer et al. 2011; Godschalx et al. 2015). Altogether, these results demonstrate the potential of the rhizosphere microbiome to induce changes in the leaf metabolome, hence affecting plant attractiveness to insect herbivores (■ Fig. 8.1). Recent advances in environmental metabolomics will help us understand the plant metabolome response to changes in the soil microbiome and its effect on plant-associated insects (Leach et al. 2017), and achieve a better integration of the interactions existing within the phytobiome.

8.2.2 The Soil Microbiome Affects Plant Aboveground Interactions Through an Induction of Plant Defense

One of the best known beneficial effects of rhizosphere microorganisms on the plant aerial tissues is the systemic triggering of defensive responses against biotrophic and necrotrophic pathogens (Pieterse et al. 2001; Schuhegger et al. 2006), mainly induced by disease-suppressive and plant growth-promoting rhizobacteria (PGPR; Van Loon et al. 1998). This conferred resistance is called rhizobacteria-mediated induced systemic resistance (ISR; Van Loon et al. 1998), and it has been proven to be effective against different pathogens and pests, including bacteria, fungi, viruses and herbivore insects (as reviewed by Pieterse et al. 2014).

The first inducible defense mechanism described in plants that renders distant, uninfected tissues resistant to pathogens, is named systemic acquired resistance (SAR; Ross 1961). SAR was initially described as a response to viruses and necrotizing pathogens (Kuć 1982; Malamy et al. 1990); since then, this state of pathogen-induced resistance has been described for a broad spectrum of plant-pathogen interactions (Klessig et al. 2018; Maruri-López et al. 2019). The triggering of immunity by SAR involves the early contact of the host plant with pathogen-derived molecules (pathogen-associated molecular patterns; PAMPs), and the accumulation of the plant hormone salicylic acid (SA), which orchestrates the concomitant amplification of the signal through the whole plant and the activation of downstream defense responses, including the systemic transcriptional activation of pathogenesis-related genes; these genes encode products that frequently possess some anti-microbial activity, i.e. chitinases and glucanases (Pieterse et al. 2001, 2012). It has been extensively reported that the SA-dependent SAR signaling suppresses the expression of genes that are responsive to another defense-related plant hormone, the jasmonic acid (JA; Caarls et al. 2015); however, this hormonal signaling crosstalk is complex, and recent evidences showed that SA and JA could either act in antagonistic or synergistic ways (Pieterse et al. 2014; Caarls et al. 2015), depending on the taxa involved in the plant-pathogen interactions.

The ISR triggered by beneficial microbes such as PGPR and *Trichoderma* spp., requires the systemic induction of JA-responsive signaling pathways to confer resistance against leaf pathogens and insect herbivores (Pineda et al. 2010; Contreras-Cornejo et al. 2021). The induction of ISR is elicited not only by Microbe-Associated Molecular Patterns (MAMPs) and soil diffusible molecules derived from microbes, but also by microbial volatile compounds that stimulate the synergistic effect of the JA and the ethylene signaling-dependent pathways (JA/ethylene; Pineda et al. 2017). Similarly, foliar herbivores such as generalist caterpillars and phloem-sucking insects activate differential molecular responses that are dependent on JA/ethylene and SA signaling pathways, respectively (◘ Fig. 8.1; Rodriguez-Saona et al. 2005).

Bacillus and *Pseudomonas* spp. comprise most of the PGPR studied species that are able to induce ISR and prime leaf tissues for anti-herbivore responses (Pineda et al. 2010). *Bacillus* spp. have been shown to be effective at enhancing parasitoid attraction to caterpillars, but they have also been found to increase tomato susceptibility to the phloem-feeding white fly *Bemisia tabaci* (Shavit et al. 2013). Contrastingly, *Bacillus subtilis* has also been reported to induce tomato resistance against *B. tabaci* through a JA-independent ISR response (Valenzuela-Soto et al. 2010). The PGPR *Pseudomonas simiae* (*P. fluorescens*) induced an ISR response in Arabidopsis, conferring resistance against the leaf-chewing herbivore *Mamestra brassicaea* through JA/ethylene-dependent signaling responses (Pangesti et al. 2016), and increased the attraction of the wasp parasitoid *Microplitis mediator* (Pangesti et al. 2015). Conversely, *Pseudomonas fluorescens* was shown to interfere with the attraction of parasitoids to the aphid *Myzus persicae,* also in *A. thaliana* (Pineda et al. 2013). It is important to point out that these reports have been focusing on the impact of individual microbial species or strains on aboveground insects, and have been mainly using plant model species to investigate bottom-up ecological interactions. Although such studies have been fundamental for establishing the mechanisms that govern the influence of belowground microorganisms on the plant defense responses against aboveground insects, it is now necessary to broaden our research scope and consider the complexity of the highly diverse soil microbiomes. This shift of focus is required in order to redirect the paradigm of the protective role of PGPR towards an extended microbiome-mediated immunity approach (as elegantly discussed in Pineda et al. 2017, Heinen et al. 2018a and Vannier et al. 2019).

Beneficial fungal species have also been described to modify plant interactions with aboveground insects. Different root-colonizing *Trichoderma* species have been reported to act as plant growth-promoting fungi (PGPF) and to mitigate the damage caused by herbivore insects under controlled experimental conditions, through the induction of volatile emissions and JA accumulation by the plant (Battaglia et al. 2013; Contreras-Cornejo et al. 2018). In tomato plants, for example, *Trichoderma longibrachiatum* affected the performance of the aphid *Macrosiphum euphorbiae* and increased its attractiveness for its natural enemies, the parasitoid *Aphidius ervi* and the predator *Macrolophus pygmaeus* (Battaglia et al. 2013). *Trichoderma atroviridae* promoted foliar herbivory resistance against the fall armyworm *Spodoptera frugiperda* in maize (*Zea mays*), by inducing the accumulation of JA and the emission of volatile terpenes (Contreras-Cornejo et al. 2018). Interestingly, a field study performed in an agroecosystem showed that the interaction of maize roots with *T. harzianum* modified the foliar community of arthropods, increasing the abundance of pest-regulating insects and decreasing the number of piercing-sucking herbivores (Contreras-Cornejo et al. 2021). In conclusion, rhizosphere inoculation with *Trichoderma* spp. impacts positively the plant responses to herbivore arthropods throughout the whole food web; however, it is important to extend our understanding of aboveground insects - *Trichoderma* interactions under natural conditions and to deeply explore the ecological functions of PGPF within the rhizosphere microbiome.

8.2.3 The Soil Microbiome Affects Plant Aboveground Interactions Through Changes in Plant Performance

The recent scientific literature is full of reports demonstrating the importance of soil microbial communities for plant performance and provides growing evidence that plants can modulate their root microbiome to sustain their growth, health and productivity (Berendsen et al. 2012; Kwak et al. 2018; Liu et al. 2019a). Diverse soil microbial communities have been shown to improve plant growth, vigor and fitness traits such as chlorophyll content, flower size, flowering time and seed production (Marschner and Rumberger 2004; Lau and Lennon 2011; Chaney and Baucom 2020). These traits are key in mediating plant-insect interactions aboveground. For example, complex soil microbial communities were associated to larger plant size and larger flower numbers in *Ipomoea purpurea* and influenced the selection on flowering time, evidencing that the soil microbiome affects both plant phenotype and fitness (Chaney and Baucom 2020). As flowering time is tightly linked with pollinator visits, changes in flowering time may decrease plant fitness through pollen limitation or a reduction in gene flow (Elzinga et al. 2007). Plant – herbivore interactions also depend on plant performance. Plant vigor, which is also influenced by belowground microbial communities (Middleton et al. 2015), has been generally associated with an increased abundance of herbivores (Price 1991; Cornelissen et al. 2008).

The importance of pollinators for plant reproduction, population dynamics and evolution of floral traits (Rosas-Guerrero et al. 2014; De Santiago-Hernández et al. 2019), added to the major decline that pollinators have been suffering worldwide (Potts et al. 2010), have prompted the need to investigate the effects of soil organisms on plant - pollinators interactions. Studies have mainly focused on the influence of root herbivores on pollinator visitation, as root herbivores strongly impact plant growth and fitness (Barber et al. 2011, 2015) and, in some cases, can induce a reduction in flower production (Barber and Soper Gorden 2015). Such negative effects on floral traits suggest that root herbivory may decrease pollinator visitation, which was confirmed by Ghyselen et al. (2016) who determined that root herbivory of *Cynoglossum officinale* by the root-feeding weevil *Mogulones cruciger* negatively affected flower numbers and pollinator visitation. However, this negative effect of root herbivory did not lead to pollen limitation. Positive effects of root herbivory on plant-pollinator interactions have also been reported. For example, root herbivory by wireworms and larvae of the striped cucumber beetle paradoxically increased *Apis mellifera* attraction to wild mustard and cucumber respectively, although the underlying mechanisms of this attraction are still unknown (Poveda et al. 2003; Barber et al. 2015). The effects of AMF on plant - pollinators interactions have also been widely studied and will be addressed below.

Soil microbes can also enhance plant performance through the so-called soil legacy effect, whereby a plant specific microbiome influences the performance of plants growing later in the same soil (van der Putten et al. 2013; Heinen et al. 2018a). Through this plant-soil feedback, a plant induces shifts in the soil microbial community structure and composition which will later affect the growth and development of other plants in that same soil, and ultimately the interactions that these plants will have with aboveground insects (Kos et al. 2015a). Since soil organisms are able to modify the composition and concentration of plant metabolites and defense compounds in the foliage (◘ Fig. 8.1; Etalo et al. 2018; Huberty et al. 2020), changes in soil microbial communities produced by plant-soil feedbacks are likely to alter the performance of aboveground organisms feeding on these newly-established plants (Kos et al. 2015a; Heinen et al. 2018a). An ecological loop can therefore take place through which herbivory will impact the soil microbiome and, via a soil legacy effect, will affect the resistance of plants subsequently growing in that soil (Kostenko et al. 2012; Howard et al. 2020). Several examples of soil legacy effects on aboveground herbivore feeding can be found in the literature. For instance, Heinen et al. (2018b) showed that distinct plant species left different soil legacies, with subsequent contrasting effects on the chewing herbivore *Mamestra brassicae*. Further evidence was provided by Bezemer et al. (2006) who showed that ragwort (*Senecio jacobaea*) plants growing in soils with a ragwort legacy were less prone to be attacked by stem borers, leaf miners and flower feeders than ragwort plants growing in other soils. Using ragwort as well, Kos et al. (2015a, b) demonstrated that the performance of the specialist *Aphis jacobaea* was influenced by the type of plant used to condition the soil, while the generalist aphid *Brachycaudus cardui* appeared to be less sensitive to soil conditioning, regardless of the plant species used to produce the plant-soil feedback. No effect of soil legacy on generalist snail performance and on plant tolerance to herbivory by banded snails was also reported by Schittko and Wurst (2014). These contrasting findings indicate that the effects of plant-soil feedbacks on aboveground plant-insect interactions will depend on the combination of plants, soils and insects involved in the interaction (Kaplan et al. 2018).

8.2.4 Arbuscular Mycorrhizal Fungi: Crucial Players in Modulating Plant Aboveground Interactions

The AMF have been amongst the most studied belowground plant symbionts due to their obligate biotrophic associations with the roots of 80% of terrestrial plants, and thus with those of most economically important crops (Smith and Read 2008). AMF are known to provide key ecosystem services such as soil fertility and resistance to erosion, and to boost plant health, fitness and nutrition (Larsen et al. 2003; Gianinazzi et al. 2010). As crucial members of the plant phytobiome, AMF are able to induce metabolic changes in their host plant, modifying primary and secondary metabolisms, and nutrient and defense compound concentrations (Korenblum and Aharoni 2019; Real-Santillán et al. 2019), which subsequently influences interactions between plants and aboveground insects at several trophic levels (Bell et al. 2019).

The plant-mediated effect of AMF on pollinating insects has been reported to be positive in several studies, as AMF inoculation increased pollinator visitation in three species of annual plants (Gange and Smith 2005) and in the perennial *Chamerion angustifolium* (Wolfe et al. 2005). Moreover, AMF suppression at the field level, in a Canadian grassland, reduced flower visitation and induced shifts in pollinator populations from large-bodied bees to small-bodied bees and flies (Cahill et al. 2008). This observed positive influence of AMF on pollinator visitations is most likely mediated by AMF-induced changes in floral traits, such as increased pollen and nectar production (Gange and Smith 2005; Pereyra et al. 2019), increased flower size and number due to a greater access to phosphorus (Gange and Smith 2005; Varga and Kytöviita 2010; Aguilar-Chama and Guevara 2012; Barber and Soper Gorden 2015), and increased seed production and germination (Poulton et al. 2002; Bennett and Meek 2020), although these latter variables should be more consistently evaluated. However, although most studies report a positive effect of AMF on pollinator visitation, some neutral or negative impacts have also been described (Varga and Kytöviita 2010; Barber et al. 2013), showing that the influence of AMF on pollination largely depends on the pollinator taxa. For example, Barber et al. (2013) reported that bumblebees and Lepidoptera were attracted to AMF-inoculated *Cucumis sativus* (Cucurbitaceae) while honey bees tended to prefer AMF-free control plants. The host plant species, gender and level of dependency on AMF also seem to be crucial in mediating the effect of AMF on pollinators (Gange and Smith 2005; Barber and Soper Gorden 2015; Heinen et al. 2018a; Bennett and Meek 2020). Overall, while AMF inoculation appears to promote pollinator visitation through an increase in attractive floral traits, further studies are required to understand the impact on AMF on plant fitness (Barber and Soper Gorden 2015). These future studies should consider the contribution of different AMF species to plant fitness, their effect on different pollinator taxa, and their influence on parameters directly associated with plant reproduction, such as pollen germination, pollen tube growth, fertilization, and seed germination (Bennett and Meek 2020). Furthermore, as increases in nectar production can be associated, in some cases, with increases in herbivory (Adler and Bronstein 2004), the AMF effect on pollinators should be studied in the wider context of aboveground plant-insect interactions (◘ Fig. 8.2).

AMF are also known to influence the fitness and behavior of herbivores. By enhancing water and nutrient acquisition by their host plant, AMF modify plant biomass, vigor and nutrient concentration in the plant tissues, hence affecting plant quality for aboveground insect herbivores (Bennett et al. 2006; Real-Santillán et al. 2019). Moreover, AMF have been shown to promote plant defense responses to biotic stresses such as those induced by herbivores or plant pathogens (Campos-Soriano et al. 2012; Selvaraj et al. 2020). Interestingly, AMF have also been reported to alleviate the existing trade-off between plant growth and defense (Vannette et al. 2013). However, studies aiming at assessing the effects of the plant

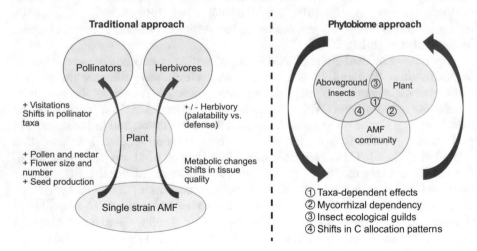

◘ **Fig. 8.2** Traditional vs. phytobiome approaches for the study of plant-mediated AMF and aboveground insect interactions. **a** The traditional approach usually considers single strains of AMF and their impact on either herbivores or pollinators. Most studies report positive effects of AMF on pollinators, mediated by AMF-induced changes in floral traits. AMF effects on herbivores, however, may either be positive or negative, as AMF can modify the nutritional status or the concentration of defense compounds in plants. **b** Our suggested phytobiome approach integrates bottom-up and top-down interactions, which could result in a feedback loop between aboveground and belowground organisms. The outcome of interactions within the phytobiome largely depends on the interacting insect, plant and AMF taxa, on the plant gender and level of dependency on AMF, on the insect ecological guild and feeding strategy, and on the insect-induced shifts in belowground C allocation by the plant

mycorrhizal status on insect herbivores have reported varying results, depending on the identity of the AMF, host plant and insects that were involved in the interaction. For example, several reports and a meta-analysis demonstrated an overall negative effect of AMF on generalist herbivores, most likely as a result of the activation of plant defenses by AMF (Pozo and Azcón-Aguilar 2007; Hartley and Gange 2009; Koricheva et al. 2009; Kaur et al. 2020), although some exceptions were reported (Bernaola et al. 2018; Real-Santillán et al. 2019). Specialist herbivores, however, may be immune to such defense compounds or may necessitate them, as they generally are positively impacted by AMF inoculation (Koricheva et al. 2009; Bernaola et al. 2018). Herbivore feeding strategies also seem to determine the outcome of the plant-mediated AMF influence. Sucking insects feeding on phloem are globally positively affected by AMF, probably due to the absence of defense compounds in the phloem, whilst the performance of chewing and sucking insects feeding on cell contents, where such compounds usually concentrate, is known to be reduced in AMF-inoculated plants (Pozo and Azcón-Aguilar 2007; Koricheva et al. 2009; Heinen et al. 2018a). The negative effect of AMF on chewing insects is also attributed to the susceptibility of these insects to JA-dependent defense responses, which are primed by AMF (Jung et al. 2012). Resource availability may be another factor influencing the balance of AMF effects on aboveground herbivores, as suggested by Aguilar-Chama and Guevara (2016). These authors demonstrated that under light-limited conditions, no benefits were obtained by *Datura stramonium* from its mycorrhizal symbionts but that the plant tolerated the cost of mycorrhizal interactions. The higher cost of mycorrhizal interactions in resource-limited environments may thus affect the outcome of plant-insect interactions aboveground, by preventing the plant to invest in defense mechanisms against herbivores for example. Several studies also showed that different AMF mixtures, species or even strains may produce different effects on herbivory (Bennett et al. 2009; Barber et al. 2013; Roger et al. 2013). Finally, future works should focus on how different AMF species, used as single inoculum or in consortia, may trigger distinct plant defense mechanisms, and on their subsequent impacts on the behavior of herbivore insects from different feeding guilds. Furthermore, top-down effects should also be integrated into the study of ecological interactions involving AMF and aboveground insects within the phytobiome (▣ Fig. 8.2), as recent evidence suggests that insect herbivory may drastically impact AMF through a decrease in C allocation (Charters et al. 2020). Nevertheless, the impact of insect herbivory on AMF communities still remains largely unknown.

8.2.5 How to Reshape Soil Microbiomes to Enhance Crop Resistance to Aboveground Insects?

Beyond the generation of scientific knowledge, the understanding of the ecological interactions occurring within the phytobiome could also provide innovative tools for integrated pest management. Managing soil microbiomes to enhance plant resistance to soil-borne diseases and obtain disease-suppressive soils has already been proposed (Mendes et al. 2011) and is the subject of many recent studies (Ou et al. 2019; Shen et al. 2019; Barelli et al. 2020; Zhang et al. 2020). Introducing complex synthetic microbiomes to the soil, or steering the existing soil microbiomes to boost their benefits, could thus also be used to promote plant resistance to herbivory (Pineda et al. 2017). The addition of soil organic amendments such as manure or biochar has been shown to modify the soil microbiome and decrease the incidence of belowground pathogens (Mazzola et al. 2015; Gu et al. 2017), and could therefore influence aboveground herbivores by activating the plant systemic resistance.

Recent evidence shows that the soil legacy concept could be applied to steer the soil microbiome and increase the resistance of plants against aboveground insect herbivores (Pineda et al. 2020). By inoculating soils conditioned by wild grasses and forbs into a sterilized soil where chrysanthemum was grown, these authors showed that it was possible to shape the soil microbiome and induce plant resistance to damages caused by thrips (although not by mites). They concluded that the observed decrease in herbivory may be due to a priming of the plant defense response rather than a positive effect of the inoculated soil on plant performance. Manipulating the phytobiome by applying plant-soil feedbacks thus seems like a promising tool to design strategies aiming at reducing aboveground pest incidence, although further research is required to determine which conditioning plants could ensure the consistent obtention of a desirable soil microbiome (Pineda et al. 2020).

8.3 Top-Down: Aboveground Insects Influence the Rhizosphere Microbiome

The influence of aboveground herbivores on belowground organisms has traditionally been studied from the point of view of litter and detritus inputs by herbivores into the soil, and their impact on the soil decomposer community (Bardgett et al. 1997; Wardle et al. 2002; Chapman et al. 2003). Insect herbivores, by adding nutrients in the form of excrements, carcasses and green litterfall to the soil, accelerate the rates of organic matter decomposition by the soil microbial community (Ritzenthaler et al. 2018) and have been shown to enhance C and N cycling in forest ecosystems in a recent meta-analysis (Kristensen et al. 2020).

The seminal work by Bardgett et al. (1998) was one of the first to highlight the importance of plant-mediated herbivore impacts on belowground organisms and processes. The authors indicated two pathways through which these top-down plant-mediated effects could occur: (1) through changes in the patterns of root exudation and C allocation following herbivory; (2) through an induction of chemical defenses by herbivory, which would modify the quality of plant tissues, and hence of litter. In this section, we will review the current knowledge on both mechanisms, and will conclude by presenting a case study about the effects of ambrosia beetles on the avocado rhizosphere microbiome.

8.3.1 Aboveground Insects Influence the Soil Microbiome Through Changes in Root Exudation Patterns

It is well documented that plant species influence soil properties and soil microbial communities through the liberation of root exudates (Hobbie 1992; Bais et al. 2006; Sasse et al. 2018). These rhizodeposits, composed mainly by low-molecular-mass compounds, mucilage and root border cells, constitute a high-quality carbon source that sustains rhizosphere microorganisms and modulates the composition of rhizosphere microbial communities (Philippot et al. 2013; Carvalhais et al. 2015). Their quantity and composition vary greatly depending on environmental conditions, on the plant species, growth rate and nutrient status, and whether the plant is under attack by a pest or pathogen (Cesco et al. 2012; Reverchon et al. 2015; Vázquez-Ochoa et al. 2020). As reviewed in 1998 by Bardgett et al., early evidences showed that herbivory influences root exudation patterns. For example, grasshopper herbivory increased the release of organic acids from the roots of blue grama grass (*Bouteloua gracilis*) and C allocation to the roots of maize and of the African grass *Panicum coloratum* (Dyer and Bokhari 1976; Dyer et al. 1991; Holland 1996). Most of these early studies used C isotope labelling techniques to explain the mechanisms underlying the effects of herbivory on plant C allocation and root exudation, and hypothesized that plants may increase the amount of carbon allocated to root tissues after foliar damage as an investment in root growth (Bardgett et al. 1998). Contrastingly, Smith (1972) showed that defoliated *Acer saccharum* trees exuded less total carbon than intact trees, most likely due to their reduced photosynthetic capacity. Smith (1972) also demonstrated that complete defoliation affected the quantity but not the quality of sugar maple root exudates.

More recent works confirmed the alteration of root exudates following shoot herbivory and showed ensuing effects on root herbivores, such as nematodes. Without directly studying root exudate composition, Kim et al. (2016) showed that aphid infestation in pepper plants modified the community structure of rhizobacteria, suggesting that the observed shifts were due to the modulation of root exudation patterns by the insects. Wang et al. (2017) described an increase in root N concentration following defoliation by clipping in the grass *Holcus lanatus*, which subsequently induced an increase in the abundance of two species of root-feeding nematodes (Wang et al. 2017). Hoysted et al. (2018) demonstrated that herbivory by phloem-sucking aphids (*Myzus persicae*) affected the root exudates of potato, reducing their glucose and fructose contents, and induced a negative effect on egg hatching of the endoparasitic nematode *Globodera pallida*. Ultimately, the alteration of root exudates or root-emitted volatile compounds by aboveground herbivory may also modify the attraction of entomopathogenic nematodes and other predator organisms, thus disrupting the belowground food web (Bezemer and van Dam 2005). Aboveground herbivory has also been shown to promote the quantity of specific compounds such as malic acid, benzoxazinoids and strigolactones in the root exudates (Pineda et al. 2017); these compounds have been associated with the recruitment of beneficial microbial taxa at the rhizosphere level such as mycorrhizal fungi and the PGPR *Bacillus subtilis* and *Pseudomonas putida* (◻ Fig. 8.1;

Akiyama and Hayashi 2006; Rudrappa et al. 2008; Neal et al. 2012). The exudation of these compounds could thus constitute a defense mechanism through which the plant might reinforce its own protection. Mechanistical studies aiming at elucidating the effects of aboveground herbivory on root exudates are, however, still scarce. The advent of metabolomics now provides us with new tools to answer questions regarding the influence of herbivory on root exudate composition and could thus help us achieve a more thorough understanding of the complex interactions existing between plant roots and the soil microbiota (van Dam and Bouwmeester 2016).

8.3.2 Aboveground Insect Herbivores Influence the Soil Microbiome Through an Induction of Plant Defenses

The modification in the levels of defense compounds belowground as a result of aboveground herbivory has been reported by several authors, as reviewed by Bezemer and van Dam (2005). For example, foliar feeding by cabbage moth (*Mamestra brassicae*) reduced pyrrolizidine alkaloid levels in ragwort roots, which are compounds involved in the plant resistance against generalist herbivores and pathogenic root fungi (Hol et al. 2004). Defoliation of rye seedlings was shown to induce an increase in the root concentration of hydroxamic acid, an allelopathic compound involved in plant resistance to herbivores and pathogens and in the synthesis of siderophores for iron uptake, which may constitute a compensation mechanism of the defoliated plants to counteract their loss of fitness (Collantes et al. 1999). Contrastingly, the same authors previously reported a lack of effect of defoliation in maize on the root concentration of hydroxamic acid, suggesting maize might invest more in compensatory root growth than in defense upon an herbivore attack (Collantes et al. 1998). These differing responses from rye and maize were attributed to the C3 nature of rye and C4 nature of maize, as C4 plants represent a less attractive food source for herbivores than C3 plants, thus requiring a lesser investment in defense strategies. Similarly, no effect of herbivory by *S. exigua* larvae on cotton leaves was found on the root concentration of terpenoid aldehydes (Bezemer et al. 2004).

As previously discussed in this chapter, plant systemic defense mechanisms triggered by the rhizosphere microbiome affect plant aboveground interactions. The reverse is also true, as foliar-feeding insects induce shifts in the concentration of defense compounds at the root level, thereby influencing the root-associated microbiome and fauna (◘ Fig. 8.1; Bezemer and van Dam 2005; Pangesti et al. 2013). The systemic effect of aboveground herbivory on root hormone levels and defense responses has been reported in several studies and seems to be largely dependent on the herbivore feeding guild (Stam et al. 2014). For example, the chewing caterpillar *S. exigua* increased the belowground synthesis of lipids involved in the JA signaling pathway (Mbaluto et al. 2020) and decreased root SA content (Kafle et al. 2017) in tomato. The leaf chewer *Plutella xylostella* was also reported to slightly induce the primary root jasmonate pathway in cabbage plants (Karssemeijer et al. 2020). These herbivore-induced belowground plant responses subsequently influenced the performance of root feeders such as plant parasitic nematodes (Machado et al. 2018) and could thus also impact that of beneficial or detrimental rhizobacteria. This was described by Song et al. (2015), who reported an increase in SA content in the root exudates of whitefly-infested tobacco plants which consequently inhibited the virulence of the soil-pathogenic bacterium *Agrobacterium tumefaciens*. These findings show that aboveground plant damage by herbivores not only produces a host defense response but has also the potential to influence root-herbivore or root-pathogen induced defense mechanisms, hence modifying root-associated organisms. Further research on interactions between aboveground herbivores and belowground organisms is thus needed to improve our understanding of the ecological processes occurring within the phytobiome.

8.3.3 Case Study: Ambrosia Beetles and the Avocado Rhizosphere Microbiome

An interesting case of aboveground pest impact on the belowground plant-associated microbiome is that of ambrosia beetles. These beetles form a symbiosis with different fungi, which they carry in structures called mycangia, farm in the galleries they form within the trunk of their host plants and use as a food source for their larvae (Carrillo et al. 2014). Some of these fungal symbionts are important plant pathogens and therefore induce a systemic defense response which affects the composition and functions of the root-associated microbiota (Hulcr and Dunn 2011). Of particular interest are the invasive ambrosia beetles *Euwallacea* sp. nr. *fornicatus* (also known as Polyphagous Shot Hole Borer, PSHB) and *Euwallacea kuroshio* (Kuroshio Shot Hole Borer, KSHB), which are native of South-East Asia but have caused great damage to tree crops and forest ecosystems in North America, Israel and South Africa, among other countries (Freeman et al. 2013; Lira-Noriega et al. 2018; van den Berg et al. 2019). Their wide host range, approximately 58 plant families, represents a significant threat to natural forests, urban landscapes and fruit crops such as avocado (*Persea americana*) (O'Donnell et al. 2016; Guevara-Avendaño et al. 2018). Their fungal symbionts are known to cause Fusarium dieback, as the fungi inoculated by the beetles invade the host vascular tissues, blocking nutrient transport within the plant and leading to wilting, branch dieback, and in severe cases, tree mortality (Eskalen et al. 2013). As mentioned earlier in this chapter, the attack of the plant by a pest or pathogen produces a chain of systemic reactions which are likely to impact the plant rhizosphere microbiome, and Fusarium dieback in avocado is no exception.

Our research group is currently aiming at elucidating the impact of Fusarium dieback on the structure and function of the avocado rhizosphere microbiome, by comparing symptomatic and asymptomatic trees from an avocado orchard in Escondido, California. Our first results indicated that Fusarium dieback caused by *Fusarium kuroshium* and vectored by KSHB decreased the richness and diversity of microbial communities associated with the avocado rhizosphere, and significantly modified rhizobacterial community structure (Bejarano-Bolívar et al.

2021). Furthermore, this observed decrease in diversity was associated with the disappearance of taxa such as Armatimonadetes (at the *phylum* level), *Sporocytophaga* and *Cellvibrio* (at the genus level), which have been related to increases in plant growth and may prevent further fungal invasions through their chitin-degrading ability (Jaiswal et al. 2017; Ma et al. 2020). Contrastingly, some bacterial taxa were exclusively found in the rhizosphere of symptomatic trees, such as bacterial genera *Myxococcus* and *Lysobacter,* which have been described as potential biocontrol agents of fungal diseases (Bull et al. 2002; Liu et al. 2019b) and may have been recruited by the plant as a protection mechanism upon infection (◘ Fig. 8.3).

These shifts in the structure of rhizosphere bacterial communities of avocado after Fusarium dieback are associated with changes in the functions that these communities perform. Metatranscriptomic analyses revealed that the composition of the active rhizobacterial community in symptomatic trees is enriched in Gammaprotobacteria and Bacilli classes, including *Pseudomonas* and *Bacillus* genera (unpublished results); moreover, the metabolism of carbohydrates and complex lipids was strongly altered in the active rhizosphere community of symptomatic trees, correlating with the gene expression profile of plant-damaged tissues (◘ Fig. 8.3). These preliminary results point out the necessity to investigate the impact of aboveground pests using a phytobiome approach, as the alterations they

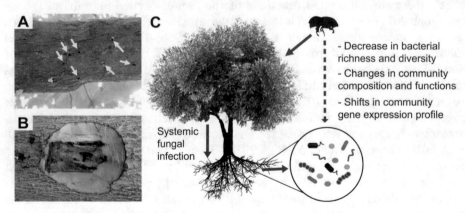

◘ **Fig. 8.3** Effect of Fusarium dieback on the avocado rhizosphere microbiome. **a** Entry points of Kuroshio Shot Hole Borer (KSHB) in the bark of an avocado branch (yellow arrows signal holes in the bark). **b** Brownish coloration signaling the systemic fungal infection caused by symbiont *Fusarium kuroshium*. **c** Plant-mediated impact (indicated by the intermittent arrow) of KSHB/*F. kuroshium* on the rhizosphere microbiome. Fusarium dieback caused by *F. kuroshium* decreases bacterial richness and diversity at the rhizosphere level, changes community structure and composition as evidenced by taxa exclusively found on the symptomatic condition, and induces functional changes as shown by the gene expression profile and identity of the active rhizobacterial community

Concluding Remarks

Over the last 30 years, a growing body of evidence has highlighted that plant aboveground and belowground interactions are closely intertwined and should consequently be studied within a phytobiome perspective. Although our understanding of the factors and mechanisms determining the outcomes of such interactions has greatly improved, some questions still remain to be addressed. Findings reported by studies on aboveground-belowground community ecology are usually correlative, which calls for experiments aiming at elucidating causal and mechanistic effects, at the genomic, transcriptomic, metabolomic and ecological levels. Furthermore, most studies investigating plant interactions with above- and belowground organisms have been carried out with a single insect species and a single strain of soil microbe. It is thus critical to go beyond individual interactions and consider the effects of more complex communities, both above (multi-species of pollinators and herbivores) and belowground (soil microbiome). Manipulating such complex communities is challenging, especially considering their dynamic nature, but crucially needed in order to elucidate which plant traits are important in mediating biotic interactions within the phytobiome. Scaling up the study of plant-mediated above-belowground interactions is also required for us to be able to reliably predict their effect on plant fitness, community structure and ecosystem services under natural settings.

produce may be found in the plant and in the soil microbiome, at molecular and ecological levels.

Key Points
- Plant-associated above- and belowground organisms influence each other through changes in the host plant metabolome and chemical communication.
- Soil microbes affect the performance of aboveground insects through shifts in plant tissue quality, through the induction of plant systemic resistance, and through alterations of plant functional traits.
- Aboveground herbivores modify the concentration of nutrient and defense compounds in plant tissue and alter root exudation patterns, thus affecting plant-associated soil organisms.
- Plant-soil feedbacks create ecological loops through which the performance of aboveground herbivores is affected, thereby further impacting the soil microbiome and the resistance of plants subsequently growing in that soil.
- The advent of new technologies such as metatranscriptomics and metabolomics will help us provide mechanistical explanations for the complex interactions existing within the phytobiome.

❓ Questions
- How could we use plant-soil feedbacks to manipulate aboveground plant-insect interactions?

- What is the concurrent impact of the rhizosphere microbiome on pollinators and aboveground herbivores?
- How can we design field experiments to elucidate causal relationships between aboveground and belowground plant interactions?

Acknowledgements We thank Kleber Del Claro and Helena Maura Torezan-Silingardi for inviting us to write this chapter and Wesley Dáttilo for recommending our contribution. We are grateful to Roger Guevara for his insightful comments that greatly improved this manuscript. We also thank Alejandro Méndez Bravo for his help with image editing. Finally, we thank the UC MEXUS –CONACYT program for funding some of our experimental research through a collaborative research grant (2015).

References

Adler LS, Bronstein JL (2004) Attracting antagonists: does floral nectar increase leaf herbivory? Ecology 85:1519–1526

Aguilar-Chama A, Guevara R (2012) Mycorrhizal colonization does not affect tolerance to defoliation of an annual herb in different light availability and soil fertility treatments but increases flower size in light-rich environments. Oecologia 168:131–139

Aguilar-Chama A, Guevara R (2016) Resource allocation in an annual herb: effects of light, mycorrhizal fungi, and defoliation. Acta Oecol 71:1–7

Akiyama K, Hayashi H (2006) Strigolactones: chemical signals for fungal symbionts and parasitic weeds in plant roots. Ann Bot 97:925–931

Badri DV, Zolla G, Bakker MG, Manter DK, Vivanco JM (2013) Potential impact of soil microbiomes on the leaf metabolome and on herbivore feeding behavior. New Phytol 198:264–273

Báez-Vallejo N, Camarena-Pozos DA, Monribot-Villanueva JL, Ramírez-Vázquez M, Carrión-Villarnovo GL, Guerrero-Analco JA, Partida-Martínez LP, Reverchon F (2020) Forest tree associated bacteria for potential biological control of *Fusarium solani* and of *Fusarium kuroshium*, causal agent of Fusarium dieback. Microbiol Res 235:126440

Bais HP, Weir TL, Perry LG, Gilroy S, Vivanco JM (2006) The role of root exudates in rhizosphere interactions with plants and other organisms. Annu Rev Plant Biol 57:233–266

Barber NA, Soper Gorden NL (2015) How do belowground organisms influence plant–pollinator interactions? J Plant Ecol 8:1–11

Barber NA, Adler LS, Bernardo HL (2011) Effects of above- and belowground herbivory on growth, pollination, and reproduction in cucumber. Oecologia 165:377–386

Barber NA, Kiers ET, Hazzard RV, Adler LS (2013) Context-dependency of arbuscular mycorrhizal fungi on plant-insect interactions in an agroecosystem. Front Plant Sci 4:338

Barber NA, Milano NJ, Kiers ET, Theis N, Bartolo V, Hazzard RV, Adler LS (2015) Root herbivory indirectly affects above-and below-ground community members and directly reduces plant performance. J Ecol 103:1509–1518

Bardgett RD, Leemans DK, Cook R, Hobbs PJ (1997) Seasonality in the soil biota of grazed and ungrazed hill grasslands. Soil Biol Biochem 29:1285–1294

Bardgett RD, Wardle DA, Yeates GW (1998) Linking above-ground and below-ground interactions: how plant responses to foliar herbivory influence soil organisms. Soil Biol Biochem 30:1867–1878

Barelli L, Waller AS, Behie SW, Bidochka MJ (2020) Plant microbiome analysis after *Metarhizium* amendment reveals increases in abundance of plant growth-promoting organisms and maintenance of disease-suppressive soil. PLoS One 15:e0231150

Battaglia D, Bossi S, Cascone P, Digilio MC, Duran Prieto J, Fanti P, Guerrieri E, Iodice L, Lingua G, Lorito M, Maffei ME, Massa N, Ruocco M, Sasso R, Trotta V (2013) Tomato below ground-above ground interactions: *Trichoderma longibrachiatum* affects the performance of *Macrosiphum euphorbiae* and its natural antagonists. Mol Plant Microbe Interact 26:1249–1256

Bell TH, Hockett KL, Alcalá-Briseño RI, Barbercheck M, Beattie GA, Bruns MA, Carlson JE, Chung T, Collins A, Emmett B, Esker P (2019) Manipulating wild and tamed phytobiomes: challenges and opportunities. Phytobiomes 3:3–21

Bennett AE, Meek HC (2020) The influence of arbuscular mycorrhizal fungi on plant reproduction. J Chem Ecol 46:707–721

Bennett AE, Alers-Garcia J, Bever JD (2006) Three-way interactions among mutualistic mycorrhizal fungi, plants, and plant enemies: hypotheses and synthesis. Am Nat 167:141–152

Bennett AE, Bever JD, Bowers MD (2009) Arbuscular mycorrhizal fungal species suppress inducible plant responses and alter defensive strategies following herbivory. Oecologia 160:771–779

Berendsen RL, Pieterse CMJ, Bakker PAHM (2012) The rhizosphere microbiome and plant health. Trends Plant Sci 17:478–486

Bernaola L, Cosme M, Schneider RW, Stout M (2018) Belowground inoculation with arbuscular mycorrhizal fungi increases local and systemic susceptibility of rice plants to different pest organisms. Front Plant Sci 9:1–16

Bejarano-Bolívar AA, Lamelas A, Aguirre von Wobeser E, Sánchez-Rangel D, Méndez-Bravo A, Eskalen A, Reverchon F (2021) Shifts in the structure of rhizosphere bacterial communities of avocado after Fusarium dieback. Rhizosphere. In Press. https://doi.org/10.1016/j.rhisph.2021.100333

Bezemer TM, van Dam NM (2005) Linking aboveground and belowground interactions via induced plant defenses. Trends Ecol Evol 20:617–624

Bezemer TM, Wagenaar R, van Dam NM, Wäckers FL (2002) Interactions between root and shoot feeding insects are mediated by primary and secondary plant compounds. Proc Section Exp Appl Entomol Neth Entomol Soc (NEV) 13:117–121

Bezemer TM, Wagenaar R, van Dam NM, van der Putten WH, Wäckers FL (2004) Above-and below-ground terpenoid aldehyde induction in cotton, *Gossypium herbaceum*, following root and leaf injury. J Chem Ecol 30:53–67

Bezemer TM, De Deyn GB, Bossinga TM, van Dam NM, Harvey JA, Van der Putten WH (2005) Soil community composition drives aboveground plant-herbivore-parasitoid interactions. Ecol Lett 8:652–661

Bezemer TM, Harvey JA, Kowalchuk GA, Korpershoek H, van der Putten WH (2006) Interplay between *Senecio jacobaea* and plant, soil, and aboveground insect community composition. Ecology 87:2002–2013

Biere A, Goverse A (2016) Plant-mediated systemic interactions between pathogens, parasitic nematodes, and herbivores above- and belowground. Annu Rev Phytopathol 54:499–527

Bull CT, Shetty KG, Subbarao KV (2002) Interactions between myxobacteria, plant pathogenic fungi, and biocontrol agents. Plant Dis 86:889–896

Caarls L, Pieterse CM, Van Wees S (2015) How salicylic acid takes transcriptional control over jasmonic acid signaling. Front Plant Sci 6:170

Cahill JF, Elle E, Smith GR, Shore BH (2008) Disruption of a belowground mutualism alters interactions between plants and their floral visitors. Ecology 89:1791–1801

Campos-Soriano L, Garcia-Martinez J, Segundo BS (2012) The arbuscular mycorrhizal symbiosis promotes the systemic induction of regulatory defence-related genes in rice leaves and confers resistance to pathogen infection. Mol Plant Pathol 13:579–592

Carrillo D, Duncan RE, Ploetz JN, Campbell AF, Ploetz RC, Peña JE (2014) Lateral transfer of a phytopathogenic symbiont among native and exotic ambrosia beetles. Plant Pathol 63:54–62

Carvalhais LC, Dennis PG, Badri DV, Kidd BN, Vivanco JM, Schenk PM (2015) Linking jasmonic acid signaling, root exudates, and rhizosphere microbiomes. Mol Plant Microbe Interact 28:1049–1058

Cesco S, Mimmo T, Tonon G, Tomasi N, Pinton R, Terzano R, Neumann G, Weisskopf L, Renella G, Landi L, Nannipieri P (2012) Plant- borne flavonoids released into the rhizosphere: impact on soil bio- activities related to plant nutrition. A review. Biol Fertil Soils 48:123–149

Chaney L, Baucom RS (2020) The soil microbial community alters patterns of selection on flowering time and fitness-related traits in *Ipomoea purpurea*. Am J Bot 107:186–194

Chaparro JM, Sheflin AM, Manter DK, Vivanco JM (2012) Manipulating the soil microbiome to increase soil health and plant fertility. Biol Fertil Soils 48:489–499

Chapman SK, Hart SC, Cobb NS, Whitham TG, Koch GW (2003) Insect herbivory increases litter quality and decomposition: an extension of the acceleration hypothesis. Ecology 84:2867–2876

Charters MD, Sait SM, Field KJ (2020) Aphid herbivory drives asymmetry in carbon for nutrient exchange between plants and an arbuscular mycorrhizal fungus. Curr Biol 30:1801–1808.e5

Collantes HG, Gianoli E, Niemeyer HM (1998) Changes in growth and chemical defences upon defoliation in maize. Phytochemistry 49:1921–1923

Collantes HG, Gianoli E, Niemeyer HM (1999) Defoliation affects chemical defenses in all plant parts of rye seedlings. J Chem Ecol 25:491–499

Contreras-Cornejo HA, Macías-Rodríguez L, del-Val E, Larsen J (2018) The root endophytic fungus *Trichoderma atroviride* induces foliar herbivory resistance in maize plants. Appl Soil Ecol 124:45–53

Contreras-Cornejo HA, Viveros-Bremauntz F, del-Val E, Macías-Rodríguez L, López-Carmona DA, Alarcón A, González-Esquivel CE, Larsen J (2021) Alterations of foliar arthropod communities in a maize agroecosystem induced by the root-associated fungus *Trichoderma harzianum*. J Pest Sci 94:363–374

Cornelissen T, Wilson Fernandes G, Vasconcellos-Neto J (2008) Size does matter: variation in herbivory between and within plants and the plant vigor hypothesis. Oikos 117:1121–1130

De Deyn GB, van der Putten WH (2005) Linking aboveground and belowground diversity. Trends Ecol Evol 20:625–633

De Santiago-Hernández MH, Martén-Rodríguez S, Lopezaraiza-Mikel M, Oyama K, González-Rodríguez A, Quesada M (2019) The role of pollination effectiveness on the attributes of interaction networks: from floral visitation to plant fitness. Ecology 100:e02803

Dyer MI, Bokhari UG (1976) Plant–animal interactions: studies of the effects of grasshopper grazing on blue grama grass. Ecology 57:762–772

Dyer MI, Acra MA, Wang GM, Coleman DC, Freckman DW, McNaughton SJ, Strain B (1991) Source-sink carbon relations in two *Panicum coloratum* ecotypes in response to herbivory. Ecology 72:1472–1483

Elzinga JA, Atlan A, Biere A, Gigord L, Weis AE, Bernasconi G (2007) Time after time: flowering phenology and biotic interactions. Trends Ecol Evol 22:432–439

Eskalen A, Stouthamer R, Lynch SC, Rugman-Jones PF, Twizeyimana M, Gonzalez A, Thibault T (2013) Host range of Fusarium dieback and its ambrosia beetle (Coleoptera: Scolytinae) vector in southern California. Plant Dis 97:938–951

Etalo DW, Jeon JS, Raaijmakers JM (2018) Modulation of plant chemistry by beneficial root microbiota. Nat Prod Rep 35:398–409

Freeman S, Sharon M, Maymon M, Mendel Z, Protasov A, Aoki T, Eskalen A, O'Donnell K (2013) *Fusarium euwallaceae* sp. nov.—a symbiotic fungus of *Euwallacea* sp., an invasive ambrosia beetle in Israel and California. Mycologia 105:1595–1606

Gange AC, Smith AK (2005) Arbuscular mycorrhizal fungi influence visitation rates of pollinating insects. Ecol Entomol 30:600–606

Ghyselen C, Bonte D, Brys R (2016) Is there a missing link? Effects of root herbivory on plant–pollinator interactions and reproductive output in a monocarpic species. Plant Biol 18:156–163

Gianinazzi S, Gollotte A, Binet MN, van Tuinen D, Redecker D, Wipf D (2010) Agroecology: the key role of arbuscular mycorrhizas in ecosystem services. Mycorrhiza 20:519–530

Godschalx AL, Schädler M, Trisel JA, Balkan MA, Ballhorn DJ (2015) Ants are less attracted to the extrafloral nectar of plants with symbiotic, nitrogen-fixing rhizobia. Ecology 96:348–354

Gu Y, Hou Y, Huang D, Hao Z, Wang X, Wei Z, Jousset A, Tan S, Xu D, Shen Q, Xu Y (2017) Application of biochar reduces *Ralstonia solanacearum* infection via effects on pathogen chemotaxis, swarming motility, and root exudate adsorption. Plant Soil 415:269–281

Guevara-Avendaño E, Carrillo JD, Ndinga-Muniania C, Moreno K, Méndez-Bravo A, Guerrero Analco JA, Eskalen A, Reverchon F (2018) Antifungal activity of avocado rhizobacteria against

Fusarium euwallaceae and *Graphium* spp., associated with *Euwallacea* spp. nr. *fornicatus*, and *Phytophthora cinnamomi*. Antonie van Leeuwenhoek 111:563–572

Hartley SE, Gange AC (2009) Impacts of plant symbiotic fungi on insect herbivores: mutualism in a multitrophic context. Annu Rev Entomol 54:323–342

Heil M (2011) Plant-mediated interactions between above- and below-ground communities at multiple trophic levels. J Ecol 99:3–6

Heinen R, Biere A, Harvey JA, Bezemer TM (2018a) Effects of soil organisms on aboveground plant-insect interactions in the field: patterns, mechanisms and the role of methodology. Front Ecol Evol 6:106

Heinen R, Sluijs M, Biere A, Harvey JA, Bezemer TM (2018b) Plant community composition but not plant traits determine the outcome of soil legacy effects on plants and insects. J Ecol 106:1217–1229

Hobbie SE (1992) Effects of plant species on nutrient cycling. Trends Ecol Evol 7:336–339

Hol WHG, Macel M, van Veen JA, van der Meijden E (2004) Root damage and aboveground herbivory change concentration and composition of pyrrolizidine alkaloids of *Senecio jacobaea*. Basic Appl Ecol 5:253–260

Hol WHG, de Boer W, Termorshuizen AJ, Meyer KM, Schneider JHM, van Dam NM, van Veen JA, van der Putten WH (2010) Reduction of rare soil microbes modifies plant-herbivore interactions. Ecol Lett 13:292–301

Holland JN (1996) Effects of above-ground herbivory on soil microbial biomass in conventional and no-tillage agroecosystems. Appl Soil Ecol 2:275–279

Howard MM, Kao-Kniffin J, Kessler A (2020) Shifts in plant–microbe interactions over community succession and their effects on plant resistance to herbivores. New Phytol 226:1144–1157

Hoysted GA, Bell CA, Lilley CJ, Urwin PE (2018) Aphid colonization affects potato root exudate composition and the hatching of a soil borne pathogen. Front Plant Sci 9:1–10

Huberty M, Choi YH, Heinen R, Bezemer TM (2020) Above-ground plant metabolomic responses to plant–soil feedbacks and herbivory. J Ecol 108:1703–1712

Hulcr J, Dunn RR (2011) The sudden emergence of pathogenicity in insect–fungus symbioses threatens naive forest ecosystems. Proc R Soc B 278:2866–2873

Jaiswal AK, Elad Y, Paudel I, Graber ER, Cytryn E, Frenkel O (2017) Linking the belowground microbial composition, diversity and activity to soilborne disease suppression and growth promotion of tomato amended with biochar. Sci Rep 7:44382

Jung SC, Martinez-Medina A, Lopez-Raez JA, Pozo MJ (2012) Mycorrhiza-induced resistance and priming of plant defenses. J Chem Ecol 38:651–664

Kafle D, Hänel A, Lortzing T, Steppuhn A, Wurst S (2017) Sequential above-and belowground herbivory modifies plant responses depending on herbivore identity. BMC Ecol 17:5

Kaplan I, Pineda A, Bezemer M (2018) Application and theory of plant–soil feedbacks on aboveground herbivores. In: Ohgushi T, Wurst S, Scott NJ (eds) Aboveground–belowground community ecology. Springer, Cham, pp 319–343

Karssemeijer PN, Reichelt M, Gershenzon J, van Loon J, Dicke M (2020) Foliar herbivory by caterpillars and aphids differentially affects phytohormonal signalling in roots and plant defence to a root herbivore. Plant Cell Environ 43:775–786

Kaur J, Chavana J, Soti P, Racelis A, Kariyat R (2020) Arbuscular mycorrhizal fungi (AMF) influences growth and insect community dynamics in Sorghum-sudangrass (*Sorghum x drummondii*). Arthropod Plant Interact 14:301–315

Kim B, Song GC, Ryu CM (2016) Root exudation by aphid leaf infestation recruits root-associated *Paenibacillus* spp. to lead plant insect susceptibility. J Microbiol Biotechnol 26:549–557

Klessig DF, Choi HW, Dempsey DMA (2018) Systemic acquired resistance and salicylic acid: past, present, and future. Mol Plant Microbe Interact 31:871–888

Korenblum E, Aharoni A (2019) Phytobiome metabolism: beneficial soil microbes steer crop plants' secondary metabolism. Pest Manag Sci 75:2378–2384

Koricheva J, Gange AC, Jones T (2009) Effects of mycorrhizal fungi on insect herbivores: a meta-analysis. Ecology 90:2088–2097

Kos M, Tuijl MAB, de Roo J, Mulder PPJ, Bezemer TM (2015a) Species-specific plant-soil feedback effects on above-ground plant-insect interactions. J Ecol 103:904–914

Kos M, Tuijl MA, de Roo J, Mulder PP, Bezemer TM (2015b) Plant–soil feedback effects on plant quality and performance of an aboveground herbivore interact with fertilisation. Oikos 124:658–667

Kostenko O, van de Voorde TFJ, Mulder PPJ, van der Putten WH, Bezemer MT (2012) Legacy effects of aboveground-belowground interactions. Ecol Lett 15:813–821

Kristensen JÅ, Rousk J, Metcalfe DB (2020) Below-ground responses to insect herbivory in ecosystems with woody plant canopies: a meta-analysis. J Ecol 108:917–930

Kuć J (1982) Induced immunity to plant disease. Bioscience 32:854–860

Kwak MJ, Kong HG, Choi K, Kwon SK, Song JY, Lee J, Lee PA, Choi SY, Seo M, Lee HJ, Jung EJ, Park H, Roy N, Kim H, Lee MM, Rubin EM, Lee SW, Kim JF (2018) Rhizosphere microbiome structure alters to enable wilt resistance in tomato. Nat Biotechnol 36:1100–1116

Larsen J, Ravnskov S, Jakobsen I (2003) Combined effect of an arbuscular mycorrhizal fungus and a biocontrol bacterium against *Pythium ultimum* in soil. Folia Geobot 38:145–154

Lau JA, Lennon JT (2011) Evolutionary ecology of plant–microbe interactions: soil microbial structure alters selection on plant traits. New Phytol 192:215–224

Leach JE, Triplett LR, Argueso CT, Trivedi P (2017) Communication in the Phytobiome. Cell 169:587–596

Lira-Noriega A, Soberón J, Equihua J (2018) Potential invasion of exotic ambrosia beetles *Xyleborus glabratus* and *Euwallacea* sp. in Mexico: a major threat for native and cultivated forest ecosystems. Sci Rep 8:10179

Liu H, Macdonald CA, Cook J, Anderson IC, Singh BK (2019a) An ecological loop: host microbiomes across multitrophic interactions. Trends Ecol Evol 34:1118–1130

Liu Y, Qiao J, Liu Y, Liang X, Zhou Y, Liu J (2019b) Characterization of *Lysobacter capsici* strain NF87–2 and its biocontrol activities against phytopathogens. Eur J Plant Pathol 155:859–869

Ma HK, Pineda A, Hannula SE, Kielak AM, Setyarini SN, Bezemer TM (2020) Steering root microbiomes of a commercial horticultural crop with plant-soil feedbacks. Appl Soil Ecol 150:103468

Machado RAR, Arce CCM, McClure MA, Baldwin IT, Erb M (2018) Aboveground herbivory induced jasmonates disproportionately reduce plant reproductive potential by facilitating root nematode infestation. Plant Cell Environ 41:797–808

Malamy J, Carr JP, Klessig DF, Raskin I (1990) Salicylic acid: a likely endogenous signal in the resistance response of tobacco to viral infection. Science 250:1002–1004

Marschner P, Rumberger A (2004) Rapid changes in the rhizosphere bacterial community structure during re-colonization of sterilized soil. Biol Fertil Soils 40:1–6

Maruri-López I, Aviles-Baltazar NY, Buchala A, Serrano M (2019) Intra and extracellular journey of the phytohormone salicylic acid. Front Plant Sci 10:423

Mazzola M, Hewavitharana SS, Strauss SL (2015) *Brassica* seed meal soil amendments transform the rhizosphere microbiome and improve apple production through resistance to pathogen reinfestation. Phytopathol 105:460–469

Mbaluto CM, Ahmad EM, Fu M, Martínez-Medina A, van Dam NM (2020) The impact of *Spodoptera exigua* herbivory on *Meloidogyne incognita*-induced root responses depends on the nematodes' life cycle stages. AoB Plants 12:plaa029

Medina-Sauza RM, Álvarez-Jiménez M, Delhal A, Reverchon F, Blouin M, Guerrero-Analco JA, Cerdán CR, Guevara R, Villain L, Barois I (2019) Earthworms building up soil microbiota, a review. Front Environ Sci 7:81

Mendes R, Kruijt M, De Bruijn I, Dekkers E, van der Voort M, Schneider JH, Piceno YM, DeSantis TZ, Andersen GL, Bakker PA, Raaijmakers JM (2011) Deciphering the rhizosphere microbiome for disease-suppressive bacteria. Science 332:1097–1100

Méndez-Bravo A, Cortazar-Murillo EM, Guevara-Avendaño E, Ceballos-Luna O, Rodríguez-Haas B, Kiel-Martínez AL, Hernández-Cristóbal O, Guerrero-Analco JA, Reverchon F (2018) Plant growth-promoting rhizobacteria associated with avocado display antagonistic activity against *Phytophthora cinnamomi* through volatile emissions. PLoS One 13:e0194665

Middleton EL, Richardson S, Koziol L, Palmer CE, Yermakov Z, Henning JA, Schultz PA, Bever JD (2015) Locally adapted arbuscular mycorrhizal fungi improve vigor and resistance to herbivory of native prairie plant species. Ecosphere 6:1–6

Mosleh Arany A, De Jong TJ, Kim HK, Van Dam NM, Choi YH, Verpoorte R, Van der Meijden E (2008) Glucosinolates and other metabolites in the leaves of *Arabidopsis thaliana* from natural populations and their effects on a generalist and a specialist herbivore. Chemoecology 18:65–71

Neal AL, Ahmad S, Gordon-Weeks R, Ton J (2012) Benzoxazinoids in root exudates of maize attract *Pseudomonas putida* to the rhizosphere. PLoS One 7:e35498

O'Donnell K, Libeskind-Hadas R, Hulcr J, Bateman C, Kasson MT, Ploetz RC, Konkol JL, Ploetz JN, Carrillo D, Campbell A, Duncan RE, Liyanage PNH, Eskalen A, Lynch SC, Geiser DM, Freeman S, Mendel Z, Sharon M, Aoki T, Cossé AA, Rooney AP (2016) Invasive Asian *Fusarium—Euwallacea* ambrosia beetle mutualists pose a serious threat to forests, urban land-scapes and the avocado industry. Phytoparasitica 44:435–442

Ou Y, Penton CR, Geisen S, Shen Z, Sun Y, Lv N, Wang B, Ruan Y, Xiong W, Li R, Shen Q (2019) Deciphering underlying drivers of disease suppressiveness against pathogenic *Fusarium oxysporum*. Front Microbiol 10:2535

Pangesti N, Pineda A, Pieterse CM, Dicke M, Van Loon JJ (2013) Two-way plant mediated interactions between root-associated microbes and insects: from ecology to mechanisms. Front Plant Sci 4:414

Pangesti N, Pineda A, Dicke M, Van Loon JJA (2015) Variation in plant-mediated interactions between rhizobacteria and caterpillars: potential role of soil composition. Plant Biol 17:474–483

Pangesti N, Reichelt M, van de Mortel JE, Kapsomenou E, Gershenzon J, van Loon JJA, Dicke M, Pineda A (2016) Jasmonic acid and ethylene signaling pathways regulate glucosinolate levels in plants during rhizobacteria-induced systemic resistance against a leaf-chewing herbivore. J Chem Ecol 42:1212–1225

Pereyra M, Grilli G, Galetto L (2019) Root-associated fungi increase male fitness, while high simu-lated herbivory decreases indirect defenses in *Croton lachnostachyus* plants. Plant Ecol 220:29–39

Philippot L, Raaijmakers JM, Lemanceau P, van der Putten WH (2013) Going back to the roots: the microbial ecology of the rhizosphere. Nat Rev Microbiol 11:789–799

Pieterse CMJ, Dicke M (2007) Plant interactions with microbes and insects: from molecular mecha-nisms to ecology. Trends Plant Sci 12:564–569

Pieterse CMJ, Van Pelt JA, Van Wees SCM, Ton J, León-Kloosterziel KM, Keurentjes JJB, Verhagen BWM, Knoester M, Van der Sluis I, Bakker PAHM, Van Loon LC (2001) Rhizobacteria-mediated induced systemic resistance: triggering, signalling and expression. Eur J Plant Pathol 107:51–61

Pieterse CMJ, Van der Does D, Zamioudis C, Leon-Reyes A, Van Wees SCM (2012) Hormonal mod-ulation of plant immunity. Annu Rev Cell Dev Biol 28:489–521

Pieterse CMJ, Zamioudis C, Berendsen RL, Weller DM, Van Wees SCM, Bakker PAHM (2014) Induced systemic resistance by beneficial microbes. Annu Rev Phytopathol 52:347–375

Pineda A, Zheng SJ, Van Loon JJ, Pieterse CM, Dicke M (2010) Helping plants to deal with insects: the role of beneficial soil-borne microbes. Trends Plant Sci 15:507–514

Pineda A, Soler R, Weldegergis BT, Shimwela MM, Van Loon JJ, Dicke M (2013) Non-pathogenic rhizobacteria interfere with the attraction of parasitoids to aphid-induced plant volatiles via jasmonic acid signalling. Plant Cell Environ 36:393–404

Pineda A, Kaplan I, Bezemer TM (2017) Steering soil microbiomes to suppress aboveground insect pests. Trends Plant Sci 22:770–778

Pineda A, Kaplan I, Hannula SE, Ghanem W, Bezemer TM (2020) Conditioning the soil microbiome through plant–soil feedbacks suppresses an aboveground insect pest. New Phytol 226:595–608

Potts SG, Biesmeijer JC, Kremen C, Neumann P, Schweiger O, Kunin WE (2010) Global pollinator declines: trends, impacts and drivers. Trends Ecol Evol 25:345–353

Poulton JL, Bryla D, Koide RT, Stephenson AG (2002) Mycorrhizal infection and high soil phospho-rus improve vegetative growth and the female and male functions in tomato. New Phytol 154:255–264

Poveda K, Steffan-Dewenter I, Scheu S, Tscharntke T (2003) Effects of below-and above-ground herbivores on plant growth, flower visitation and seed set. Oecologia 135:601–605

Poveda K, Steffan-Dewenter I, Scheu S, Tscharntke T (2005) Effects of decomposers and herbivores on plant performance and aboveground plant-insect interactions. Oikos 108:503–510

Pozo MJ, Azcón-Aguilar C (2007) Unraveling mycorrhiza-induced resistance. Curr Opin Plant Biol 10:393–398

Price PW (1991) The plant vigor hypothesis and herbivore attack. Oikos 62:244–251

Real-Santillán RO, del-Val E, Cruz-Ortega R, Contreras-Cornejo HÁ, González-Esquivel CE, Larsen J (2019) Increased maize growth and P uptake promoted by arbuscular mycorrhizal fungi coincide with higher foliar herbivory and larval biomass of the Fall Armyworm *Spodoptera frugiperda*. Mycorrhiza 29:615–622

Reverchon F, Hosseini Bai S, Liu X, Blumfield TJ (2015) Tree plantation systems influence nitrogen retention and the abundance of nitrogen functional genes in the Solomon Islands. Front Microbiol 6:1439

Ritzenthaler CA, Maloney CE, Maran AM, Moore EA, Winters A, Pelini SL (2018) The feedback loop between aboveground herbivores and soil microbes via deposition processes. In: Ohgushi T, Wurst S, Scott NJ (eds) Aboveground–belowground community ecology. Springer, Cham, pp 201–221

Rodriguez-Saona C, Chalmers JA, Raj S, Thaler JS (2005) Induced plant responses to multiple damagers: differential effects on an herbivore and its parasitoid. Oecologia 143:566–577

Roger A, Gétaz M, Rasmann S, Sanders IR (2013) Identity and combinations of arbuscular mycorrhizal fungal isolates influence plant resistance and insect preference. Ecol Entomol 38:330–338

Rosas-Guerrero V, Aguilar R, Martén-Rodríguez S, Ashworth L, Lopezaraiza-Mikel M, Bastida JM, Quesada M (2014) A quantitative review of pollination syndromes: do floral traits predict effective pollinators? Ecol Lett 17:388–400

Ross AF (1961) Systemic acquired resistance induced by localized virus infections in plants. Virology 14:340–358

Rudrappa T, Czymmek KJ, Paré PW, Bais HP (2008) Root-secreted malic acid recruits beneficial soil bacteria. Plant Physiol 148:1547–1556

Sasse J, Martinoia E, Northen T (2018) Feed your friends: do plant exudates shape the root microbiome? Trends Plant Sci 23:25–41

Schittko C, Wurst S (2014) Above- and belowground effects of plant-soil feedback from exotic *Solidago canadensis* on native *Tanacetum vulgare*. Biol Invasions 16:1465–1479

Schuhegger R, Ihring A, Gantner S, Bahnweg G, Knappe C, Vogg G, Hutzler P, Schmid M, van Breusegem F, Eberl L, Hartmann A, Langebartels C (2006) Induction of systemic resistance in tomato by N-acyl-L-homoserine lactone-producing rhizosphere bacteria. Plant Cell Environ 29:909–918

Selvaraj A, Thangavel K, Uthandi S (2020) Arbuscular mycorrhizal fungi (*Glomus intraradices*) and diazotrophic bacterium (*Rhizobium* BMBS) primed defense in blackgram against herbivorous insect (*Spodoptera litura*) infestation. Microbiol Res 231:126355

Shavit R, Ofek-Lalzar M, Burdman S, Morin S (2013) Inoculation of tomato plants with rhizobacteria enhances the performance of the phloem-feeding insect *Bemisia tabaci*. Front Plant Sci 4:306

Shen Z, Xue C, Penton CR, Thomashow LS, Zhang N, Wang B, Ruan Y, Li R, Shen Q (2019) Suppression of banana Panama disease induced by soil microbiome reconstruction through an integrated agricultural strategy. Soil Biol Biochem 128:164–174

Smith WH (1972) Influence of artificial defoliation on exudates of sugar maple. Soil Biol Biochem 4:111–113

Smith SE, Read DJ (2008) Mycorrhizal symbiosis, 3rd edn. Academic, Cambridge

Song GC, Lee S, Hong J, Choi HK, Hong GH, Bae DW, Mysore KS, Park YS, Ryu CM (2015) Aboveground insect infestation attenuates belowground agrobacterium-mediated genetic transformation. New Phytol 207:148–158

Stam JM, Kroes A, Li Y, Gols R, van Loon JJ, Poelman EH, Dicke M (2014) Plant interactions with multiple insect herbivores: from community to genes. Annu Rev Plant Biol 65:689–713

Thamer S, Schädler M, Bonte D, Ballhorn DJ (2011) Dual benefit from a belowground symbiosis: nitrogen fixing rhizobia promote growth and defense against a specialist herbivore in a cyanogenic plant. Plant Soil 341:209–219

Valenzuela-Soto JH, Estrada-Hernández MG, Ibarra-Laclette E, Délano-Frier JP (2010) Inoculation of tomato plants (*Solanum lycopersicum*) with growth-promoting *Bacillus subtilis* retards whitefly *Bemisia tabaci* development. Planta 231:397–410 van Dam NM, Bouwmeester HJ (2016) Metabolomics in the rhizosphere: tapping into belowground chemical communication. Trends Plant Sci 21:256–265

van den Berg N, du Toit M, Morgan SW, Fourie G, de Beer ZW (2019) First report of *Fusarium euwallaceae* on *Persea americana* in South Africa. Plant Dis 103:1774

van der Putten WH, Bardgett RD, Bever JD, Bezemer TM, Casper BB, Fukami T, Kardol P, Klironomos JN, Kulmatiski A, Schweitzer JA, Suding KN, van de Voorde TFJ, Wardle DA (2013) Plant–soil feedbacks: the past, the present and future challenges. J Ecol 101:265–276

Van Loon LC, Bakker PAHM, Pieterse CMJ (1998) Systemic resistance induced by rhizosphere bacteria. Annu Rev Phytopathol 36:453–483

Van Wees SC, Van der Ent S, Pieterse CMJ (2008) Plant immune responses triggered by beneficial microbes. Curr Opin Plant Biol 11:443–448

Vannette RL, Hunter MD, Rasmann S (2013) Arbuscular mycorrhizal fungi alter above-and belowground chemical defense expression differentially among *Asclepias* species. Front Plant Sci 4:361

Vannier N, Agler M, Hacquard S (2019) Microbiota-mediated disease resistance in plants. PLoS Pathog 15(6):e1007740

Varga S, Kytöviita MM (2010) Gender dimorphism and mycorrhizal symbiosis affect floral visitors and reproductive output in *Geranium sylvaticum*. Funct Ecol 24:750–758

Vázquez-Ochoa F, Reverchon F, Sánchez-Velásquez L, Ruíz-Montiel C, Pineda-López R (2020) Soil chemical properties and biological processes under pine trees with and without *Dendroctonus* bark beetle infestation. J Sustain For. In press. https://doi.org/10.1080/10549811.2020.1759103

Wang M, Biere A, van der Putten WH, Bezemer TM, Brinkman EP (2017) Timing of simulated aboveground herbivory influences population dynamics of root-feeding nematodes. Plant and Soil 415:215–228

Wardle DA, Bonner KI, Barker GM (2002) Linkages between plant litter decomposition, litter quality, and vegetation responses to herbivores. Funct Ecol 16:585–595

Wolfe BE, Husband BC, Klironomos JN (2005) Effects of a belowground mutualism on an aboveground mutualism. Ecol Lett 8:218–223

Wondafrash M, Van Dam NM, Tytgat TO (2013) Plant systemic induced responses mediate interactions between root parasitic nematodes and aboveground herbivorous insects. Front Plant Sci 4:87

Wurst S, Sonnemann I, Zaller JG (2018) Soil macro-invertebrates: their impact on plants and associated aboveground communities in temperate regions. In: Ohgushi T, Wurst S, Scott NJ (eds) Aboveground–belowground community ecology. Springer, Cham, pp 175–200

Zhang J, Wei L, Yang J, Ahmed W, Wang Y, Fu L, Ji G (2020) Probiotic consortia: reshaping the rhizospheric microbiome and its role in suppressing root-rot disease of *Panaxnotoginseng*. Front Microbiol 11:701

How Plant-Arthropod Interactions Modify the Environment: Concepts and Perspectives

Eduardo Soares Calixto,
Danilo Ferreira Borges dos Santos,
Diego V. Anjos, and Eva Colberg

Contents

Electronic Supplementary Material The online version of this chapter (https://doi.org/10.1007/978-3-030-66877-8_9) contains supplementary material, which is available to authorized users. The videos can be accessed by scanning the related images with the SN More Media App.

© Springer Nature Switzerland AG 2021
K. Del-Claro, H. M. Torezan-Silingardi (eds.), *Plant-Animal Interactions*, https://doi.org/10.1007/978-3-030-66877-8_9

⌂ **Learning Objectives**
- The terms ecosystem engineers and ecosystem engineering.
- Autogenic and allogenic ecosystem engineers.
- Plant-arthropod interactions as ecosystem engineering.
- Relationship between ecosystem engineers and biodiversity.
- Predicting the outcomes of the loss or addition of an ecosystem engineer.

Concepts & Definitions

- *Ecosystem engineer* describes an organism whose presence, growth, or behavior alters the physical state or structure of their abiotic or biotic environment and thereby alters the flow of resources to other species.
- *Ecosystem engineering* is the process by which an organism affects resource flow to other organisms by modifying its environment.
- *Autogenic ecosystem engineers* change the environment directly via their own physical structures, i.e., their living and dead tissues.
- *Allogenic ecosystem engineers* modify external biotic or abiotic aspects of their environment.
- *Physical engineering* refers to a subset of ecosystem engineering in which an organism alters the physical state or structure of something in the environment, as opposed to altering chemical properties or transport processes.
- *Niche construction* refers to organismally mediated environmental changes that impact the ecological niche of the organism creating the change, and potentially that of other organisms as well. This is a term from evolutionary biology to describe the influence of phenotype on selection.

9.1 Introduction

The concept of ecosystem engineering (Jones et al. 1994), although more often used to describe the effects of a single organism on its environment, can also be applied to plant-animal interactions in order to better understand their non-trophic effects on ecosystems. To this end, we will use this chapter to provide justification for and examples of how the ecosystem engineering concept has been and can be further applied to plant-animal interactions. Although plant-animal interactions can take place in many environments between many types of organisms, we will focus on the physical ecosystem impacts of plant-arthropod interactions. Within this focus, we will differentiate between autogenic and allogenic cases of ecosystem engineering, explain the implications of both for biodiversity, and conclude by pointing out some novel and underrepresented perspectives for future studies and comparisons.

9.1.1 Origins of the Ecosystem Engineering Concept

The idea that organisms have ecological effects beyond trophic and competitive is not novel—Darwin wrote a monograph on the ways in which earthworms altered the soil (Darwin 1881), and others before him had casually observed the same (Morgan 1868). Indeed, multiple authors (Buchman et al. 2007; Wilson 2007) have reviewed how ecologists discussed organismal modifications to the environment prior to the rise of the term *ecosystem engineering* (Jones et al. 1994; Cuddington 2007; Wilson 2007). Even the term "engineering" itself had already been used in an ecological context, describing the impacts of pocket gophers on soil characteristics (Huntly and Inouye 1988). However, the full term "ecosystem engineering" was officially coined and codified only in 1994 (Jones et al. 1994), and thereafter expanded upon and refined (Jones et al. 1997a; Jones and Gutiérrez 2007). As first put forth, the term *ecosystem engineer* describes an organism whose presence, growth, or behavior alters the physical state or structure of their abiotic or biotic environment and thereby alters the flow of resources to other species (Jones et al. 1994). The term overlaps with *facilitation* (*sensu* Bronstein 2009), where one organism's environmental modifications benefit another organism, but ecosystem engineering as a whole is agnostic of benefit, and thus also comprises environmental changes with negative and net neutral effects on other organisms.

The Jones et al. (1994) definition specifically referred to "*physical engineering*", although the authors recognized that non-physical state alterations, such as the modification of chemical and transport processes could also modify resource availability (Jones et al. 1994). Subsequent definitions have incorporated some of these processes into the overall ecosystem engineering framework (Berke 2010). Although the original concept and its definition have been accompanied by debate and alternative interpretations (Wright and Jones 2006), use of the term ecosystem engineering has greatly increased since its conception (Berke 2010; Pearce 2011).

The wording, definition, and utility of ecosystem engineering have all been subject to critique. To some, the word "engineering" connotes purposeful action, which some have argued is reason to limit (Power 1997) or reject (Reichman and Seabloom 2002a) use of the term, although the original authors plead against such a value-driven interpretation (Jones et al. 1997b). Confusion has also stemmed from conflation with *niche construction*, a term from evolutionary biology subject to its own swathe of debate, which some have used interchangeably with ecosystem engineering (Laland et al. 1999; Boogert et al. 2006; Odling-Smee et al. 2013). Although some argue that ecosystem engineering is most interesting when an organism modulates the environment for its own benefit (Wilson 2007), the Jones et al. (Jones et al. 1994) definition limits ecosystem engineering to environmental changes that have repercussions for organisms beyond the agent of change, whereas *niche construction* can apply to organisms whose environmental alterations affect only themselves (Odling-Smee et al. 2013).

More extensive critiques involve the generality of ecosystem engineering, which in its original conception does not include any threshold of impact level, and thus could arguably be applied to most if not all organisms (Reichman and Seabloom 2002b; Wilson 2007). However, limiting the definition to organismally driven abiotic modifications with significant impacts (a term also subject to multiple definitions) can render ecosystem engineering a subset of Power et al.'s (1996) keystone species concept (Wilson 2007), and limits the ubiquity of the term (Jones and Gutiérrez 2007). Similarly, limiting the term to environmental modifications with positive benefits on other organisms would render it a synonym of facilitation (*sensu* Bronstein 2009), when in reality the impacts of environmental modification can also be negative, neutral, or variable. Therefore, we side with the view that it is more useful to focus on the *process* of ecosystem engineering before quantifying its impacts and outcomes (Wright and Jones 2006; Jones and Gutiérrez 2007). In focusing on the process first, we can consider how the outcomes will change under different conditions, as variability of outcomes can itself be a subject of interest and evolutionary implication (Pearce 2011).

9.1.2 Autogenic vs. Allogenic Engineers

From its conception, physical ecosystem engineering was recognized as having multiple origins (Jones et al. 1994), with a distinction between cases in which an organism modifies its own structure (*autogenic ecosystem engineering*) and cases in which an organism modifies some external biotic or abiotic structure (*allogenic ecosystem engineering*). Examples of arthropods as allogenic engineers abound at multiple scales. On an individual level, many lepidopterans manipulate leaf structures to create shelters that other arthropods can then use (Lill and Marquis 2003). On a larger scale, ants and termites alter soil structure, which increases aeration and nutrient cycling and availability (Folgarait 1998; Bignell 2006). Arthropod examples of autogenic engineers are comparatively rare. Some authors include the creation of biomaterials as autogenic engineering, implying that social spiders (Keiser and Pruitt 2014) and silk moths (Raath et al. 2017) can act as autogenic engineers by creating silken structures that directly or indirectly alter resources for other organisms. From the plant perspective, plants autogenically modify the environment in myriad ways, casting shade and altering the microclimate with their entire growth form (Arredondo-Núñez et al. 2009) or forming specialized structures (e.g. phytotelmata) that hold water and create microclimatic conditions suitable for shelter of other organisms (Kitching 2000). However, plants can also act as allogenic engineers as their roots change the porosity and density of soil structure, or even simultaneously act as autogenic and allogenic engineers, with their roots forming part of the structure that itself provides substrate stability while also changing already-existing substrate (Jones et al. 1997a).

Jones et al. (1994) further differentiated between several cases of both autogenic and allogenic engineering, based on whether an organism itself is or directly modifies resource flow, or whether its effects on resource flow are further modulated by the products of engineering and/or abiotic factors. Other divisions of ecosystem engineering beyond autogenic and allogenic have also been put forth, such as classification based on the broad mechanisms behind the engineering process (Berke 2010), or a distinction between obligate and facultative engineers (Cuddington et al. 2009). However, for the purposes of this chapter, we will limit our divisions to these first two cited categories, and will provide more in-depth discussion of each below.

9.1.3 Plant-Arthropod Interactions as Ecosystem Engineering

The original definition of ecosystem engineers treated individual organisms as the agents of engineering (Jones et al. 1994). However, plant-animal interactions inherently involve at least two different organisms: the plant and the animal. As such, there are several different ways in which the ecosystem engineering framework can be applied to plant-animal interactions, depending on which organism is the engineer, which organism is subject to engineering, and whether the engineering requires both organisms in order to occur. We embrace this variety and consider the ways in which plant ecosystem engineers affect arthropods, how arthropod ecosystem engineers affect plants, and how interactions between plants and arthropods can also affect other organisms through environmental modulation.

9.2 Autogenic Ecosystem Engineers

Autogenic ecosystem engineers are organisms that modify the environment through their own physical structure, creating, modifying and maintaining habitats through their growth, death or maintenance (Jones et al. 1994, 1997a). The clearest and most common examples of autogenic ecosystem engineers are plants. As they grow, the development of their above and belowground parts, such as trunks, branches, leaves, and roots, create habitats for a wide range of other living organisms, including mammals (e.g., squirrels and small marsupials), birds, herptiles, arthropods, or even other plants, such as epiphytes (Reyes-López et al. 2003; Yasuda et al. 2009; Mitchell et al. 2010). For instance, in tropical environments, lianas (woody vines) grow through the forest canopy connecting trees and forming arboreal pathways by which different animals (e.g., monkeys, snakes, and arthropods) can travel without having to descend to the ground (Charles-Dominique 1971; Charles-Dominique et al. 1981). Although less common, arthropods can also act as autogenic engineers. For instance, recently ants were observed using empty cocoons of the economically important silk moth, *Gonometapostica* (Lasiocampidae), as shelter and/or nesting sites (Raath et al. 2017). In this context, the moth larva physically changes the habitat by using its own biomaterials to create a structure, and this structure becomes a possible resource for ants.

9.2.1 Trees as Biodiversity Amplifiers in Terrestrial Ecosystems

Trees influence a wide range of niches above and belowground (Reyes-López et al. 2003; Yasuda et al. 2009; Mitchell et al. 2010). Berke (2010) has included most trees in three functional classes (structural, light and chemical engineers) out of four possible classes of ecosystem engineers (the fourth being bioturbators). These autogenic engineers create new suitable habitats for species to colonize and reproduce in (Jones et al. 1997a). A single tree can be essential for maintaining biodiversity in forest and urban environments (Reyes-López et al. 2003; Yasuda et al. 2009; Mitchell et al. 2010), and in restoring ecosystems (Byers et al. 2006). For instance, a single tree can create shade that favors certain ant species, leading to higher ant densities under the treetops than in more exposed areas. Increases in richness and diversity of the ground ant community are influenced mainly by the less extreme temperatures and higher humidity found under tree canopies (Perfecto and Vandermeer 1996; Reyes-López et al. 2003). Similar to tree shade, many other plant features from endogenous processes (e.g., tree growth, development) increase the diversity of niches for animals and plants (Yasuda et al. 2009; Mitchell et al. 2010).

The forest canopy, consisting of overlapping leaves and branches of trees, shrubs, or both, houses a great diversity of species and is considered the top layer of biodiversity in terrestrial ecosystems (Ozanne et al. 2003). At large scales, this forest layer helps to regulate the flow of water and carbon (Calder 2001). At a smaller scale, the canopy creates micro-habitats for a huge number of species. The canopy harbors about 40% of living species, with a quarter of these species being canopy-specialists (Hammond 1997; Novotny et al. 2002).

Mainly in the tropics, the canopy can support tank bromeliads, epiphytes whose overlapping vase-shaped leaves modify the environment by forming phytotelmata (pools of water contained in plants or parts of plants) (Benzing 2000). Part of the largest family (Bromeliaceae) of flowering plants in the neotropics, with >2600 species, tank bromeliads autogenically provide habitats and modulate resources (e.g. litter and dead invertebrates) for aquatic and terrestrial animals, especially arthropods. Both biotic (e.g., competition and predation) and abiotic factors (e.g., water volume and plant physical structure) determine the species richness of bromeliad-inhabiting communities (Armbruster et al. 2002). For instance, in the lowland forests of Ecuador, Armbruster et al. (2002) found 354 animal species in three Bromeliaceae genera, of which more than half were considered rare. While some organisms, such as frogs, mosquitoes, and dragonflies, spend part of their development on bromeliads, other species are adapted to living their entire life cycles in these habitats, such as rotifers, crustaceans, and diving beetles (Dytiscidae) (Balke et al. 2008). The loss of these tank bromeliads due to deforestation or invasive species can thus decrease biological diversity by reducing habitat and altering the availability of water and nutrients in the canopy (Cooper et al. 2014). Conversely, the reintroduction of epiphytic bromeliads during forest restoration can increase the diversity of arthropods by providing habitat and providing environmental buffering at a microclimatic scale (Fernandez Barrancos et al. 2017).

Just as the canopy alters microhabitat conditions, so too can tree bark. Bark is defined as any tissue outside the vascular cambium whose function involves mechanical support, photosynthesis, storage of water and other components, protection and recovery after fire, and protection against pathogenic attacks and herbivory (Evert 2006). Jackson (1979) considered tree bark to be a "bedroom community" where arthropods lay eggs or overwinter, but this term understates the potential of bark to beget and maintain biodiversity. Bark morphology differs based on a tree species' ecological and physiological needs, and this high morphological diversity suggests a great number of niches for small animals, which has been corroborated since Jackson's perspective. Factors such as microstructure (e.g., thickness, hardness), microclimate and even bark color can determine the associated fauna and their behavior (Nicolai 1986). For instance, *Paratemnoides nidificator* (Pseudoscorpiones) uses bark openings to determine its predation strategies and capture prey of different sizes (Moura et al. 2018) (◙ Fig. 9.1a). Many species use bark for egg laying and larval development, mating and nesting areas, and overwintering and migration zones (Majer et al. 2003; Yasuda et al. 2009; Moura et al. 2018). Some species of Psocoptera and Hemiptera (e.g., *Empicoris vagabundus*) live exclusively in tree bark (Nicolai 1986). Deep bark furrow sare used by soft-bodied arthropods, mainly spiders. Regarding spider fauna, 57% of species in a central European forest were found only in tree bark (Albert 1976). Spider abundance may determine the abundance of predators, as with *Certhia americana* (Passeriformes) in southern Washington's Cascade Range (Mariani and Manuwal 1989). Many other predators, including woodpeckers, can take advantage of bark-living fauna. Hooper (1996) demonstrated that arthropod biomass changes according to temperature and age of longleaf pine trees (*Pinus palustris*) (Pinaceae), influencing the foraging of endangered red-cockaded woodpecker (*Picoides borealis*) (Passeriformes). This endangered bird may feed on up to 400 genera of arthropods, mainly spiders, ants, hemiptera and bark beetles that live in the bark of longleaf pines (Hanula and Franzreb 1998).

Autogenic engineering is not limited to living and growing organisms, either. Dead plant parts (e.g., leaves and branches) that fall on the soil form the litter layer, an above-ground layer composed of organic material from mainly vegetal but also animal and fungal sources that serves asa potential microhabitat for many species (◙ Fig. 9.1b). Plant litter changes raindrop impact, drainage, nutrient cycling, andheat and gas exchange, in addition to offering physical protection for organisms such as seeds, animals, and microbes (Facelli and Pickett 1991; Hättenschwiler et al. 2005; Stinchcombe and Schmitt 2006). Litter properties also depend on leaf size, leaf morphology, and leaf moisture content, so the more diverse the forest, the more diverse the litter. For instance, Kaneko and Salamanca (1999) found higher abundance of microarthropods in mixed-species litter than in litter from a single plant species, and Hansen and Coleman (1998) found similar results for oribatid mite (Acari: Oribatida) species richness.

Below the litter layer, tree roots autogenically create suitable conditions for fauna in the soil through physical and biochemical processes, along with many allogeneic ecosystem engineers (e.g., earthworms, ants, termites) (Angers and Caron 1998) (◙ Fig. 9.1b). This is particularly important because the belowground environment is extremely complex and pivotal for global nutrient cycling (Bardgett and van der Putten 2014). Tree roots and other woody parts also play an important role in engineering the riparian zone by trapping sediment and influencing water

◘ **Fig. 9.1** Examples of the types of ecosystem engineering that can be performed by or in a single tree: **a** Tree bark facilitates book scorpions (Pseudoscorpiones) preying on a Tettigoniidae (Orthoptera). **b** Plant roots create niches for species that live below ground; and aboveground litter, created mainly by organic vegetable material such as dead leaves and branches, provides nutrients, habitat, and other resources. **c** Caterpillars manipulate leaves to create a tent shelter. **d** Another example of arthropods manipulating leaves, a leaf roller on *Miconia* sp. (Melastomataceae). **e** Shelter created by *Cerconota achatina* in leaves of *Byrsonima intermedia*. **f** Multiple galls produce a leaf distortion and create new habitats. For more details see online supplementary material (QR Code support)

flow and retention (Gurnell and Petts 2006). Since Lavelle's (1986) "first link" hypothesis showing that mutualistic interactions between roots and other soil organisms are the first link in a suite of processes that increase biodiversity, vegetation dynamics have been shown to be strongly associated directly or indirectly with these mutualistic interactions (Wardle et al. 2004; Eisenhauer and Scheu 2008). Changes induced by root biological functions such as respiration, uptake, and exudation may affect soil and water quality and even the growth of other plants (Ehrenfeld 2003; Hinsinger et al. 2005). Several organisms in these habitats (or microhabitats) depend on the physical conditions created by these autogenic engineers (Jones et al. 1997a). Soil animals indirectly benefit from root resources when they feed on metazoans, protozoans, fungi, and bacteria that obtain their resources from roots (Pollierer et al. 2007). In addition, root exudates may attract specific groups of microbes and thereby establish underground trophic webs (Ayres et al. 2007; Henry et al. 2008).

9.2.2 Mutualisms Mediated by Autogenic Engineers

The presence of food corpuscles and sugary solutions (e.g., extrafloral nectaries) offered by plants facilitates mutualistic interactions with arthropod visitors and even symbionts (Beattie 1985; Calixto et al. 2018). However, as the role of plants as ecosystem engineers goes beyond their roles in the food supply and trophic interactions (Jones et al. 1994, 1997a), we will not consider these food-based examples in this session, although ▶ Chap. 5 of this book addresses this topic in much greater detail. Nonetheless, there are other plant-arthropod mutualisms that depend upon resources beyond food, which plants can also provide.

As mentioned above, plants are the main autogenic engineers of terrestrial ecosystems. However, plants may also directly or indirectly mediate mutualistic interactions, in some cases acting as autogenic engineers and in other cases being subjected to allogenic engineering by their mutualists. The classical example of a mutualism directly mediated by plants is the myrmecophyte–ant relationship. Myrmecophytes are plants that offer cavities for ant species to nest in, in exchange for protection against herbivores and competing plants (Beattie 1985). Myrmecophytic plants are common in tropical areas, and belong to more than 100 genera (Heil and McKey 2003). These plants have specific structures, called *domatia* (singular *domatium*), which are often formed by hypertrophy of the internal tissue at specific locations (i.e., branches, petiole, stipule and thorns). These cavities appear with the plant ontogeny and may harbor different ant species at a time within the same individual (Fiala and Maschwitz 1992). Domatia directly facilitate and promote ant nesting by providing primarily open chambers, such as hollow spines in *Acacia* (Fabaceae) plants (Janzen 1966) (▶ Box 9.1), or indirectly by providing closed cavities that need to be excavated by ants themselves, including leaf rachises in *Tachigali* (Fabaceae), nodes of *Cordia* (Boraginaceae), and internodes of *Macaranga* (Euphorbiaceae) (Fiala and Maschwitz 1992; Fonseca 1994; Longino 1996; Pacheco and Del-Claro 2018). In the former case, when plant parts are naturally hollow and available for use as domatia, the plants are considered autogenic engineers whose growth directly forms these cavities that then alter environmental conditions and add to the habitat available for ants and other arthropods. In the latter case, ants are considered to be allogenic engineers because they modify a living plant structure (Jones et al. 1997a), and the domatia are the result of this modification (Benson 1985).

There is a great diversity of domatia regarding the position of swelling in the stem, size and shape, and whether the cavities are discrete and separate or continuous throughout the branch and other plant features (Brouat and McKey 2000).

Box 9.1 Small Guests But Great Defenders: David and Goliath's Struggle for *Acacia*

The mutualism between ants and plants of the *Acacia* genus (*sensu* lato) has been one of the most investigated study systems of the past 50 years. Janzen's pioneering work (1966) has served as an inspiration for hundreds of ecologists and evolutionists. Put simply, in this system the plants offer food and housing for ants, which in return protect the plants against herbivory. *Acacia* plants act as autogenic engineers by developing hollow spines, which are modified leaves that prevent water loss and serve as shelters for several ant species. An amazing version of this story occurs in African savannas. While studying *Acacia drepanolobium,* the myrmecophytic "whistling thorn" trees in Kenya, and nonmyrmecophytic congeners, Goheen and Palmer (2010) demonstrated that autogenic engineering by plants had indirect effects beyond the ant community.

Crematogaster mimosa in *Acacia drepanolobium* in Kenya. Photo credit: Roberth Pringle. As part of a long-term experiment to exclude elephants, the authors showed with satellite images that elephants decrease tree cover. However, the elephants preferred to browse on plants without ants, and did not significantly impact the ant-inhabited *A. drepanolobium.*

To directly test whether ants were reducing elephant browsing in *A. drepanolobium*, the authors experimentally removed ants from the plants, and then assessed the subsequent level of damage by elephants. The results showed that the greater the activity of ants, the smaller the number of branches browsed. Finally, a manipulative food preference experiment was conducted in which four types of branches were offered to elephants: *A. drepanolobium* with and without ants and *A. mellifera* (an *Acacia* that has no associations with ants, nonmyrmecophytic) with and without ants. Elephants clearly preferred branches of both species of *Acacia* that had no ants — despite their thick dermis, elephants are highly sensitive around their eyes and inner membranes of their trunks, and ants tend to attack these areas of thin skin by biting and firmly holding on with their mandibles. Thus, the mutualism facilitated by the autogenic ecosystem engineering of *A. drepanolobium* can affect elephant browsing behavior by attracting ants that then prevent elephant herbivory. This story shows how ecosystem engineering can create a situation reminiscent of David's victory against Goliath, in which tiny insects are able to protect *Acacia* against an herbivore that has a weight and body size millions of times greater than these little guests.

Elephant eating *Acacia* trees in Kenya. Photo credit: Roberth Pringle.

The specific environmental conditions offered by plants, through plant anatomy and even ant behavior, allow individual myrmecophytes to harbor a single or several species of ants at a time, of which a large part are only able to live in these domatia (Beattie 1985). It is worth noting that a similar relationship also occurs in some plant species that have acarodomatia (or leaf domatia for mites, Acari) (Romero and Benson 2004). Leaf domatia are tiny cavities or capillary tufts often found on the leaf undersides, and can harbor beneficial mites which protect their host plant against phytophagous mites and other herbivorous arthropods (O'Dowd and Willson 1991).

Another example of a positive ant-plant interaction indirectly mediated by trees are ant-gardens (AGs). AGs are associations between epiphytes and arboreal ants that build carton nests in the tree canopy. Carton nests are dry and sturdy, made from a mixture of plant and soil material, vertebrate feces, and ant secretions (Davidson 1988; Yu 1994). Some of these carton nests can share more than one ant species (an association known as parabiosis), and thus the creation of these nests is already a form of allogenic engineering (Orivel and Leroy 2011). Then, as canopy ants build their carton nests, they collect and incorporate seeds and fruits of epiphytes that then germinate and grow on these carton nests. As the plants grow, the ants also protect and feed (through the nutrient-rich carton) them, further providing services to the plants. In return, epiphytic roots offer support to the ant nests, and thus the plants begin to impact the nest environment and act as autogenic engineers. Additionally, several AG plants provide raindrop protection and food resources (e.g. extrafloral and floral nectar, elaiosomes, and fruit pulp) to the ants (Orivel and Leroy 2011; Calixto et al. 2018). The complexity of these interactions comes from the specificity of these associations, without which several species (both ants and epiphytes) cannot survive (Orivel and Leroy 2011). Moreover, some ant species (e.g., *Odontomachus hastatus*) do not build carton nests like AGs, but nest exclusively on the roots of epiphytes (bromeliads) (Camargo and Oliveira 2012), in which case the epiphytes act as autogenic ecosystem engineers.

The above examples demonstrate how plant-arthropods mutualisms can modify resource availability for the mutualist partners, as well as for other organisms outside the mutualism (see ▶ Box 9.1). Although plant-arthropod mutualisms can involve autogenic ecosystem engineering, usually by the plant, they are also often accompanied by allogenic ecosystem engineering on behalf of the arthropod or plant, demonstrating that both types of ecosystem engineering can operate at the same time or in coordination. In the following section, we will provide more examples of allogenic engineering in the context of plant-arthropod interactions.

9.3 Allogenic Ecosystem Engineers

Allogenic ecosystem engineers are organisms capable of changing the environment by creating, maintaining or modifying living or non-living materials from one physical state or condition to another (Jones et al. 1994, 1997a). As a result, these modifications change the flux of available resources to other species such as living space, shelter, resting locations, food and energy, that will subsequently affect other organisms (Jones et al. 1997a). However, it is important to highlight that ecosystem engineers do not themselves directly provide the resources, but rather they alter the quality, quantity, and distribution of available biotic and abiotic structures (Jones et al. 1994). These modifications can trigger several impacts on animal community composition, and an extensive literature has shown the positive impacts of ecosystem engineers in structuring arthropod communities (Jones et al. 1994, 1997a; Martinsen et al. 2000; Bailey and Whitham 2003; Fournier et al. 2003; Lill and Marquis 2003; Buse et al. 2008; Cornelissen et al. 2016). For instance, Lill and Marquis (2003) showed that leaf shelters constructed by caterpillars from the genus *Pseudolteulphusa* (Gelechiidae) are an important factor for the species richness of associated herbivores on their host plant, white oak (*Quercus alba*). The caterpillar binds pairs of leaves together with its silk, creating and maintaining shelter structures called leaf ties. These constructs can act as a resource for several other organisms beyond their original caterpillar architects, creating conditions that help to ameliorate harsh abiotic conditions, prevent desiccation, increase food quality, and provide shelter from rain, extreme temperatures, and natural enemies (Fukui 2001), thereby influencing herbivore distributions throughout the growing season. In only one season, researchers recorded up to nine species of chewing insects within a single leaf tie (Lill and Marquis 2003), showing and quantifying the importance of these constructs for secondary residents. However, to some extent the impacts of leaf shelters on the herbivore community can also be negative. For example, generalist predators can also use leaf constructs as shelter or hunting grounds, with negative impacts for the herbivore community (Martinsen et al. 2000; Fournier et al. 2003; Buse et al. 2008). These different results illustrate how the impacts of an ecosystem engineer on species richness and abundance can vary greatly, and do not necessarily have to be positive (Jones et al. 1997a). Indeed, the effect of any environmental modification will be positive for some species and negative for others, especially if the modified material comes from a source already in use by another animal, such as branches and leaves (Jones et al. 1997b).

Another common material modified by ecosystem engineers is the wood of trees, which can be excavated to create chambers or galleries (Buse et al. 2008). For instance, larvae of the beetle *Cerambyx cerdo* (Cerambycidae) feed on trunks and thick branches of oak trees (*Quercus* spp.), creating new spaces and galleries which can act as entrances, habitats and refuges for other species. A whole community of saprophilic insects can then benefit from the space cleared by the feeding habits of cerambycid larvae, including *Euglenes culatus* (Aderidae), *Dorcatoma flavicornis* (Anobiidae), *Globicornis nigripes* (Dermestidae), and *Procraerus tibialis* (Elateridae) (Buse et al. 2008). Additionally, species that do not inhabit these galleries but use them to find their prey also benefit from these new spaces created, as is the case of *Brachygonus megerlei* and *Ampedus cardinalis*, both Elateridae. These spaces and holes can persist over many years or even decades, and therefore can act as habitats for many generations.

To sum up, the benefits that arthropod ecosystem engineers offer to other organisms can extend to the whole ecological community. By modifying living and non-living materials, allogenic ecosystem engineers can ameliorate the harsh abiotic conditions of their immediate environment, providing other organisms with a favorable microclimate, increased food quality, and shelter from harsh abiotic events (Fukui et al. 2002; Lill and Marquis 2003). However, some predators and parasites have also learned how to identify these constructs and environmental modifications as potential resources, either as shelter for themselves or as indicators of prey presence. When the structures created by ecosystem engineers prevail in the environment for an extended period of time (Lill and Marquis 2007), some generalist predators can move in as secondary occupants that then prey upon unsuspecting shelter-seekers (Martinsen et al. 2000; Fournier et al. 2003). Additionally, visual cues displayed by some types of shelters, as well as the general ability of predators to identify and target shelters, can increasing encounter rates between predators and prey. For example, birds can actively forage on leaf rolls or dead curled leaves, using the visual appearance of the structure to locate their prey (Remsen and Parker 1984; Murakami 1999). As another example, social wasps (*Polistes* spp.) can associate certain aspects of shelters with the presence of prey, and then use these cues to locate and attack skipper larvae inside their shelters (Weiss et al. 2004). Therefore, simply occupying shelters does not guarantee protection from predators.

The next two following sections we will discuss in more detail specific types of allogenic modification, such as shelter-building insects, focusing on leaf tiers and rollers, and gall-inducing insects. We will also point out the ways in which these systems increase the abundance and richness of their surrounding communities.

9.3.1 Shelter-Building Insects as Allogenic Engineers

Arthropods, including spiders, caterpillars, beetles, and ants, are widely known for their ability to manipulate plant structures to build shelters (Lill and Marquis 2007). By covering, tying, cutting, folding (◻ Fig. 9.1c), or rolling (◻ Fig. 9.1d, e) plant parts along with their own additional structures (e.g., silk), these organisms build external constructs on host plants (Carroll and Kearby 1978; Cappuccino and Martin 1994), which they can then use for much of their lives. For instance, caterpillars of *Rhopabota naevana* (Lepidoptera: Totricidae) feed, develop and pupate all within the rolled-leaf shelters that they construct in lilac trees (*Syringa reticulata*) (Fukui et al. 2002). The same occurs with *Pseudotelphusa* sp. (Gelechiidae), a leaf-tying caterpillar that modifies leaves of white oak (*Quercus alba*) (Lill and Marquis 2003), usually by sewing two or more mature leaves together with silk to form a leaf tie (Carroll and Kearby 1978). In all of these cases, when the shelter-building organism leaves its shelter for any reason (e.g., emergence, predation, death, unfavorable environmental changes), the vacant structure can remain on the plant as a shelter for many other organisms. Indeed, leaf shelters in the Brazilian Cerrado plant *Byrsonima intermedia* (Malpighiaceae) built by the caterpillar *Cerconota achatina* (◻ Fig. 9.2) provide shelter from intense temperatures, and the addition of experimental shelters provided space for 153 arthropod morphospecies while experimentally-added extended "leaves" of the same material had nearly no effect on the arthropod community (Velasque and Del-Claro 2016). Thus, by modifying leaves and other parts of the environment to create shelters, shelter-building insects also increase structural complexity and vacant spaces available to other arthropods. Additionally, shelter-building activities can also alter food resources by decreasing leaf toughness and phenolic content, ultimately

◘ Fig. 9.2 **a** *Cerconota achatina* caterpillar rolling a leaf of *Byrsonima intermedia* (Malpighiaceae) in the Brazilian Cerrado. **b, c** Examples of shelter constructs made by *C. achatina* rolled and tied with a vacant space inside. **d** An example of *B. intermedia* infested with *C. achatina* and the display change caused by the shelter constructs

impacting the diversity of other arthropods (Jones et al. 1994; Jones et al. 1997a; Lill and Marquis 2003, 2007; Cornelissen et al. 2016).

An interesting example of an environmental modification that influences abiotic factors is the yellow ant, *Lasius flavus* (Formicidae), known for its mound-building habit. These ants build soil mounds to control the temperature inside the nest, and as vegetation around the mound becomes taller, the ants increase the size of their mounds (Blomqvist et al. 2000). Yellow ants disturb the soil and surrounding environment by building these mounds, affecting soil physical and chemical properties and also cutting roots or other organs of nearby plants (Dauber et al. 2008). However, ant mounds can also increase colonization by arbuscular mycorrhizal fungi (Dauber et al. 2008), which form mutualistic symbiotic associations with plant roots to the benefit of nearby plants (Rillig 2004). Thus, mound-building ant activities alter multiple aspects of the environment, with impacts on resource availability and habitat suitability for plants and fungi.

Termites are also recognized as allogenic ecosystem engineers. Their mound-building and other activities affect nutrients, moisture, texture, and chemical composition of soils, contributing to the micro-topographical and spatial distribution of nutrients available to other organisms (Watson 1977; Kang 1978; Jones et al. 1994; Bignell and Eggleton 1995). Among the nutrients most influenced by termites are Ca, K and Mg. Since the soil and nutrient modifications around termite mounds affect plant development, termite constructs have consequences for plant diversity. For instance, Beaudrot et al. (2011) found a positive correlation between different termite mound traits and plant composition. They observed a higher diversity of seedling and juvenile trees near termite mounds than in areas without mounds. Therefore, by altering soil nutrients and composition, termite mounds contribute to the spatial heterogeneity of plants (Joseph et al. 2013). Termite mounds can persist for 20–25 years, providing long-term benefits to plants and ultimately to the ecosystem as a whole (Lee and Wood 1971; McComie and Dhanarajan 1993; Wood 1996), as suggested by Bonachela et al. (2015), in which termite mounds can help stabilize dry ecosystems under global change.

9.3.2 Gall-Inducing Insects as Allogenic Ecosystem Engineers

Gall-inducing insects prompt cellular differentiation of plant host tissues, creating a physical structure in which their larvae can grow (Price et al. 1987; Cornelissen et al. 2016). These physical structures can provide their inhabitants with several advantages such as habitat, protection from natural enemies, and metabolism manipulation in order to increase the nutritional quality of surrounding plant tissues (Price et al. 1987; Stone and Schönrogge 2003). Since they modify plant tissues to create physical structures, which are often used by other organisms after abandoned, gall-inducing insects can be considered allogenic ecosystem engineers (Jones et al. 1994; Cornelissen et al. 2016). Among the main arthropods inhabiting abandoned galls, ants stand out. For instance, eight species of ants were found in vacant galls made by the mealy oak gall wasp, *Disholcaspis cinerosa* (Wheeler and Longino 1988). The ants often occupied galls with greater diameter and volume (dos Santos et al. 2017).

Similar to other allogenic engineers, gall-inducing insects also positively influence the abundance and richness of other arthropod species by providing habitat, protection, and food resources (Fukui 2001). Waltz and Whitham (1997) showed that the removal of the gall-forming aphid *Pemphigus betae* from cotton woods decreased the richness and abundance of associated arthropods by 32% and 55%, respectively. Likewise, gall-forming wasps can indirectly affect arthropod communities in the surrounding unmodified foliage (▶ Box 9.2). In an experimentally-driven study based on the removal of senesced galls from trees, Wetzel et al. (2016) showed that gall removal resulted in 59% and 26% greater herbivore density and richness, and 27% greater arthropod density after leaves sprouted in the spring. Finally, Wheeler and Longino (1988) found 16 non-ant arthropod species inhabiting *D. cinerosa* galls, showing the potential that these structures have on the arthropod biodiversity. Therefore, the effects of allogenic ecosystem engineers' modifications on environment are not limited to providing habitat for organisms that specialize in exploiting these changes, but that these effects can go further and influence a whole community of organisms, even if they are not directly linked to the habitat modification.

Box 9.2 Cascading Indirect Effects from a Gall-Inducing Wasp on Arthropod Diversity

The impacts of an engineer can go beyond direct regulations mainly if the engineered constructs prevail for an extended period of time in nature (Wetzel et al. 2016). The gall wasp *Andricusquercus californicus* forms an "apple gall" on valley oak trees, *Quercus lobate,* can act as a type of allogenic ecosystem engineering that indirectly affects the foliage-dwelling arthropod community. The formation of a woody gall is the first step in a cascade of effects that can eventually impact members of the community that do not interact directly with the galls themselves, but rather are mediated by jumping spiders (Salticidae) that are generalist predators of herbivorous arthropods

A dissected senesced oak apple gall showing spider silk in several chambers (left), and an oak apple gall showing an emergence tunnel excavated by a California gall wasp (right). Photo credits: W. C. Wetzel.

The gall-inducing wasp *A. quercuscalifornicus* can manipulate plant tissues to create a swollen physical structure called a gall. In the late spring or summer, *A. quercuscalifornicus* induces a spherical, multi-chambered gall in the host plant. Then, in the autumn after the gall-inducer oviposits and leaves the structure, jumping spiders will often move in to occupy the senesced galls.

In order to understand how this type of gall-former influences the surrounding communities, Wetzel et al. (2016) set up a field experiment involving trees in three different treatments: control trees with galls, trees with galls removed, and naturally gall-free trees. To investigate the impacts of galls on the community, they sampled the pre-treatment arthropod community in the spring (16–23 March 2013) and again 2 and 5 weeks later (11–14 April and 2–7 May).

The dominant group found in galls were the jumping spiders, with one or more adult spiders found in 49.5% of the galls dissected. Jumping spiders are generalist predators of herbivorous arthropods, thus arthropod richness and abundance would be expected to be lower in trees where jumping spiders could seek refuge in galls. Indeed, during the final sampling period, there were 27.3% more arthropods, 58.8% more herbivores, and 25.6% greater herbivore richness on removal trees than on control trees. These results suggest that gall removal had a positive effect on herbivore density and richness, potentially by removing space that would otherwise be occupied by jumping spiders. However, beta diversity was significantly higher among control trees than among removal trees, perhaps indicating more variability in communities between galled trees than gall-free trees.

The indirect effects caused by the gall wasp *A. quercuscalifornicus* can shape patterns of abundance and diversity at the community level by increasing top-down predation, influencing beta diversity at large scale. Gall structures result in differences in community composition among trees within both the engineered microhabitat (galls) and the surrounding unmodified microhabitat (leaves), hence playing an important role in community dynamics. In this specific case, the authors suggest that these results were possible due to the galls persisting on trees for several years after their engineers died. Thus, a gall can impact the arthropod community well beyond the gall-making engineer's lifetime.

9.4   Ecosystem Engineers and Their Impact on Biodiversity

Most of the effects associated with ecosystem engineering result in the increase of environmental heterogeneity, both at individual and community scales. The increase of ecosystem complexity resulting from ecosystem engineering may directly and indirectly influence biodiversity (Crooks 2002). Ecosystem engineering generally facilitates the presence and coexistence of different species due to greater diversity of habitats and resources (Jones et al. 1997a; Wright et al. 2002; Wright and Jones 2004), although the specific outcomes may vary with environmental stressors (Crain and Bertness 2005, 2006). With the creation of new habitats and spaces, ecosystem engineering can lead to niche differentiation and diversification and then coexistence of species in the same or other trophic levels (Laland et al. 1999; Erwin 2008). In this context, ecosystem engineering can contribute to the stable coexistence of species, enhancing species diversity (Chesson 2000).

Ecosystem engineering can alter the distribution and abundance of species, and significantly modify biodiversity patterns (Jones et al. 1994, Jones et al. 1997a; Wright et al. 2002). For instance, shelter-building caterpillars make different types of leaf shelters such as leaf rolls, ties, folds and tents (☐ Fig. 9.1e). These leaf modifications beget new microhabitats which are concurrently and subsequently used by many other animals, especially other arthropods (☐ Fig. 9.1f). In our previous examples, we have also shown that ecosystem engineering can also have effects on higher trophic levels, as increased arthropod abundance can provide more food for vertebrate and invertebrate predators. The impacts of arthropods as ecosystem engineers go beyond the examples mentioned here, however. In a meta-analysis including 122 studies, Romero et al. (2015) observed that invertebrate ecosystem engineers of aquatic and terrestrial ecosystems significantly increased species richness across temperate and tropical regions. Additionally, given the ubiquity of plants in most ecosystems, and the many ways in which plants can act as both autogenic and allogenic engineers, it very well might be impossible to quantify the global impacts of plant ecosystem engineers. Nonetheless, the impacts of ecosystem engineering, and the roles of plants and arthropod engineers in particular, are clearly important to our understanding of biodiversity.

Abiotic and biotic changes caused by ecosystem engineers mediate interactions between species, an attribute that can aid in the restoration of degraded areas (Byers et al. 2006). Understanding the factors that determine environmental dynamics allows us to predict the impacts of the removal or addition of ecosystem engineers, assisting in decision-making and management (Byers et al. 2006). Knowing that some environmental conditions can be modified through ecosystem engineering, ecosystem engineers might be used to assist in the transition between degraded and healthy areas (Mayer and Rietkerk 2004). For instance, Wigginton et al. (2014) showed that the invasive weed *Lepidium latifolium* (Brassicaceae) can directly impact soil-dwelling and canopy-dwelling invertebrate communities in a brackish marsh ecosystem. The authors observed that the presence of *L. latifolium* increased abundance of soil-dwelling invertebrates and decreased the species richness of canopy-dwelling invertebrates in specific regions within the brackish marsh ecosystem. Thus, invasive organisms which act as ecosystem engineers might alter community and trophic dynamics, resulting in a negative-to-positive range of effects within the ecosystem. Within this focus, we believe that studies evaluating the direct and indirect impacts of ecosystem engineers are essential to understanding the consequences that these organisms will bring to the ecosystem. Thus, conservation, reforestation, and impact mitigation projects need to consider the presence of this environmental modification process (ecosystem engineering), especially as it relates to plants and arthropods, which occupy mostly the first trophic levels.

9.5 Perspectives

The study of ecosystem engineers has increased in recent years, with a focus on describing processes and providing mechanisms for a better understanding of community dynamics and structure (Romero et al. 2015). Although ecosystem engineering is seen as a key concept and widely accepted in different ecological areas, its impacts at individual and community levels are underestimated in *tropical* regions (◘ Fig. 9.3, ◘ Table 9.1). For instance, Cornelissen et al. (2016) showed that most studies of shelter-building insect ecosystem engineers were conducted in temperate areas, with long-lived host plants, and with leaf rollers (◘ Fig. 9.1g). Tropical regions present different biotic and abiotic characteristics and therefore are very important places for understanding the impacts of ecosystem engineers on biodiversity patterns.

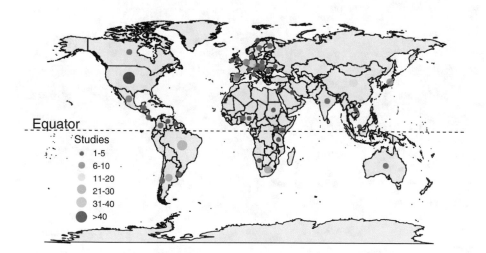

◘ **Fig. 9.3** Number of studies by country considering plant-arthropod interaction as an ecosystem engineering process. Circles represent the country centroid. There were 174 studies (between 1900 - August 2020) distributed in 47 countries, highlighting countries with temperate environments. For absolute and relative numbers, see ◘ Table 9.1. Studies were retrieved by using different combinations of expressions, namely "ecosystem engineer*", "insect*", "arthropod*", and "plant*" in the Web of Science database platform

◩ **Table 9.1** Number (%) of studies by country considering plant-arthropod interaction as an ecosystem engineering process for a total of 174 studies distributed across 47 countries between 1900- August 2020

Country	Studies	Country	Studies
United States of America	43(24.7)	Uruguay	2(1.1)
Brazil	24(13.8)	Czech Republic	1(0.5)
China	11(6.3)	Denmark	1(0.5)
Germany	9(5.1)	Guyana	1(0.5)
Argentina	7(4.0)	Hungary	1(0.5)
South Africa	7(4.0)	Ireland	1(0.5)
Japan	6(3.4)	Israel	1(0.5)
Netherlands	5(2.8)	Italy	1(0.5)
Spain	5(2.8)	Kuwait	1(0.5)
Mexico	4(2.2)	Malaysia	1(0.5)
Slovakia	3(1.7)	Namibia	1(0.5)
Australia	2(1.1)	Nigeria	1(0.5)
Austria	2(1.1)	Norway	1(0.5)
Belize	2(1.1)	Poland	1(0.5)
Benin	2(1.1)	Romania	1(0.5)
Canada	2(1.1)	Sudan	1(0.5)
Chile	2(1.1)	Switzerland	1(0.5)
Colombia	2(1.1)	United Republic of Tanzania	1(0.5)
Costa Rica	2(1.1)	Uganda	1(0.5)
Finland	2(1.1)	Ukraine	1(0.5)
France	2(1.1)	United Kingdom	1(0.5)
India	2(1.1)	Venezuela	1(0.5)
Kenya	2(1.1)	Vietnam	1(0.5)
Sweden	2(1.1)		

Furthermore, given the utility of the ecosystem engineering concept in understanding the impacts of species on their environment and other organisms, ecosystem engineering will be an increasingly useful consideration for conservation, restoration, and applied ecology. For instance, considering the roles and impacts of different ecosystem engineers can provide a fuller understanding of the impacts of various conservation and restoration approaches (Boogert et al. 2006). An understanding of the ecosystem engineering effects of a species can also help in predicting and minimizing the impacts of species loss due to climate change and other anthropogenic pressures. Additionally, *invasive species* are one of the greatest threats to biodiversity and the functioning of ecosystems (Strayer 2012). Since some of these invasive species are also considered ecosystem engineers (e.g. most plants) (Crooks 2002), or can directly interact with them, understanding and predicting how invasive species can impact the environment is still a challenge (Lodge 1993; Simberloff et al. 2013). Notwithstanding, since the belowground environment is poorly studied compared to the aboveground environment, investigating how plants modulate resources and create belowground micro-habitats through their root growth is still a current and major challenge. Finally, besides these many future directions pointed out, studies related to ecosystem engineering can also consider biogeochemical heterogeneity, indirect interactions, key species, facilitation, creation, maintenance and destruction of habitats, human impact and use of species for conservation, and restoration of degraded habitats (Cuddington et al. 2011).

Conclusions

As we have shown, multiple aspects of plant-arthropod interactions can be considered ecosystem engineering. Through their own presence, growth, and death, plants autogenically modify the availability of resources for arthropods, providing suitable microhabitats and shelter for some arthropods while enhancing food supply and facilitating hunting for others. In some cases, this modification is part of a specialized relationship (ant gardens, domatia), and in other cases, it is aseemingly more generalized side effect of plant growth and development (shade, root structure, litter layer, bark, phytotelmata). Plants are also subject to modification by allogenic arthropod engineers, as with shelter-building and gall-forming insects, whose modifications of plant structures have implications for herbivore abundance and diversity. In other cases, arthropod activities can modulate the availability of resources for plants, as in the case of termite and ant mound-building. The vast majority of examples that we have provided show positive impacts ofecosystem engineering on biodiversity, and although not a universal outcome (Crain and Bertness 2005, 2006), this trend has been supported in the literature (Romero et al. 2015).

❓ Questions

1. What are the two main divisions of ecosystem engineers and what is the difference between them?
2. Think of a plant-insect interaction—which aspects of this interaction could be considered autogenic or allogenic engineering, and how might they modify the environment?
3. What are the implications of ecosystem engineering for biodiversity?

Acknowledgment   We thank Kleber Del-Claro, Roberth Pringle, Todd Palmer, Everton Tizo Pedroso and Escobar Studios for kindly handing over their photographs and videos.

References

Albert VR (1976) Zusammensetzung und vertikalverteilung der spinnenfauna in Buchenwäldern des solling. Faun-Ökol Mitt 5:65–80

Angers DA, Caron J (1998) Plant-induced changes in soil structure: processes and feedbacks. Biogeochemistry 42:55–72

Armbruster P, Hutchinson RA, Cotgreave P (2002) Factors influencing community structure in a South American tank bromeliad fauna. Oikos 96:225–234. https://doi.org/10.1034/j.1600-0706.2002.960204.x

Arredondo-Núñez A, Badano E, Bustamante R (2009) How beneficial are nurse plants? A meta-analysis of the effects of cushion plants on high-Andean plant communities. Community Ecol 10:1–6

Ayres E, Dromph KM, Cook R et al (2007) The influence of below-ground herbivory and defoliation of a legume on nitrogen transfer to neighbouring plants. Funct Ecol 21:256–263. https://doi.org/10.1111/j.1365-2435.2006.01227.x

Bailey JK, Whitham TG (2003) Interactions among elk, aspen, galling sawflies and insectivorous birds. Oikos 101:127–134. https://doi.org/10.1034/j.1600-0706.2003.12185.x

Balke M, Gó mez-Zurita J, Ribera I et al (2008) Ancient associations of aquatic beetles and tank bromeliads in the Neotropical forest canopy. Proc Natl Acad Sci 105:6356–6361

Bardgett RD, van der Putten WH (2014) Belowground biodiversity and ecosystem functioning. Nature 515:505–511

Beattie AJ (1985) The evolutionary ecology of ant-plant mutalisms. Cambridge University Press, New York

Beaudrot L, Du Y, Rahman Kassim A et al (2011) Do epigeal termite mounds increase the diversity of plant habitats in a tropical rain forest in peninsular malaysia? PLoS One 6. https://doi.org/10.1371/journal.pone.0019777

Benson WW (1985) Amazon ant-plants. In: Amazonia, pp 239–266

Benzing DH (2000) Bromeliaceae: profile of an adaptive radiation. Cambridge University Press, Cambridge

Berke S (2010) Functional groups of ecosystem engineers: a proposed classification with comments on current issues. Integr Comp Biol 50:147–157

Bignell DE (2006) Termites as soil engineers and soil processors. In: König H (ed) Intestinal microorganisms of termites and other invertebrates. Springer, Berlin/New York, pp 183–220

Bignell DE, Eggleton P (1995) On the elevated intestinal pH of higher termites (Isoptera: Termitidae). Insect Soc 42:57–69. https://doi.org/10.1007/BF01245699

Blomqvist MM, Olff H, Blaauw MB et al (2000) Interactions between above- and belowground biota: importance for small-scale vegetation mosaics in a grassland ecosystem. Oikos 90:582–598. https://doi.org/10.1034/j.1600-0706.2000.900316.x

Bonachela JA, Pringle RM, Sheffer E et al (2015) Termite mounds can increase the robustness of dryland ecosystems to climatic change. Science (80-) 347:651–655

Boogert NJ, Paterson DM, Laland KN (2006) The implications of niche construction and ecosystem engineering for conservation biology. Bioscience 56:570–578

Bronstein JL (2009) The evolution of facilitation and mutualism. J Ecol 97:1160–1170

Brouat C, McKey D (2000) Origin of caulinary ant domatia and timing of their onset in plant ontogeny: evolution of a key trait in horizontally transmitted ant-plant symbioses. Biol J Linn Soc 71:801–819. https://doi.org/10.1111/j.1095-8312.2000.tb01292.x

Buchman N, Cuddington K, Lambrinos J (2007) A historical perspective on ecosystem engineering. In: Cuddington K, Byers JE, Wilson WG, Hastings A (eds) Ecosystem engineers: plants to protists. Academic Press, Amsterdam/Boston, pp 25–46

Buse J, Ranius T, Assmann T (2008) An endangered longhorn beetle associated with old oaks and its possible role as an ecosystem engineer. Conserv Biol 22:329–337. https://doi.org/10.1111/j.1523-1739.2007.00880.x

Byers JJE, Cuddington K, Jones CCG et al (2006) Using ecosystem engineers to restore ecological systems. Trends Ecol Evol 21:493–500. https://doi.org/10.1016/j.tree.2006.06.002

9

Calder IR (2001) Canopy processes: implications for transpiration, interception and splash induced erosion, ultimately for forest management and water resources. Plant Ecol 153:203–214

Calixto ES, Lange D, Del-Claro K (2018) Protection mutualism: an overview of ant-plant interactions mediated by extrafloral nectaries. Oecol Aust 22:410–425. https://doi.org/10.4257/oeco.2018.2204.05

Camargo RX, Oliveira PS (2012) Natural history of the Neotropical arboreal ant, *Odontomachus hastatus*: nest sites, foraging schedule, and diet. J Insect Sci 12:1–9. https://doi.org/10.1673/031.012.4801

Cappuccino N, Martin M-A (1994) Eliminating early-season leaf-tiers of paper birch reduces abundance of mid-summer species. Ecol Entomol 19:399–401

Carroll RM, Kearby WH (1978) Microlepidopterous oak leaftiers (Lepidoptera: Gelechioidea) in central Missouri. J Entomol Soc 51:457–471

Charles-Dominique P (1971) Eco-éthologie et vie sociale des Prosimiens du Gabon. Biol gabonica 7:121–228

Charles-Dominique P, Atramentowicz M, Charles-Dominique M et al (1981) Les mammiferes frugivores arboricoles nocturnes d'une foret guyanaise: inter-relations plantes-animaux. Rev d'écologie 35:341–435

Chesson P (2000) Mechanisms of maintenance of species diversity. Annu Rev Ecol Syst 31:343–366. https://doi.org/10.1146/annurev.ecolsys.31.1.343

Cooper T, Frank J, Oecologica RC-A, 2014 U (2014) Loss of phytotelmata due to an invasive bromeliad-eating weevil and its potential effects on faunal diversity and biogeochemical cycles. Acta Oecol 54:51–56

Cornelissen T, Cintra F, Santos JC (2016) Shelter-building insects and their role as ecosystem engineers. Neotrop Entomol 45:1–12. https://doi.org/10.1007/s13744-015-0348-8

Crain CM, Bertness MD (2005) Community impacts of a tussock sedge: is ecosystem engineering important in benign habitats? Ecology 86:2695–2704

Crain CM, Bertness MD (2006) Ecosystem engineering across environmental gradients: implications for conservation and management. Bioscience 56:211–218

Crooks JA (2002) Characterizing ecosystem-level consequences of biological invasions: the role of ecosystem engineers. Oikos 97:153–166. https://doi.org/10.1034/j.1600-0706.2002.970201.x

Cuddington K (2007) Ecosystem engineering: utility contention, and progress. In: Cuddington K, Byers J, Wilson W, Hastings A (eds) Ecosystem engineers: plants to protists. Academic Press, Cambridge, pp 69–74

Cuddington K, Wilson WG, Hastings A (2009) Ecosystem engineers: feedback and population dynamics. Am Nat 173:488–498

Cuddington K, Byers JE, Wilson WG, Hastings A (2011) Ecosystem engineers: plants to protists. Academic Press, London

Darwin C (1881) The formation of vegetable mould, through the action of worms, with observations on their habits. Murray, London

Dauber J, Niechoj R, Baltruschat H, Wolters V (2008) Soil engineering ants increase grass root arbuscular mycorrhizal colonization. Biol Fertil Soils 44:791–796. https://doi.org/10.1007/s00374-008-0283-5

Davidson DW (1988) Ecological studies of neotropical ant gardens. Ecology 69:1138–1152. https://doi.org/10.2307/1941268

dos Santos LR, dos Feitosa R SM, Carneiro MAA (2017) The role of senescent stem-galls over arboreal ant communities structure in Eremanthus erythropappus (DC.) MacLeish (Asteraceae) trees. Sociobiology 64:7–13

Ehrenfeld JG (2003) Effects of exotic plant Iivasions on soil nutrient cycling processes. Ecosystems 6:503–523. https://doi.org/10.1007/s10021-002-0151-3

Eisenhauer N, Scheu S (2008) Invasibility of experimental grassland communities: the role of earthworms, plant functional group identity and seed size. Oikos 117:1026–1036. https://doi.org/10.1111/j.0030-1299.2008.16812.x

Erwin DH (2008) Macroevolution of ecosystem engineering, niche construction and diversity. Trends Ecol Evol 23:304–310. https://doi.org/10.1016/j.tree.2008.01.013

Evert R (2006) Esau's plant anatomy: meristems, cells, and tissues of the plant body: their structure, function, and development. Wiley, Hoboken

Facelli JM, Pickett STA (1991) Plant litter: light interception and effects on an old-field plant community. Ecology 72:1024–1031. https://doi.org/10.2307/1940602

Fernandez Barrancos EP, Reid JL, Aronson J (2017) Tank bromeliad transplants as an enrichment strategy in southern Costa Rica. Restor Ecol 25:569–576. https://doi.org/10.1111/rec.12463

Fiala B, Maschwitz U (1992) Domatia as most important adaptations in the evolution of myrmecophytes in the paleotropical tree genus Macaranga (Euphorbiaceae). Plant Syst Evol 180:53–64. https://doi.org/10.1007/BF00940397

Folgarait PJ (1998) Ant biodiversity and its relationship to ecosystem functioning: a review. Biodivers Conserv 7:1221–1244

Fonseca C (1994) Herbivory and the long-lived leaves of an Amazonian ant-tree. J Ecol 82:833–842

Fournier V, Rosenheim JA, Brodeur J et al (2003) Herbivorous mites as ecological engineers: indirect effects on zarthropods inhabiting papaya foliage. Oecologia 135:442–450. https://doi.org/10.1007/s00442-003-1212-9

Fukui A (2001) Indirect interactions mediated by leaf shelters in animal-plant communities. Popul Ecol 43:31–40. https://doi.org/10.1007/PL00012013

Fukui A, Murakami M, Konno K et al (2002) A leaf-rolling caterpillar improves leaf quality. Entomol Sci 5:263–266

Goheen JR, Palmer TM (2010) Defensive plant-ants stabilize megaherbivore-driven landscape change in an African savanna. Curr Biol 20:1768–1772

Gurnell A, Petts G (2006) Trees as riparian engineers: the Tagliamento River, Italy. Earth Surf Process Landf 31:1558–1574. https://doi.org/10.1002/esp.1342

Hammond PM (1997) Tree-crown beetles in context: a comparison of canopy and other ecotone assemblages in a lowland tropical forest in Sulawesi. In: Stork NE, Adis J, Didham RK (eds) Canopy arthropods. Chapman & Hall, London, pp 184–223

Hansen RA, Coleman DC (1998) Litter complexity and composition are determinants of the diversity and species composition of oribatid mites (Acari: Oribatida) in litterbags. Appl Soil Ecol 9:17–23

Hanula J, Franzreb K (1998) Source, distribution and abundance of macroarthropods on the bark of longleaf pine: potential prey of the red-cockaded woodpecker. For Ecol Manage 102:89–102

Hättenschwiler S, Tiunov AV, Scheu S (2005) Biodiversity and litter decomposition in terrestrial ecosystems. Annu Rev Ecol Evol Syst 36:191–218. https://doi.org/10.1146/annurev.ecolsys.36.112904.151932

Heil M, McKey D (2003) Protective ant-plant interactions as model systems in ecological and evolutionary research. Annu Rev Ecol Evol Syst 34:425–553. https://doi.org/10.1146/annurev.ecolsys.34.011802.132410

Henry S, Texier S, Hallet S et al (2008) Disentangling the rhizosphere effect on nitrate reducers and denitrifiers: insight into the role of root exudates. Environ Microbiol 10:3082–3092. https://doi.org/10.1111/j.1462-2920.2008.01599.x

Hinsinger P, Gobran GR, Gregory PJ, Wenzel WW (2005) Rhizosphere geometry and heterogeneity arising from root-mediated physical and chemical processes. New Phytol 168:293–303. https://doi.org/10.1111/j.1469-8137.2005.01512.x

Hooper RG (1996) Arthropod biomass in winter and the age of longleaf pines. For Ecol Manage 82:115–131

Huntly N, Inouye R (1988) Pocket gophers in ecosystems: patterns and mechanisms. Bioscience 38:786–793

Jackson J (1979) Tree surfaces as foraging substrates for insectivorous birds. In: Dickson JG, Connor RN, Fleet RR, Jackson JA, Kroll J (eds) The role of insectivorous birds in Forest Ecosystems. Academic Press, New York, pp 69–93

Janzen DH (1966) Coevolution of mutualism between ants and Acacias in Central America. Evolution (N Y) 20:249–275. https://doi.org/10.2307/2406628

Jones CG, Gutiérrez JL (2007) On the purpose, meaning, and usage of the physical ecosystem engineering concept. In: Cuddington K, Byers JE, Wilson WG, Hastings A (eds) Ecosystem engineers: plants to protists. Elsevier/Academic Press, Burlington, pp 3–20

9

Jones CG, Lawton JH, Shachak M (1994) Organisms as ecosystem engineers. Oikos 69:373–386. https://doi.org/10.2307/3545850

Jones CG, Lawton JH, Shachak M (1997a) Positive and negative effects of organisms as physical ecosystem engineers. Ecology 78:1946–1957. https://doi.org/10.1890/0012-- 9658(1997)078[1946:PANEOO]2.0.CO;2

Jones CG, Lawton JH, Shachak M (1997b) Ecosystem engineering by organisms: why semantics matters. Trends Ecol Evol 12:275

Joseph GS, Seymour CL, Cumming GS et al (2013) Termite mounds as islands: woody plant assemblages relative to termitarium size and soil properties. J Veg Sci 24:702–711

Kaneko N, Salamanca EF (1999) Mixed leaf litter effects on decomposition rates and soil microarthropod communities in an oak-pine stand in Japan. Ecol Res 14:131–138. https://doi.org/10.1046/j.1440-1703.1999.00292.x

Kang BT (1978) Effect of some biological factors on soil variability in the tropics - III. Effect of macrotermes mounds. Plant and Soil 50:241–251. https://doi.org/10.1007/BF02107175

Keiser CN, Pruitt JN (2014) Spider aggressiveness determines the bidirectional consequences of host-inquiline interactions. Behav Ecol 25:142–151

Kitching RL (2000) Food webs and container habitats: the natural history and ecology of phytotelmata. Cambridge University Press, Cambridge

Laland KN, Odling-Smee FJ, Feldman MW (1999) Evolutionary consequences of niche construction and their implications for ecology. Proc Natl Acad Sci U S A 96:10242–10247. https://doi.org/10.1073/pnas.96.18.10242

Lavelle P (1986) Associations mutualistes avec la microflore du sol et richesse spécifique sous les tropiques: l'hypothèse du premier maillon. Comptes rendus l'Académie des Sci 302:11–14

Lee KE, Wood TG (1971) Termites and soils. Academic Press, London

Lill JT, Marquis RJ (2003) Ecosystem engineering by caterpillars increases insect herbivore diversity on white oak. Ecology 84:682–690. https://doi.org/10.1890/0012-9658(2003)084[0682:EEBCII]2.0.CO;2

Lill JT, Marquis RJ (2007) Microhabitat manipulation: ecosystem engineering by shelter-building insects. In: Cuddington K, Byers J, Wilson W, Hastings A (eds) Ecosystem engineers: plants to protists. Academic Press, Cambridge, pp 107–138

Lodge DM (1993) Biological invasions: lessons for ecology. Trends Ecol Evol 8:133–137

Longino J (1996) Taxonomic characterization of some live-stem inhabiting *Azteca* (Hymenoptera: Formicidae) in Costa Rica, with special reference to the ants of *Cordia* (Boraginaceae) and *Triplaris* (Polygonaceae). J Hymenopt Res 5:131–156

Majer JD, Recher HF, Graham R, Gupta R (2003) Trunk invertebrate faunas of Western Australian forests and woodlands: influence of tree species and season. Austral Ecol 28:629–641. https://doi.org/10.1046/j.1442-9993.2003.01320.x

Mariani J, Manuwal D (1989) Factors influencing brown creeper (*Certhia americana*) abundance patterns in the southern Washington Cascade range. Stud Avian Biol 13:53–57

Martinsen GD, Floate KD, Waltz AM et al (2000) Positive interactions between leafrollers and other arthropods enhance biodiversity hybrid cottonwoods. Occologia 123:82–89. https://doi.org/10.1007/s004420050992

Mayer AL, Rietkerk M (2004) The dynamic regime concept for ecosystem management and restoration. Bioscience 54:1013. https://doi.org/10.1641/0006-3568(2004)054[1013:tdrcfe]2.0.co;2

McComie LD, Dhanarajan G (1993) The physical and chemical composition of mounds of *Macrotermes carbonarius* (Hagen) (Termitidae, Macrotermitinae), in Penang, Malaysia. J Soil Sci 44:427–433

Mitchell RJ, Campbell CD, Chapman SJ, Cameron CM (2010) The ecological engineering impact of a single tree species on the soil microbial community. J Ecol 98:50–61

Morgan LH (1868) The American beaver and his works. TheClassics.us

Moura RF, Tizo-Pedroso E, Del-Claro K (2018) Colony size, habitat structure, and prey size shape the predation ecology of a social pseudoscorpion from a tropical savanna. Behav Ecol Sociobiol 72:103. https://doi.org/10.1007/s00265-018-2518-2

Murakami M (1999) Effect of avian predation on survival of leaf-rolling lepidopterous larvae. Res Popul Ecol (Kyoto) 41:135–138. https://doi.org/10.1007/s101440050015

Nicolai V (1986) The bark of trees: thermal properties, microclimate and fauna. Oecologia 69:148–160. https://doi.org/10.1007/BF00399052

Novotny V, Basset Y, Miller SES et al (2002) Low host specificity of herbivorous insects in a tropical forest. Nature 416:841–844. https://doi.org/10.1038/416841a

O'Dowd D, Willson M (1991) Associations between mites and leaf domatia. Trends Ecol Evol 6:179–182

Odling-Smee FJ, Laland KN, Feldman MW (2013) Niche construction: the neglected process in evolution (MPB-37). Princeton university press, New Jersey

Orivel J, Leroy C (2011) The diversity and ecology of ant gardens (Hymenoptera: Formicidae; Spermatophyta: Angiospermae). Myrmecological News 14:73–85

Ozanne CHP, Anhuf D, Boulter SL et al (2003) Biodiversity meets the atmosphere: a global view of forest canopies. Science (80-) 301:183–186. https://doi.org/10.1126/science.1084507

Pacheco PSM, Del-Claro K (2018) *Pseudomyrmex concolor* Smith (Formicidae: Pseudomyrmecinae) as induced biotic defence for host plant *Tachigali myrmecophila* Ducke (Fabaceae: Caesalpinioideae). Ecol Entomol 43:782–793. https://doi.org/10.1111/een.12665

Pearce T (2011) Ecosystem engineering, experiment, and evolution. Biol Philos 26:793–812

Perfecto I, Vandermeer J (1996) Microclimatic changes and the indirect loss of ant diversity in a tropical agroecosystem. Oecologia 108:577–582. https://doi.org/10.1007/BF00333736

Pollierer MM, Langel R, Körner C et al (2007) The underestimated importance of belowground carbon input for forest soil animal food webs. Ecol Lett 10:729–736. https://doi.org/10.1111/j.1461-0248.2007.01064.x

Power ME (1997) Ecosystem engineering by organisms: why semantics matters Reply from M. Power. Trends Ecol Evol 12:275–276

Power ME, Tilman D, Estes JA et al (1996) Challenges in the quest for keystones: identifying keystone species is difficult—but essential to understanding how loss of species will affect ecosystems. Bioscience 46:609–620

Price PW, Fernandes GW, Waring GL (1987) Adaptive nature of insect galls. Environ Entomol 16:15–23. https://doi.org/10.1086/330311

Raath MJ, Le Roux PC, Veldtman R et al (2017) Empty *Gonometa postica* cocoons function as nest sites and shelters for arboreal ants. J Arid Environ 144:42–47

Reichman OJ, Seabloom EW (2002a) Ecosystem engineering: a trivialized concept?: Response from Reichman and Seabloom. Trends Ecol Evol 17:308

Reichman OJ, Seabloom EW (2002b) The role of pocket gophers as subterranean ecosystem engineers. Trends Ecol Evol 17:44–49

Remsen JV, Parker TA (1984) Arboreal dead-leaf-searching birds of the neotropics. Condor 86:36–41. https://doi.org/10.2307/1367341

Reyes-López J, Ruiz N, Fernández-Haeger J (2003) Community structure of ground-ants: the role of single trees in a Mediterranean pastureland. Acta Oecol 24:195–202

Rillig MC (2004) Arbuscular mycorrhizae and terrestrial ecosystem processes. Ecol Lett 7:740–754. https://doi.org/10.1111/j.1461-0248.2004.00620.x

Romero GQ, Benson WW (2004) Leaf domatia mediate mutualism between mites and a tropical tree. Oecologia 140:609–616. https://doi.org/10.1007/s00442-004-1626-z

Romero GQ, Gonçalves-Souza T, Vieira C, Koricheva J (2015) Ecosystem engineering effects on species diversity across ecosystems: a meta-analysis. Biol Rev 90:877–890. https://doi.org/10.1111/brv.12138

Simberloff D, Martin J, Genovesi P et al (2013) Impacts of biological invasions: what's what and the way forward. Trends Ecol Evol 28:58–66

Stinchcombe JR, Schmitt J (2006) Ecosystem engineers as selective agents: the effects of leaf litter on emergence time and early growth in Impatiens capensis. Ecol Lett 9:255–267. https://doi.org/10.1111/j.1461-0248.2005.00872.x

Stone GN, Schönrogge K (2003) The adaptive significance of insect gall morphology. Trends Ecol Evol 18:512–522. https://doi.org/10.1016/S0169-5347(03)00247-7

Strayer DL (2012) Eight questions about invasions and ecosystem functioning. Ecol Lett 15:1199–1210

Velasque M, Del-Claro K (2016) Host plant phenology may determine the abundance of an ecosystem engineering herbivore in a tropical savanna. Ecol Entomol 41:421–430. https://doi.org/10.1111/een.12317

9

Waltz AMYM, Whitham TG (1997) Plant development affects arthropod communities: opposing impacts of species removal. Ecology 78:2133–2144

Wardle D, Bardgett R et al (2004) Ecological linkages between aboveground and belowground biota. Nature 304:1629–1633

Watson JP (1977) The use of mounds of the termite *Macrotermes falciger* (Gerstäcker) as a soil amendment. J Soil Sci 28:664–672. https://doi.org/10.1111/j.1365-2389.1977.tb02273.x

Weiss MR, Wilson EE, Castellanos I (2004) Predatory wasps learn to overcome the shelter defences of their larval prey. Anim Behav 68:45–54. https://doi.org/10.1016/j.anbehav.2003.07.010

Wetzel WC, Screen RM, Li I et al (2016) Ecosystem engineering by a gall-forming wasp indirectly suppresses diversity and density of herbivores on oak trees. Ecology 97:427–438

Wheeler J, Longino JT (1988) Arthropods in live oak galls in Texas. Entomol News 99:25–29

Wigginton RD, Pearson J, Whitcraft CR (2014) Invasive plant ecosystem engineer facilitates community and trophic level alteration for brackish marsh invertebrates. Ecosphere 5:1–17. https://doi.org/10.1890/ES13-00307.1

Wilson WG (2007) A new spirit and concept for ecosystem engineering. In: Cuddington K, Byers JE, Wilson WG, Hastings A (eds) Ecosystem engineers: plants to protists. Academic Press, London, pp 47–59

Wood TG (1996) The agricultural importance of termites in the tropics. In: Evans K (ed) Agricultural zoology reviews. Intercept Ltd, Andover, pp 117–155

Wright JP, Jones CG (2004) Emphasizing new ideas to stimulate research in ecology predicting effects of ecosystem engineers on patch-scale species richness from primary productivity. Ecology 85:2071–2081

Wright JP, Jones CG (2006) The concept of organisms as ecosystem engineers ten years on: progress, limitations, and challenges. Bioscience 56:203–209. https://doi.org/10.1641/0006--3568(2006)056[0203:tcooae]2.0.co;2

Wright JP, Jones CG, Flecker AS (2002) An ecosystem engineer, the beaver, increases species richness at the landscape scale. Oecologia 132:96–101. https://doi.org/10.1007/s00442-002-0929-1

Yasuda M, Planning FK-L and U (2009) The contribution of the bark of isolated trees as habitat for ants in an urban landscape. Landsc Urban Plan 92:276–281

Yu DW (1994) The structural role of epiphytes in ant gardens. Biotropica 26:222–226

Additional Resources
Boxes

https://www.sciencedirect.com/science/article/pii/S0960982210010055
https://esajournals.onlinelibrary.wiley.com/doi/full/10.1890/15-1347.1

Figure 1

A. https://link.springer.com/article/10.1007/s00265-018-2518-2
B. https://link.springer.com/article/10.1023/A:1005944025343#:~:text=By%20penetrating%20the%20soil%2C%20roots,and%20strengthening%20of%20the%20soil
C,D,F. https://link.springer.com/article/10.1007/s13744-015-0348-8
E. https://onlinelibrary.wiley.com/doi/abs/10.1111/een.12317

Videos

https://www.youtube.com/watch?v=ccpDJvGMVuw
https://www.youtube.com/watch?v=4YoKaR9AMhc
https://www.youtube.com/watch?v=cSzTGmX7KaY
Ants swarming out of a swollen thorn of *Acacia drepanolobium*.

Disentangling Plant-Animal Interactions into Complex Networks: A Multi-view Approach and Perspectives

Pedro Luna and Wesley Dáttilo

Contents

© Springer Nature Switzerland AG 2021
K. Del-Claro, H. M. Torezan-Silingardi (eds.), *Plant-Animal Interactions*, https://doi.org/10.1007/978-3-030-66877-8_10

🔲 **Learning Objectives**

After completing this chapter, you should have an understanding of the following:

- The terms: species interaction networks, nestedness, modularity, individual-based networks, multilayer networks.
- How stochastic and deterministic factors shape the organization of plant-animal networks.
- The ways in which interaction networks vary over space and time and are affected by environmental perturbations.
- The potential of ecological networks for understanding plant-animal interactions and their evolutionary and coevolutionary processes.

10.1 Introduction

Since the seminal book *On the Origin of Species* by Charles Darwin, modern ecologists have been aware that the origin, maintenance and function of biodiversity are products of biotic interactions among species (Darwin 1859; Ings and Hawes 2018). Particularly, the majority of animal and plant species are embedded in highly diverse trophic systems which include different types of organisms within a large and complex web of life (Herrera and Pellmyr 2002). The idea of a web of life has been used as a theoretical framework to study the complexity of these plant-animal relationships in species interaction networks (Dáttilo and Rico-Gray 2018). In these interaction networks, plant and animal species are denoted by nodes and their interactions by links describing the use of plants as resources by animals (Ings et al. 2009). By studying plant-animal interaction networks, researchers have provided new insights on how interactive communities are structured and the ecological and evolutionary dynamics of these systems (Bascompte et al. 2003; Bascompte and Jordano 2007; Vázquez et al. 2009).

In general, plant-animal interactions range from negative (e.g., parasitism and herbivory; Poisot et al. 2017; López-Carretero et al. 2018) to positive (e.g., pollination and seed dispersal; Vizentin-Bugoni et al. 2018; Anjos et al. 2020) interactions, all of which directly contribute to the functioning of ecosystems. For example, in the tropics, 90% of the woody plant species depend on animals to move their pollen and to disperse their seeds (Jordano 2000; Ollerton 2017). Without such mutualistic interactions, these plants would be unable to complete their life cycles, and their animal partners would be unable to feed themselves (Herrera and Pellmyr 2002). Recent evidence supports the idea that ecological interactions are the driving force of natural selection in nature, where all species reciprocally affect each other across their populations and communities (Guimarães et al. 2011; Medeiros et al. 2018). For example, a study performed by Galetti et al. (2013) showed that the loss of large-gape avian frugivores in forest fragments led to a reduction in seed size, highlighting the role of ecological interactions in the evolutionary trajectories and compositions of tropical forests. Despite the important role of biotic interactions in both ecological and evolutionary processes, there is still only limited use of biotic interaction theory to address ecological hypotheses.. Therefore, the incorporation of the diversity of biotic interactions into species interaction networks remains an important gap in our knowledge of biodiversity and ecosystem functioning (Luna et al. 2020).

Even though plant-animal interactions can be studied under several approaches and facets, here we aim to explain how plants and animals interact by forming non-random ecological networks, and how both stochastic and deterministic factors shape the organization of these networks across environmental gradients. Moreover, in this chapter, we will explain how ecological networks can be used to study not only how groups of individuals from different species and trophic levels interact with each other (i.e., interspecific networks) but also how individuals within groups vary in their interactions (i.e., individual-based networks). Finally, we will introduce a multi-layer view of ecological networks, and we will conclude by discussing the future of ecological networks and their robustness and stability to different environmental disturbances.

10.2 The Non-random Organization of Plant-Animal Interaction Networks

Historically most studies have focused on describing and explaining the ecological and evolutionary dynamics only between pairs of species (see Herrera and Pellmyr 2002). For this reason, ecological interactions among *groups* of species are one of the less understood and studied components of biodiversity (Luna et al. 2020). In fact, only in the past 20 years have we started to understand the role of ecological interactions in shaping biodiversity and ecosystem functioning (Ings and Hawes 2018). Now, ecological theory recognizes that ecological communities are groups of species that not only coexist but also depend on each other (Vellend 2016). For example, plants provide food resources for many animal species in exchange for positive services (e.g., pollination or seed dispersal; Escribano-Avila et al. 2018; Vizentin-Bugoni et al. 2018), but are also strongly affected by antagonistic partners (e.g., herbivores and seed predators; López-Carretero et al. 2018; Luna et al. 2018a). These plant-animal relationships are part of the evolutionary history of natural ecosystems, in which groups of species from both trophic levels have developed optimal strategies that maintain the complexity and functionality of biodiversity (Bascompte et al. 2003; Bascompte and Jordano 2007). Under a complex network approach, functionality arises from organization, and therefore disordered (i.e., random) networks might not be functional. To better understand the relationship between organization and functionality, imagine that you have a disassembled car: if you put all of its parts together in a random manner, you will end up with many important parts in the wrong places, and the final product will not be a functional car. However, if you assemble all the pieces of the car in an ordered manner, it will be much more likely to run and be functional. The same idea applies to plant-animal interactions. Species play different functional roles that allow some species to establish multiple interactions and maintain cohesive networks, providing functionality and robustness to the system (Jordano et al. 2003; Olesen et al. 2007; Dáttilo et al. 2016). As plant-animal interactions are key to maintaining the functionality of environments, the maintenance of functionality depends on spe-

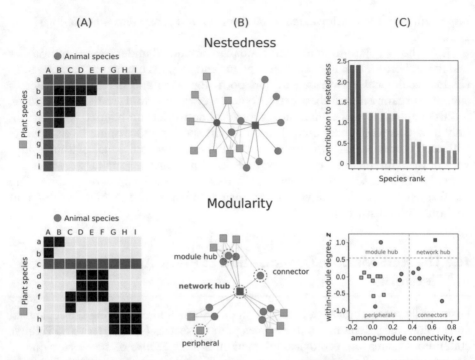

■ **Fig. 10.1** **a** Matrices denoting the nested (top) and modular pattern of interactions (bottom) between plants (rows) and animals (columns). Black cells represent species pairs that interaction, gray cells represent species pairs that do not interact, and red cells represent interactions involving highly connected species. **b** Plant-animal interaction networks representing the interactions in **a**. Blue and green nodes represent animal and plant species, respectively. Red nodes denote highly connected species within both nested and modular pattern of interactions. **c** Plant and animal species contributions to nested and modular patterns of interactions (see text for detailed information)

cies interactions having an organized structure (i.e., interactions are not randomly assembled; Bascompte et al. 2003; Olesen et al. 2007).

Two main non-random structural properties have been frequently reported in interaction networks: nestedness and modularity (Dehling 2018). In nested networks, species that engage in few interactions (specialists) are connected to the rest of the community by their interactions with a subset of highly interactive species (generalists), while interactions between specialists rarely occur in nature (■ Fig. 10.1, Bascompte et al. 2003). Moreover, a core group of generalist plants and animals interact with virtually all other species in the network, performing similar roles and thereby lending high functional redundancy to the system (Bascompte and Jordano 2007). The other non-random structural property, modularity, describes a pattern in which a subgroup of species in one trophic level interacts more frequently with a group of species from another trophic level (■ Fig. 10.1, Olesen et al. 2007). Both nestedness and modularity occur along a continuum, and a network can be nested and modular at the same time, or even have nested modules (Dehling 2018).

In the classical literature, nestedness has been called the *"architecture of biodiversity"* since this organization promotes species coexistence by reducing competition and increasing biodiversity (Bascompte and Jordano 2007; Bastolla et al. 2009). But what generates nestedness in species interaction networks? Well, one of the main explanations for the emergence of nestedness in plant-animal networks is attributed to the natural structure of ecological communities. We know that about 60–70% of nested patterns can be attributed to differences in relative species abundances (i.e., few abundant and many rare species; Vázquez et al. 2007; Krishna Jr et al. 2008). In this case, abundant species should interact most frequently with each other as well as with less abundant species, while rare species are more likely to interact with more abundant species rather than with other rare species. This abundance-based process shaping plant-animal networks means that interactions with rare species become subsets of the interactions of the most abundant species and, consequently, result in nested interaction networks (Krishna Jr et al. 2008). However, stochastic factors structuring ecological communities (i.e., dispersal, drift and selection as proposed in the Unified Neutral Theory of Biodiversity; Hubbell 2001) also influence the ways in which species interact. The variability in nested networks can also be explained by deterministic factors, and elucidating which of and how these factors affect ecological networks has become one of the main challenges for field ecologists. One study performed by Dáttilo et al. (2014a) showed that ant dominance hierarchies can determine about 50% of the nestedness patterns found in ant-plant networks mediated by extrafloral nectaries. Specifically, these authors showed that an ant species' position within the nested network could be predicted only by differences among numerical dominance and recruitment of ant species. Therefore, both stochastic (i.e., neutral theory) and deterministic (i.e., niche-based theory) processes underlying nestedness have been proposed to explain the origin of structural patterns in ecological networks.

For modular networks, the presence of semi-independent groups of highly interactive species has been shown to be the result of multiple drivers of organization and functioning (Hintze and Adami 2008). Some of the most studied factors driving modularity are pollination and seed dispersal syndromes and spatio-temporal variation, which are well documented ecological filters (Vazquez et al. 2009; Donatti et al. 2011; Tylianakis and Morris 2017). The idea of pollination syndromes is based on observations suggesting that floral phenotypes reflect specialization toward certain groups of floral visitors (Fenster et al. 2004). In the case of seed dispersal syndromes, the hypothesis is based on the fact that fruits have a heterogeneous set of traits (i.e., different shapes, sizes and colors) and, therefore, different groups animals (e.g., birds, bats or ants) interact with particular sets of plant species (Donatti et al. 2011). Following this idea, ornithophilous flowers should only rarely be visited by insects or mammals, so humming birds and ornithophilous flowers should form a cohesive group within a larger plant-pollinator network (Olesen et al. 2007; Vazquez et al. 2009). This same reasoning can be extended to networks that include multiple groups of frugivores, in which large frugivores (e.g., mammals) eat bigger fruits while smaller frugivores (e.g. birds) eat smaller fruits, generating modularity (Donatti et al. 2011). In addition, the temporal and spatial co-occurrence of species is another factor affecting the presence of an interaction, and may also generate modularity. In a study performed by Araujo et al. (2018) in the Neotropical savanna, the authors found that plant species compositions along a spatial gradient were not the same (i.e., high species turnover) and that their floral traits and floral visitors also were exclusive to certain regions, resulting in isolated/modular networks. Moreover, modular networks may reflect phylogenetic clustering and divergent evolutionary histories (Schleuning et al. 2014), or even the simple spatial foraging of pollinators as demonstrated by Dupont et al. (2014). These findings indicate that, as with species occurrences, plant-animal interaction networks are context and scale dependent (Dáttilo et al. 2019). Despite the importance of considering the drivers of interaction network patterns, we are only beginning to understand the main mechanisms and processes behind the organization of plant-animal interaction networks.

Since interaction frequencies are highly heterogeneous and differ among species within a network, they also reflect variation in trophic specialization (i.e., trophic niche breadth; Devictor et al. 2010). Under a network approach, specialization has two components: (i) niche breadth, which is the number of other species with which an individual species interacts, and (ii) niche overlap, which is the similarity of interactions between species in a network (Blüthgen et al. 2006). Thereby, trophic specialization at the network level is a measure of the total number of interactions in a network and the similarity in interactions between species, which gives a complete overview of how the group of species interact (Blüthgen et al. 2006; Vázquez et al. 2007). Note that this approach based on network theory is different from the common interpretation of trophic specialization, in which if a bird just eats fruit from a few plant species it will be considered a specialist, but if it eats fruits from a large set of plant species it will be considered a generalist, without considering other species with which the plant or bird interacts. However, in ecological networks, the fact that an animal eats a small set of food resources might not reflect a true specialization, since such an animal might be a rare species, or those resources might also be eaten by many other species of animals. To avoid assuming that all rare interactions are specialized ones, measures of species specialization within interaction networks tend to consider the availability of partners (Blüthgen et al. 2006).

Until now we have explained network structure considering the network as a whole, but we can also explain network organization with a species-level framework. As the building blocks of ecological networks are individuals from different species, we can measure the contribution of each species to the overall structure previously described, namely nestedness or modularity (Olesen et al. 2007; Saavedra et al. 2011). For nestedness, we can measure how much the contribution of a given species to nestedness differs from that expected randomly (Saavedra et al. 2011). Such an approach is based on the idea that if species interactions are randomly sorted, one would not expect a network to remain cohesive and functional. Species play different roles in maintaining network structure, and studies have shown that only a few species contribute positively to maintaining nested structures (◘ Fig. 10.1c, Saavedra et al. 2011). This is a paradox, as in some cases the species that most contribute to the network structure are also those most vulnerable to extinction (Vidal et al. 2014). For modular networks, species also play different roles in holding the network together. To elucidate the role of each species within a modular network, we can compute the extent to which each species is connected to the other species in its module (i.e., within-module degree, z_i), as well as the degree to which the interactions of a given species are evenly distributed across modules (i.e., among-module connectivity, c_i. Olesen et al. 2007; Dehling 2018, ◘ Fig. 10.1c). Species can then be classified as peripherals (i.e., species with few interactions with other species), connectors (i.e., species connecting several modules to each other), module hubs (i.e., species with many interactions only within their own modules), or network hubs (i.e., species who have both many interactions within their own modules but also connect several modules to each other). Moreover, we also can think about the structural role of a species in maintaining a cohesive and resilient network. This species-level framework is extremely useful in obtaining additional information on how roles within plant-animal networks may act in complementary ways, ultimately allowing identification of potential network collapses and extinction cascades (Dáttilo et al. 2016).

10.3 Plant-Animal Networks Across Natural Gradients

Just as the composition, abundance and richness of species vary over environmental gradients, ecological interactions also vary over space and time (Poisot et al. 2015). The variation of ecological interactions is mainly due to the highly dynamic nature of species interactions (CaraDonna et al. 2017). For instance, species interactions can be assembled and disassembled over short periods of time (e.g., through the course of a day and even between day and night) (Luna et al. 2018b). Moreover, although two species can occur in the same place, simple co-occurrence does not automatically determine that two species will interact, as species must pass through ecological filters in order to establish an interaction (Dormann et al. 2017; Tylianakis and Morris 2017). In addition, with the current situation of global change, the introduction of an exotic species into an ecosystem may alter local trophic chains and create additional variation in ecological network dynamics (de M Santos et al. 2012). An important point we need to consider is that the high dynamism of species interactions has shaped their evolutionary and co-evolutionary history over natural gradients (Medeiros et al. 2018). For this reason, the study of interaction networks has attracted the attention of researchers working with the role of biotic interactions in maintaining biodiversity.

In the last section we pointed out that the encounter probability of a partner (i.e., stochastic factors) is one factor that determines how species interact. However, ecological filters related to environmental conditions can also change the way species interact, independently of species composition. For instance, landscapes with higher forest cover and landscape heterogeneity hold higher diversity of ant-plant interactions than landscapes with lower forest cover and landscape heterogeneity (Corro et al. 2019). Similar trends have been reported for plant-pollinator networks, where the number of interactions and network nestedness increase with increasing landscape heterogeneity (Moreira et al. 2015). In addition, Dalsgaard et al. (2011) showed that the specialization level in plant-hummingbird networks is

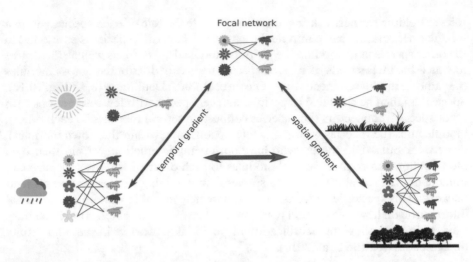

◘ Fig. 10.2 Hypothetical representation highlighting how species and their ecological interactions within a focal pollination network observed in the field can change over temporal (here represented as dry and rainy seasons) and spatial (here represented as pasture and forest environments) gradients

associated with contemporary precipitation across the Americas. In these cases, the ways in which ecological networks vary are driven by changes in environmental conditions over space (◘ Fig. 10.2). Furthermore, to better understand how environmental factors affect ecological networks at large spatial scales, we also need to consider elevational and latitudinal gradients.

Historically, both elevational and latitudinal gradients have been used as natural experiments to elucidate which environmental factors explain the distribution of biodiversity around the globe. We know that with increasing elevation, environmental conditions (e.g., lower temperature and higher solar radiation) change over very short distances, providing an excellent framework for studying ecological networks along these elevation gradients. As this field of study is quite new to involving ecological networks, only a few studies have assessed how environmental factors affect the organization of species interactions across elevations. However, we know that as elevation increases, the nestedness of plant-pollinator networks decreases, making them more susceptible to random extinctions (Ramos-Jiliberto et al. 2010). More recent evidence has also revealed that the trophic specialization of plant-pollinator networks decreases with increasing elevation in Mt. Kilimanjaro (Africa), highlighting the importance of environmental gradients in structuring species interactions (Classen et al. 2020). In the case of latitudinal gradients, one the most influential studies showed that trophic specialization of plant-pollinator and seed-dispersal networks is lower at tropical than at temperate latitudes (Schleuning et al. 2012). This trend is due to the greater climatic stability in the tropics, which generates higher resource diversity and, therefore, higher generalization of consumer species (Schleuning et al. 2012). Another study found that Quaternary climate-change (climatic shifts in the last 2.6 million years) is negatively associated with modularity and positively associated with nestedness (Dalsgaard et al. 2013). These findings indicate that both current and historical climate together offer a complementary explanation for the organization of biotic interactions that we find in nature today.

In addition to spatial variation, ecological networks are also driven by seasonal and temporal changes in environmental conditions. One study dealing with ants that visit plants bearing extrafloral nectaries showed that in periods of the year when plant extrafloral nectar production peaked, ant-plant interaction networks tended to be more nested and less specialized (Lange et al. 2013). Similarly, Ramos-Robles et al. (2016) found that in periods of the year with higher fruit availability, plant-frugivore networks also presented a more nested and generalized pattern of interactions (Ramos-Robles et al. 2016). For antagonistic networks, a recent study found that in the season with higher precipitation, harvester ants collected fewer seed species which led to less specialized networks (Anjos et al. 2019). It is important to mention that studies on the variation of plant-animal networks throughout time are not yet common when compared to studies on plant-animal networks along spatial gradients and, therefore, new studies should test the generality of these conclusions.

As environmental conditions fluctuate and affect plant-animal network properties, they also affect both species and biotic interactions. But how can we measure network variation over space and time? Poisot et al. (2012) proposed a new framework that considers another dimension of ecological networks: the interaction beta diversity. By mentioning the term *beta*, we refer to the turnover of species interactions and two additive components: interaction turnover (i.e., changes in the identities of interactions due to spatial/temporal changes in species composition) and interaction rewiring (i.e., changes in the identities of the interactions generated by the reassembly of interactions between the same species in different sampling sites/times; Poisot et al. 2012) (◘ Fig. 10.3). Studies dealing with interaction beta diversity have shown that despite evident changes in species composition over time or space, network properties such as nestedness or modularity do not change over both environmental gradients (Dáttilo et al. 2013a; Kemp et al. 2017). Therefore, to elucidate another aspect of network variation in relation to time or space we can consider the turnover of interactions. In recent years, network researchers have observed that species interactions tend to vary more frequently than species composition, and that abiotic factors can be a main driver of this interaction turnover (Carstensen et al. 2014; Poisot et al. 2017; Dáttilo and Vasconcelos 2019). Moreover, we now know that when geographic and environmental distances increase between habitats, species turnover has been identified as an important driver of the interaction beta diversity (Dáttilo and Vasconcelos 2019). Conversely, when we compare sites with highly similar species compositions, the main driver of interaction beta diversity tends to be interaction rewiring among species (Luna et al. 2018b). In addition to studying interaction beta diversity over spatial gradients, this framework also allows us to measure how ecological networks vary over time. For example, CaraDonna et al. (2017) showed that interaction rewiring was the main driver of interaction beta diversity in plant-pollinator networks across time, in this case between weeks. This ability of species to switch partners was mainly due to constant changes in the co-occurrence and abundance of species, directly influencing who could interact with whom (CaraDonna et al. 2017). Although current knowl-

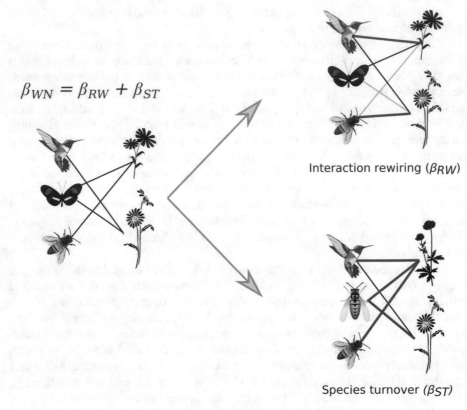

$$\beta_{WN} = \beta_{RW} + \beta_{ST}$$

Interaction rewiring (β_{RW})

Species turnover (β_{ST})

□ Fig. 10.3 Diagram illustrating the components of interaction turnover (i.e., beta diversity of interactions). The gray arrows indicate two possible paths of interaction change between the original plant-animal network on the left and the two alternative networks on the right. The top-right diagram shows an interaction rewiring, where gray links denote those interactions that are lost with respect to the network in the left side. Red links denote the new interactions between shared species and black links indicate the interactions that did not change between the two networks. In the bottom-right diagram, changes in the composition of species drive changes in species interactions as denoted by blue links, while the black link is the interaction that did not change between the two networks

edge allows us to hypothesize how environmental patterns can change ecological networks, there is still a lot of work to be done. Now, researchers working with interaction networks are seeking to understand how environmental factors affect the structure and the ecological and evolutionary dynamics of plant-animal interaction networks in different ecosystems around the world.

10.4 Individual-Based Networks: Linking Communities to Populations

In the previous section, we described how groups of plants and animals establish non-random patterns of organization within "species interaction networks" and how these networks vary through space and time. In those networks at the species level, we pool all records of the interactions of individuals of a species in the field and represent them as a single node in the network. However, species do not actually interact in nature. That is, individuals of species meet and interact with each other, rather than the "species" themselves interacting. In this sense, some studies have recently used tools and concepts derived from Graph Theory to study intra-population network variation (Pires et al. 2011; Dáttilo et al. 2013b). These individual-based networks use a set of mathematical abstractions to identify and connect interactions performed by individuals within a population, such as different feeding habits. It is worth clarifying that although the use of interspecific networks is now popular and widespread, it is still uncommon to use network analysis to assess the variation in interactions among individuals of a population. In order to unravel the details of individual-based networks, in this section we will present a biological and ecological framework for studying how populations interact amongst themselves by applying an intrapopulation network approach. Moreover, we will explain how such networks are non-randomly assembled similarly to the non-random patterns displayed by interspecific networks.

The variation in individual traits (i.e., phenotype) is a property of all populations and it is a key element of evolutionary processes, since natural selection can only act on variation within populations. These intraspecific trait variations are directly related to the ways in which different individuals of the same species interact, thereby affecting both population and community dynamics (Begon et al. 2006; Bolnick et al. 2011). Imagine a population of shrubs of the same species, all of them genetically similar but not phenotypically equal. In this theoretical shrub population, each individual has a different size, age and set of traits (e.g., number of flowers or fruits). Such differences among individuals of the same species might influence how many interactions each shrub can establish. For instance, if one shrub bears a higher number of fruits it might be more attractive to frugivorous birds compared to another individual with fewer fruits (◻ Fig. 10.4). Correspondingly, this could also be applied to animals. For example, the individual traits of each individual bird in a population (e.g., beak size, age, gender, body size and ecophysiological needs) are going to affect how many interactions each individual can establish (◻ Fig. 10.3). Therefore, intra-population variation and individual specialization are the cornerstones of individual-based networks.

As shown for interspecific networks, individual-based networks also display both nestedness and modularity as non-random structures. However, both measures have to be interpreted properly for individual-based networks to avoid any confusion with their interspecific counterparts. Thereby, the nestedness of an individual-based network reflects that the use of resources by individuals has a hierarchy, in which the set of resources used by individuals with few interactions represents a subset of the resources used by individuals with many interactions (Pires et al. 2011). Intraspecific modularity can show that there are individuals that use resources in a "selective" way, establishing more interactions with a set of resources than with others. Nestedness has been shown to be a property of both mutualistic (e.g., ant-plant protective systems and primate frugivory respectively; Dáttilo et al. 2014a, b) and antagonistic individual-based networks (e.g., seed predation by harvester ants; Luna et al. 2018a, b). The empirical evidence indicates that nestedness

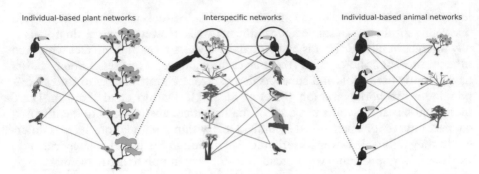

Fig. 10.4 Interspecific ecological networks involving plants and animals can be viewed as individual-based networks to assess interindividual variation. In these individual-based networks, plant (left) and animal (right) individuals are nodes, and their ecological relationships with species from the other trophic level are depicted by links (lines)

in individual-based networks also could be explained by both stochastic and deterministic factors. Under the framework of stochastic factors, Luna et al. (2018a) showed that resource abundance explains around 60% of the diet of harvester ants. Specifically, the authors found that the diets of ant nests located in sites with lower plant abundances represented subsets of the diets of ant nests located in sites with higher plant abundances. Intriguingly, the same authors observed that the variation that could not be explained by abundance was explained by the determinism of the Optimal Diet Theory (i.e., ant colonies tended to forage in more energetic resources at shorter distances) (Luna et al. 2018a). In other case, some studies have shown that for howler monkey-plant networks, the age and social role of an individual determine its role within the network. Such studies observed that older and dominant monkeys tended to access a greater number of resources, while the diets of younger and non-dominant monkeys were just subsets of the diets of dominant monkeys (Dáttilo et al. 2014c; Benitez-Malvido et al. 2016). On the other hand, modularity in individual-based networks has also been shown to be a property of mutualistic plant–pollinator (Dupont et al. 2014; Valverde et al. 2016). and antagonistic plant-herbivore interactions (Carvalho et al. 2021) For individual-based plant–pollinator networks, the modular pattern has been explained by the restricted movements of bumblebees in space, which generate isolated network compartments between patches of plants (Dupont et al. 2014). Other evidence has shown that modularity can be a consequence of plant phenology, since the alternating flowering of different plant species leads to animal individuals interacting more frequently with some plants during certain periods of the year (Valverde et al. 2016). The heuristic power of this individual-based network approach has already revealed new insights on the ecological and evolutionary dynamics of species interactions that cannot be detected using interspecific networks. For instance, one of the most interesting findings dealing with plant-pollinator networks is that plant individuals that have higher centrality in the network (i.e., nodes with high degrees that act as bridges maintaining cohesive networks) have higher fitness, and therefore should be favored by natural selection (Gómez and Perfectti 2012). Note that the previous examples show that by studying individuals we can assess the variation in the interactions of individual plants (i.e., producers) and animals, ultimately furthering our understanding of interspecific interactions (see Guerra et al. 2017). Despite the potential use of individual-based networks to evaluate the ecological and evolutionary dynamics of species interactions, the amount of theoretical and empirical information available is still limited. Therefore, we encourage researchers from around the world to employ an interspecific approach to link population dynamics to community structure.

10.5 Plant-Animal Multilayer Networks

The implementation of ecological networks to study plant-animal interactions has prompted the development of numerous scientific discoveries that have increased our understanding of the function and dynamics of species interactions (Dáttilo and Rico-Gray 2018). Despite ecological networks being a fundamental area of research in ecological theory, most of the knowledge generated by implementing such an approach only describes interaction patterns between two trophic levels (e.g., animals and the plants they use as food sources). This limited view of species interactions ignores the multiple relationships that a species can establish in an environment as well as any variation across time and space (Dáttilo et al. 2016). Thereby, when bipartite ecological networks are described through time and space, they are considered static entities isolated one from another, leading to posible biased conclusions. For example, we might consider a plant a specialist if we assess its interactions with herbivores, but the same plant could be a generalist if we assessed its interactions with pollinators. By integrating multiple interaction types and spatio-temporal variation into a single network, one could more accurately and thoroughly describe species interactions and reveal new insights into the complexity of species interactions (Genrich et al. 2017; Pilosof et al. 2017). Recent developments in the field of complex networks have provided a mathematical framework for studying networks with multiple layers, which include multiple types of vertices (e.g., herbivores, pollinators and plants) with multiple types of edges (e.g., mutualistic and antagonistic interactions). In fact, the implementation of multilayer networks in ecology has already made advances in understanding biotic interactions (Pilosof et al. 2017), and in this section we will explain how to use this multilayer-network framework to study plant-animal relationships.

A multilayer network is an approximation used to connect multiple entities/interactions into a single network, allowing the use of interlayer edges that connect different layers of a network. For instance, layers can be represented by networks of different interactions types (e.g., pollination, herbivory and seed dispersal) connected by shared species (e.g., plants). The interesting and heuristic part of this approach is that interlayer edges represent additional and more realistic ecological processes that are not considered in single bipartite networks (e.g., the effect of one interaction on the output of another interaction). However, this approach is not limited to merging multiple interaction types, since multilayer networks allow more elaborate scenarios like a network with multiple interaction types over time and space, something that we will explain in the next paragraph.

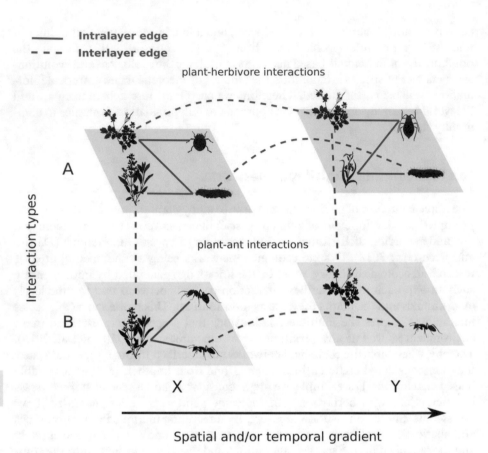

□ Fig. 10.5 A schematic plant-animal multilayer network. The first layer is represented by the interaction type (A = plant-herbivore interactions. B = plant-ant defensive mutualism). The second layer is represented by ecological interactions shared between spatial or temporal gradients (from X to Y). Intralayer edges are in blue, and interlayer edges showing different ecological and evolutionary processes across layers are in red dashed lines

Similar to bipartite networks, an ecological multilayer network (EMN) is also built by a set of nodes. However, we also need a set of layers, which can be different interaction types, networks in different periods of time or networks across space (□ Fig. 10.5). Note that these EMNs are not limited to only one set of layers. In other words, we can use an EMN representing one set of layers for different interaction types and another type of layer for periods of time. Each type of layer is called an aspect (e.g., an annual set of networks or a set of antagonistic and mutualistic interactions of a single system). Thus, to refer to each element of each aspect we refer to an elementary layer (e.g., the pollination network of a month). As species assemblages vary over space and time not all nodes can be found in all the elementary layers of an EMN. Thus, the nodes that are part of a specific elementary layer are called state nodes. Finally, to connect the network layers, we have intralayer edges (i.e., interactions within each layer, □ Fig. 10.5) and interlayer edges, which connect state nodes across elementary layers. Edges can be weighted or unweighted (binary, frequencies or abundances), which could affect interpretation of the results as previously showed for bipartite networks (Miranda et al. 2019). Moreover, it is worth mentioning that the nature of each layer is only limited by the imagination, and that there are more types of layering than those mentioned in this chapter.

But how can we describe the organization of EMNs? Initially, we can identify influential nodes or edges with three main descriptors: (i) eigenvector versatility (i.e., a species' importance depends on its connections within and across layers, and on the connections of its neighbors), (ii) betweenness versatility (i.e., how often the shortest paths between each pair of nodes go through a given node within and across layers) and, (iii) multidegree (i.e., the degree of each species across all layers). These descriptors can be used to assess how a group of species might be affected if certain species are removed or which species are relevant to main network cohesion. If species connect different layers in a non-random way, this could imply that there are nonrandom patterns of interactions across layers. Consequently, we can use a measure of modularity to identify which species are present in different layers and interacting with other species in those different layers (Pilosof et al. 2017). In addition, we can also identify motifs, which are small numbers of species/ nodes that interact in repetitive and predictable ways (e.g., one plant always interacts with the same two pollinators, or one herbivore always predates the same three plants). By searching for motifs in networks, we can assess whether certain motifs appear more often within or across layers in order to answer questions such as: Are there certain types of interactions that are predictable or constant across space or time? Or, how different are interaction motifs between interaction types? Applying the above-mentioned descriptors could reveal the ecological dynamics of species interactions and even evolutionary consequences of biotic interactions in a more integrative way.

Current evidence shows that plants connect herbivores and pollinators between layers, suggesting that the dynamics of antagonistic and mutualistic communities are interconnected via plants (Sauve et al. 2016). This more comprehensive idea of species interactions could not be visualized using bipartite networks. For example, in a study performed by Timóteo et al. (2018), the authors used the different habitats as a layer along with plants and animals, to show that animals with wide-ranging movements (e.g., birds and elephants) disperse plants across habitats. In another study, by merging highly diverse mutualistic networks (a total of 390 species including pollinators, bird seed dispersers and ants), Dáttilo et al. (2016) showed that only a few species contributed to the maintenance of the whole network structure. This finding indicates that conserving interactions of keystone mutualists in interlinked mutualistic networks is crucial to the persistence of species-rich mutualistic assemblages. However, despite the possible applications of EMN for studying plant-animal interactions, ecologists must carefully study the underlying mathematical framework in order to fully understand and correctly use this approach. Moreover, the field of multilayer networks is still growing, and many of its descriptors and tools used are still in development and discussion. Thus, this is an open area of research which promises to provide new tools for understanding nature and its complexity.

Conclusion and Future Directions

To end this chapter, we would like to highlight that animals and plants are embedded in complex and species-rich interaction networks. Moreover, both stochastic and deterministic factors explain the organization of such plant-animal interaction networks. By using the complex network approach and by studying how biotic interactions vary between individuals of the same population, we can create a bridge between population and community ecology. As also observed in interspecific networks, individual-based networks are not randomly assembled. In fact, the structure and function of interaction networks vary over time and space and can be driven by environmental conditions. However, interaction networks are only an initial approach for studying the complexity of species relationships within the web of life and, therefore, multilayer networks hold promise for elucidating the real complexity of natural systems. Our planet is mega-diverse and there are regions in which you can find thousands of species at relatively small spatial scales (e.g., the Tropical Andes or Amazon rainforest), begging the question, how we can study and identify who interacts with whom when diversity is so high? In recent years, molecular techniques have started to help to solve this problem, such as DNA barcoding allowing us to analyze the gut content of herbivores and frugivores and identify the plants in different animals' diets (García-Robledo et al. 2013; González-Varo et al. 2014). Such techniques are still slow and expensive, but we hope that in coming years molecular techniques will become more accessible and commonplace, boosting the study of plant-animal interactions in the process.

Charles Darwin was aware of the high diversity of interactions that species establish, using the image of an "entangled bank" to call to mind this vast complexity in his seminal book *On the Origin of Species* (Darwin 1859). More recently, the variety of life forms and the interactions between them that caught Darwin's attention also inspired John N. Thompson to propose the Geographic Mosaic Theory of Coevolution (GMTC). This theory postulates that many of the co-evolutionary dynamics between groups of interacting species occur at a geographic scale above the level of populations (Thompson 2005). He hypothesized that a selection mosaic favors different evolutionary trajectories between populations, allowing some regions to be co-evolutionary hotpots (i.e., where the interaction affects the fitness of both partners) and coldspots (i.e., where selection is not reciprocal or there is no selection; Thompson 2005). Thanks to ecological networks, we now know that co-evolution is a diffuse process and that changes in one trait of a species can affect another species in a network without direct interaction, favoring similar traits at a community level (Guimarães et al. 2011, 2017). In another study, it was shown that ecological interactions promote gene flow across large geographical distances, favoring trait matching and thereby co-evolutionary dynamics (Medeiros et al. 2018). Likewise, some evidence suggests that the structure of pollinator and seed-dispersal networks (i.e., nestedness and modularity) could be a result of trait matching and exploitation barriers, both co-evolutionary processes (de Andreazzi et al. 2020). Thus, it is clear that interactions between animals and plants can drive selection, as both groups reciprocally affect each other across population and communities. On the other hand, many interactions have never been observed and studied, and with the current trajectory of global changes (e.g., global warming, habitat loss & biological invasions), if human lifestyles continue in the same way we might never be able to observe such natural wonders. Thus, we invite the reader to increase their efforts and collaborate to elucidate Darwin's intriguing and inspiring entangled bank.

Key Points
- Plant-animal interactions can be studied by implementing a complex network approach.
- Complex networks that arise from plant-animal interactions are non-randomly organized, which can be explained by stochastic and deterministic factors.
- The organization of plant-animal networks varies through space and time.
- Individuals are the building blocks of ecological networks.
- Plant-animal interaction networks can help us understand the complexity of evolutionary and co-evolutionary processes in interactive communities.

Questions
- What are the main benefits and limitations of studying interaction networks?
- What is a multilayer network?
- Could the use of molecular techniques change the current understanding of plant-animal networks?

References

Anjos DV, Luna P, Borges CCR et al (2019) Structural changes over time in individual-based networks involving a harvester ant, seeds, and invertebrates. Ecol Entomol een.12764. https://doi.org/10.1111/een.12764

Anjos DV, Leal LC, Jordano P, Del-Claro K (2020) Ants as diaspore removers of non-myrmecochorous plants: a meta-analysis. Oikos. https://doi.org/10.1111/oik.06940

Araujo AC, Martín González AM, Sandel B et al (2018) Spatial distance and climate determine modularity in a cross-biomes plant–hummingbird interaction network in Brazil. J Biogeogr 45:1846–1858. https://doi.org/10.1111/jbi.13367

Bascompte J, Jordano P (2007) Plant-animal mutualistic networks: the architecture of biodiversity. Annu Rev Ecol Evol Syst 38:567–593. https://doi.org/10.1146/annurev.ecolsys.38.091206.095818

Bascompte J, Jordano P, Melián CJ, Olesen JM (2003) The nested assembly of plant-animal mutualistic networks. Proc Natl Acad Sci U S A 100:9383–9387. https://doi.org/10.1073/pnas.1633576100

Bastolla U, Fortuna MA, Pascual-García A et al (2009) The architecture of mutualistic networks minimizes competition and increases biodiversity. Nature 458:1018–1020. https://doi.org/10.1038/nature07950

Begon M, Townsend CR, Harper JL (2006) Ecology: from individuals to ecosystems. Wiley-Blackwell, Malden

Benitez-Malvido J, Martínez-Falcón AP, Dáttilo W et al (2016) The role of sex and age in the architecture of intrapopulation howler monkey-plant networks in continuous and fragmented rain forests. PeerJ 4:e1809. https://doi.org/10.7717/peerj.1809

Blüthgen N, Menzel F, Blüthgen N (2006) Measuring specialization in species interaction networks. BMC Ecol 6. https://doi.org/10.1186/1472-6785-6-9

Bolnick DI, Amarasekare P, Araújo MS et al (2011) Why intraspecific trait variation matters in community ecology. Trends Ecol Evol 26:183–192. https://doi.org/10.1016/j.tree.2011.01.009

CaraDonna PJ, Petry WK, Brennan RM et al (2017) Interaction rewiring and the rapid turnover of plant-pollinator networks. Ecol Lett. https://doi.org/10.1111/ele.12740

Carvalho, R. L., Anjos, D. V., Fagundes, R., Luna, P., & Ribeiro, S. P. (2021). Similar topologies of individual-based plant-herbivorous networks in forest interior and anthropogenic edges. Austral Ecology. https://doi.org/10.1111/aec.13001

Classen, A., Eardley, C. D., Hemp, A., Peters, M. K., Peters, R. S., Ssymank, A., & Steffan-Dewenter, I. (2020). Specialization of plant–pollinator interactions increases with temperature at Mt. Kilimanjaro. Ecology and evolution, 10(4), 2182-2195. https://doi.org/10.1002/ece3.6056

Carstensen DW, Sabatino M, Trøjelsgaard K, Morellato LPC (2014) Beta diversity of plant-pollinator networks and the spatial turnover of pairwise interactions. PLoS One 9. https://doi.org/10.1371/journal.pone.0112903

Corro EJ, Ahuatzin DA, Jaimes AA et al (2019) Forest cover and landscape heterogeneity shape ant–plant co-occurrence networks in human-dominated tropical rainforests. Landsc Ecol 34:93–104. https://doi.org/10.1007/s10980-018-0747-4

Dalsgaard B, Magård E, Fjeldså J et al (2011) Specialization in plant-hummingbird networks is associated with species richness, contemporary precipitation and quaternary climate-change velocity. PLoS One 6. https://doi.org/10.1371/journal.pone.0025891

Dalsgaard B, Trøjelsgaard K, Martín González AM et al (2013) Historical climate-change influences modularity and nestedness of pollination networks. Ecography (Cop) 36:1331–1340. https://doi.org/10.1111/j.1600-0587.2013.00201.x

Darwin CR (1859) On the origin of species. John Murray, London. pp 1–502

Dáttilo W, Rico-Gray V (2018) Ecological networks in the tropics. Springer, Switzerland, pp 1–202

Dáttilo W, Vasconcelos HL (2019) Macroecological patterns and correlates of ant – tree interaction networks in Neotropical savannas. Glob Ecol Biogeogr:1–12. https://doi.org/10.1111/geb.12932

Dáttilo W, Guimarães PR, Izzo TJ (2013a) Spatial structure of ant-plant mutualistic networks. Oikos 122:1643–1648. https://doi.org/10.1111/j.1600-0706.2013.00562.x

Dáttilo W, Izzo TJ, Vasconcelos HL, Rico-Gray V (2013b) Strength of the modular pattern in Amazonian symbiotic ant-plant networks. Arthropod Plant Interact 7:455–461. https://doi.org/10.1007/s11829-013-9256-1

Dáttilo W, Díaz-Castelazo C, Rico-Gray V (2014a) Ant dominance hierarchy determines the nested pattern in ant-plant networks. Biol J Linn Soc 113:405–414. https://doi.org/10.1111/bij.12350

Dáttilo W, Fagundes R, Gurka CAQ et al (2014b) Individual-based ant-plant networks: diurnal-nocturnal structure and species-area relationship. PLoS One 9:e99838. https://doi.org/10.1371/journal.pone.0099838

Dáttilo W, Serio-Silva JC, Chapman CA, Rico-Gray V (2014c) Highly nested diets in intrapopulation monkey-resource food webs. Am J Primatol 76:670–678. https://doi.org/10.1002/ajp.22261

Dáttilo W, Lara-Rodríguez N, Guimarães PR et al (2016) Unravelling Darwin's entangled bank: architecture and robustness of mutualistic networks with multiple interaction types. Proc Soc Biol 283:1–9. https://doi.org/10.1098/rspb.2016.1564

Dáttilo W, Vizentin-Bugoni J, Debastiani VJ et al (2019) The influence of spatial sampling scales on ant–plant interaction network architecture. J Anim Ecol 88:903–914. https://doi.org/10.1111/1365-2656.12978

de Andreazzi CS, Astegiano J, Guimarães PR (2020) Coevolution by different functional mechanisms modulates the structure and dynamics of antagonistic and mutualistic networks. Oikos 129:224–237. https://doi.org/10.1111/oik.06737

de M Santos GM, Aguiar CML, Genini J et al (2012) Invasive Africanized honeybees change the structure of native pollination networks in Brazil. Biol Invasions 14:2369–2378. https://doi.org/10.1007/s10530-012-0235-8

Dehling DM (2018) The structure of ecological networks. In: Ecological networks in the tropics. Springer, Switzerland, pp 29–42

Devictor V, Clavel J, Julliard R et al (2010) Defining and measuring ecological specialization. J Appl Ecol 47:15–25. https://doi.org/10.1111/j.1365-2664.2009.01744.x

10

Donatti CI, Guimarães PR, Galetti M et al (2011) Analysis of a hyper-diverse seed dispersal network: Modularity and underlying mechanisms. Ecol Lett 14:773–781. https://doi.org/10.1111/j.1461--0248.2011.01639.x

Dormann CF, Fründ J, Schaefer HM (2017) Identifying causes of patterns in ecological networks: opportunities and limitations. Annu Rev Ecol Evol Syst 48:annurev-ecolsys-110316-022928. https://doi.org/10.1146/annurev-ecolsys-110316-022928

Dupont YL, Trøjelsgaard K, Hagen M et al (2014) Spatial structure of an individual-based plant-pollinator network. Oikos 123:1301–1310. https://doi.org/10.1111/oik.01426

Escribano-Avila G, Lara-Romero C, Heleno R, Traveset A (2018) Tropical seed dispersal networks: emerging patterns, biases, and keystone species traits. In: Ecological networks in the tropics, pp 93–110. https://doi.org/10.1007/978-3-319-68228-0_7

Fenster CB, Armbruster WS, Wilson P et al (2004) Pollination syndromes and floral specialization. Annu Rev Ecol Evol Syst 35:375–403. https://doi.org/10.1146/annurev.ecolsys.34.011802.132347

Galetti M, Guevara R, Côrtes MC et al (2013) Functional extinction of birds drives rapid evolutionary changes in seed size. Science (80-) 340:1086–1091

García-Robledo C, Erickson DL, Staines CL et al (2013) Tropical plant-herbivore networks: reconstructing species interactions using DNA barcodes. PLoS One 8. https://doi.org/10.1371/journal.pone.0052967

Genrich CM, Mello MAR, Silveira FAO et al (2017) Duality of interaction outcomes in a plant–frugivore multilayer network. Oikos 126:361–368. https://doi.org/10.1111/oik.03825

Gómez JM, Perfectti F (2012) Fitness consequences of centrality in mutualistic individual-based networks. Proc R Soc B Biol Sci 279:1754–1760. https://doi.org/10.1098/rspb.2011.2244

González-Varo JP, Arroyo JM, Jordano P (2014) Who dispersed the seeds? The use of DNA barcoding in frugivory and seed dispersal studies. Methods Ecol Evol 5:806–814. https://doi.org/10.1111/2041-210X.12212

Guerra TJ, Dayrell RLC, Arruda AJ et al (2017) Intraspecific variation in fruit–frugivore interactions: effects of fruiting neighborhood and consequences for seed dispersal. Oecologia 185:233–243. https://doi.org/10.1007/s00442-017-3943-z

Guimarães PR, Jordano P, Thompson JN (2011) Evolution and coevolution in mutualistic networks. Ecol Lett 14:877–885. https://doi.org/10.1111/j.1461-0248.2011.01649.x

Guimarães PR, Pires MM, Jordano P et al (2017) Indirect effects drive coevolution in mutualistic networks. Nature 550:511–514. https://doi.org/10.1038/nature24273

Herrera CM, Pellmyr O (2002) Plant animal interactions: an evolutionary approach. Blackwell Publishing, New Jersey, pp 1–328

Hintze A, Adami C (2008) Evolution of complex modular biological networks. PLoS Comput Biol 4. https://doi.org/10.1371/journal.pcbi.0040023

Hubbell SP (2001) The unified neutral theory of biodiversity and biogeography. Princeton University Press, New Jersey, pp 1–375

Ings TC, Hawes JE (2018) The history of ecological networks. In: Ecological networks in the tropics. Springer, Switzerland, pp 15–28

Ings TC, Montoya JM, Bascompte J et al (2009) Ecological networks - Beyond food webs. J Anim Ecol 78:253–269. https://doi.org/10.1111/j.1365-2656.2008.01460.x

Jordano P (2000) Fruits and frugivory. In: Seeds: the ecology of regeneration in plant communities. CABI, New York, pp 125 166

Jordano P, Bascompte J, Olesen JM (2003) Invariant properties in coevolutionary networks of plant-animal interactions. Ecol Lett 6:69–81. https://doi.org/10.1046/j.1461-0248.2003.00403.x

Kemp JE, Evans DM, Augustyn WJ, Ellis AG (2017) Invariant antagonistic network structure despite high spatial and temporal turnover of interactions. Ecography (Cop) 40:1315–1324. https://doi.org/10.1111/ecog.02150

Krishna A Jr, Guimarães PR, Jordano P, Bascompte J (2008) A neutral-niche theory of nestedness in mutualistic networks. Oikos 117:1609–1618. https://doi.org/10.1111/j.1600-0706.2008.16540.x

Lange D, Dáttilo W, Del-Claro K (2013) Influence of extrafloral nectary phenology on ant-plant mutualistic networks in a neotropical savanna. Ecol Entomol 38:463–469. https://doi.org/10.1111/een.12036

López-Carretero A, del-Val E, Boege K (2018) Plant-herbivore networks in the tropics. In: Ecological networks in the tropics, pp 111–126. https://doi.org/10.1007/978-3-319-68228-0

Luna P, García-Chávez JH, Dáttilo W (2018a) Complex foraging ecology of the red harvester ant and its effect on soil seed bank. Acta Oecol 86:57–65. https://doi.org/10.1016/j.actao.2017.12.003

Luna P, Peñaloza-Arellanes Y, Castillo-Meza AL et al (2018b) Beta diversity of ant-plant interactions over day-night periods and plant physiognomies in a semiarid environment. J Arid Environ 156:69–76. https://doi.org/10.1016/j.jaridenv.2018.04.003

Luna P, Corro EJ, Antoniazzi R, Dáttilo W (2020) Measuring and linking the missing part of biodiversity and ecosystem function: the diversity of biotic interactions. Diversity 12:86. https://doi.org/10.3390/d12030086

Medeiros LP, Garcia G, Thompson JN, Guimarães PR (2018) The geographic mosaic of coevolution in mutualistic networks. Proc Natl Acad Sci U S A 115:12017–12022. https://doi.org/10.1073/pnas.1809088115

Miranda PN, Ribeiro JELDS, Luna P et al (2019) The dilemma of binary or weighted data in interaction networks. Ecol Complex 38. https://doi.org/10.1016/j.ecocom.2018.12.006

Moreira EF, Boscolo D, Viana BF (2015) Spatial heterogeneity regulates plant-pollinator networks across multiple landscape scales. PLoS One 10:1–19. https://doi.org/10.1371/journal.pone.0123628

Olesen JM, Bascompte J, Dupont YL, Jordano P (2007) The modularity of pollination networks. Proc Natl Acad Sci 104:19891–19896. https://doi.org/10.1073/pnas.0706375104

Ollerton J (2017) Pollinator diversity: distribution, ecological function, and conservation. Annu Rev Ecol Evol Syst 48:353–376. https://doi.org/10.1146/annurev-ecolsys-110316-022919

Pilosof S, Porter MA, Pascual M, Kéfi S (2017) The multilayer nature of ecological networks. Nat Ecol Evol 1. https://doi.org/10.1038/s41559-017-0101

Pires MM, Guimarães PR, Araújo MS et al (2011) The nested assembly of individual-resource networks. J Anim Ecol 80:896–903. https://doi.org/10.1111/j.1365-2656.2011.01818.x

Poisot T, Canard E, Mouillot D et al (2012) The dissimilarity of species interaction networks. Ecol Lett 15:1353–1361. https://doi.org/10.1111/ele.12002

Poisot T, Stouffer DB, Gravel D (2015) Beyond species: why ecological interaction networks vary through space and time. Oikos 124:243–251. https://doi.org/10.1111/oik.01719

Poisot T, Cynthia G-J, Fortin M-J et al (2017) Hosts, parasites and their interactions respond to different climatic variables. Glob Ecol Biogeogr 26:942–951. https://doi.org/10.1111/geb.12602

Ramos-Jiliberto R, Domínguez D, Espinoza C et al (2010) Topological change of Andean plant-pollinator networks along an altitudinal gradient. Ecol Complex 7:86–90. https://doi.org/10.1016/j.ecocom.2009.06.001

Ramos-Robles M, Andresen E, Díaz-Castelazo C (2016) Temporal changes in the structure of a plant-frugivore network are influenced by bird migration and fruit availability. PeerJ 4:e2048. https://doi.org/10.7717/peerj.2048

Saavedra S, Stouffer DB, Uzzi B, Bascompte J (2011) Strong contributors to network persistence are the most vulnerable to extinction. Nature 478:233–235. https://doi.org/10.1038/nature10433

Sauve AMC, Thébault E, Pocock MJO, Fontaine C (2016) How plants connect pollination and herbivory networks and their contribution to community stability. Ecology 97:908–917. https://doi.org/10.1890/15-0132

Schleuning M, Fründ J, Klein AM et al (2012) Specialization of mutualistic interaction networks decreases toward tropical latitudes. Curr Biol 22:1925–1931. https://doi.org/10.1016/j.cub.2012.08.015

Schleuning M, Ingmann L, Strauß R et al (2014) Ecological, historical and evolutionary determinants of modularity in weighted seed-dispersal networks. Ecol Lett 17:454–463. https://doi.org/10.1111/ele.12245

Thompson JN (2005) The geographic mosaic of coevolution. University of Chicago Press, Chicago

Timóteo S, Correia M, Rodríguez-Echeverría S et al (2018) Multilayer networks reveal the spatial structure of seed-dispersal interactions across the Great Rift landscapes. Nat Commun 9:1–11. https://doi.org/10.1038/s41467-017-02658-y

Tylianakis JM, Morris RJ (2017) Ecological networks across environmental gradients. Annu Rev Ecol Evol Syst 48:annurev-ecolsys-110316-022821. https://doi.org/10.1146/annurev-ecolsys-110316-022821

10

Valverde J, Gómez JM, Perfectti F (2016) The temporal dimension in individual-based plant pollination networks. Oikos 125:468–479. https://doi.org/10.1111/oik.02661

Vázquez DP, Melián CJ, Williams NM et al (2007) Species abundance and asymmetric interaction strength in ecological networks. Oikos 116:1120–1127. https://doi.org/10.1111/j.2007.0030-1299.15828.x

Vazquez DP, Bluthgen N, Cagnolo L, Chacoff NP (2009) Uniting pattern and process in plant-animal mutualistic networks: a review. Ann Bot 103:1445–1457. https://doi.org/10.1093/aob/mcp057

Vázquez DP, Bluthgen N, Cagnolo L, Chacoff NP (2009) Uniting pattern and process in plant-animal mutualistic networks: a review. Ann Bot 103:1445–1457. https://doi.org/10.1093/aob/mcp057

Vellend M (2016) The theory of ecological communities. Monogr Popul Biol 57:229

Vidal MM, Hasui E, Pizo MA et al (2014) Frugivores at higher risk of extinction are the key elements of a mutualistic network. Ecology 95:3440–3447

Vizentin-Bugoni J, Maruyama PK, de Souza CS et al (2018) Plant-pollinator networks in the tropics: a review. In: Ecological networks in the tropics, pp 73–91. https://doi.org/10.1007/978-3-319-68228-0_6

Further Reading/Additional Resources

Dáttilo W, Rico-Gray V (eds) (2018) Ecological networks in the tropics: an integrative overview of species interactions from some of the most species-rich habitats on earth. Springer, Cham

The Gift That Keeps on Giving: Why Does Biological Diversity Accumulate Around Mutualisms?

Judith L. Bronstein

Contents

© Springer Nature Switzerland AG 2021
K. Del-Claro, H. M. Torezan-Silingardi (eds.), *Plant-Animal Interactions*, https://doi.org/10.1007/978-3-030-66877-8_11

> **⊜ Learning Objectives**
>
> After completing this chapter, you should have an understanding of the following:
> - What a mutualistic resource is, and which common mutualisms involve the exchange of such resources and which ones involve the exchange of a resource for a service.
> - The kinds of organisms, other than one single best mutualistic partner, that utilize mutualistic resources.
> - How mutualistic resources function to link species and sets of interactions into larger networks.

11.1 Introduction

The study of mutualism had a late start, but in the past two decades has begun to thrive (Bronstein 2015). It is now widely recognized that mutualists are not only present, but are common and prominent interactions in every habitat on Earth. New mutualisms are continually being discovered and described; interactions that had been thought not to confer mutual benefits have now been convincingly demonstrated to be mutualistic. A significant shift has happened in the past 20 years, however: broad conceptual questions now guide research in this field. Key ecological topics span levels of inquiry, from the individual to populations, communities, and ecosystems, addressing both basic and applied questions using methods ranging from observations to experiments to theoretical models. Particularly pressing questions include these: Is population growth constrained by the abundance of mutualistic partners (e.g., Thomson 2019)? How do mutualisms contribute to essential ecosystem functions such as nutrient cycling (e.g., Taylor et al. 2020?) Which mutualisms are at greatest threat of anthropogenic disruption, and what are the consequences if they are lost (e.g., Lever et al. 2020; Pires et al. 2020)?

Exciting progress in our understanding of mutualism is taking place at the community level in particular. This has been spurred by the development of new tools in community analysis, notably ecological networks theory (Bascompte and Jordano 2014). One of the clearest messages of this work is that mutualisms are deeply nested within most communities: specialists commonly interact with generalists, generalists interact with generalists, and multiple mutualisms are then linked in multilayer networks (e.g., Montesinos-Navarro et al. 2018; Mello et al. 2019). This new understanding is, I believe, about to lead to a new generation of inquiry into the ecological significance of mutualism, not just for the individual species involved, but for the communities and ecosystems in which they are embedded. Consideration of these issues has already begun (Chomicki et al. 2019).

Here, I develop the hypothesis that the deep embedding of mutualism within ecological communities is not only a *consequence* of communities' rich biological diversity, but actually one of its primary *causes*. I build this argument based on four fundamental features of mutualism biology. First, most mutualisms involve a one-way or two-way exchange of resources. Second, most consumers of mutualistic resources are rather generalized: a single resource-providing species sustains many consumers, and in turn many consumers return benefits to it. Diversity therefore accumulates around individual mutualistic consumer and resource species. Third, mutualistic resources are difficult to modulate to the needs of the best mutualistic consumers, and should commonly be produced in overabundance. This opens the door to consumers that provide no benefits in return, building biodiversity further. Some of these non-mutualistic consumers function as mutualists of other resource-offering species, generating yet more cross-linkages within and among communities. Fourth and last, different types of mutualisms form links with each other, and mutualisms form links to antagonisms, generating robust multiplex structures that may further accumulate and sustain biodiversity. As a consequence, I will argue, (a) interaction networks form around mutualisms that extend well beyond the mutualists themselves; (b) non-mutualists taking advantage of one species' resources may often be mutualists of other species in the community; and (c) different mutualistic networks are also linked, because successful mutualisms often generate resources that form the base of yet other mutualisms. I develop three examples from nature to illustrate my argument. I conclude by offering a set of predictions to move these ideas further.

11.2 Four Features of Mutualism Linked to Resource Production

11.2.1 Most Mutualisms Involve Resource Exchange

Mutualisms are economic exchanges that take place in a "biological marketplace" (Noë and Hammerstein 1994; Schwartz and Hoeksema 1998; Hoeksema and Schwartz 2001). Each of two species uses its partner to obtain commodities to which it otherwise has no or limited access; in exchange, it offers commodities it can provide cheaply. Commodities can be divided into "resources" and "services". Mutualisms caninvolve the exchange of services, the exchange of resources, or the exchange of a resource for a service. A good example of the first category, service-for-service mutualisms, involves species that benefit from feeding in multispecies flocks because different species spot predators in different ways and warn the group (Goodale et al. 2020). In this case the service rendered, as well as the service received, is protection from predators. Resource-for-resource mutualisms, the second category, are notable because they are critical for ecosystem functioning (Johnson 2015). The symbiosis between leguminous plants and rhizobial bacteria is an excellent example: plants provide the bacteria with carbon (which they but not the bacteria have in excess) in exchange for nitrogen (which the bacteria convert to a usable form that would otherwise be strongly limiting to plants). Symbiotic mutualisms between plants and mycorrhizal fungi are also resource-exchange mutualisms. Resource-for-service mutualisms, the third category, are diverse and widespread. They include most pollination mutualisms, in which plants offer animals nectar in exchange for pollen transport services; most seed dispersal mutualisms, in which plants offer animals fruit in exchange for transport of seeds to better germination sites; and protection mutualisms, in which one organism offers a food reward to partner in exchange for protection from natural enemies or the abiotic environment.

Setting aside service-for-service interactions, which appear to be relatively rare in nature (but see, e.g., Brooks and Gwaltney 1993, Aubier and Elias 2020), one sees that resources (often termed rewards) are key to mutualisms. To reiterate, resources are commodities of exchange. Some resources are substances that a species either has but does not require, or substances that it has in excess to its own needs. For example, certain pairs of microbes exchange waste products, benefiting both of them and leading to stable associations between them (Hoek et al. 2016); cleaner fish remove parasites from their hosts, the parasites in this case being the (unwanted) resource that is exchanged for the service of cleaning (Grutter et al. 2019). In other mutualisms, resources must be produced, but production costs are lower than the benefit the producer would receive from offering it to the partner. Floral nectar is the best-studied example (Parachnowitsch et al. 2019).

Although it has not been explicitly recognized, resources lie at the heart of most of the general questions we ask about the ecology and evolution of mutualism. The question of when and why mutualisms evolve often centers around how organisms find the resources they require to survive and thrive. Questions about the evolution and distribution of specialization and generalization in mutualism revolve around whether resources can be selectively delivered to the best partners. Questions about the ubiquity and consequences of cheating in mutualistic systems often center on how organisms can obtain the benefits partners provide while minimizing or eliminating the costs of resource provision.

Parallels between mutualisms and antagonism are now recognized to be linked to the idea that most mutualisms are consumer-resource interactions. However, in the mutualistic cases, the consumer is obtaining a resource provided by the partner, rather than the partner itself being the resource that is consumed (Holland et al. 2005).

11.2.2 Resources Commonly Are Obtained by Many Mutualists, Not Just One

It is increasingly recognized that most mutualisms are relatively generalized: one or both "partners" is not a species, butrather a guild of species (Stanton 2003). The degree of generalization varies across mutualism types (Blüthgen et al. 2007; Chomicki et al. 2020). Generalization in pollination mutualisms is exceptionally well-documented (Waser et al. 1996), although many exceptions exist (see ▶ Sects. 11.4.2 and 11.4.3). Specialization by either partner is rare in seed dispersal mutualisms (Howe 1993; but see Yule et al. 2020). Ant-insect protective mutualisms too tend to be relatively generalized (Glasier et al. 2018). Ant-plant protective mutualisms range from highly specialized to highly generalized; specialization is relatively uncommon, but has been studied in great depth where it does occur (Guimarães et al. 2006). Symbiotic mutualisms show a higher degree of specialization than non-symbiotic ones (Fisher et al. 2017), although difficulties in distinguishing microbial strains and species make this a difficult pattern to elucidate. It is important to clarify that referring to a mutualism as generalized or specialized is problematic. Until the late 1990's, this was a common way of classifying mutualisms

(Bronstein 2015). This approach was reinforced by in-depth studies of a handful of mutualisms, including the fig pollination systems (see ▶ Sect. 11.4.3), in which there is indeed extreme specialization on both sides of the interaction. However, more recent studies have emphasized the ubiquity of asymmetries in specialization (Bascompte et al. 2003). Further, on each side of the interaction, specialization and generalization are not a dichotomy but a continuum (Ollerton 2006).

Specialization is often not very labile evolutionarily; it is sometimes retained over long periods of an interaction's history, even as other features of the interaction change greatly (e.g., Pellmyr and Thompson 1992). In an ecological sense, however, the extent to which a mutualism is specialized is usually tied closely to the provision of resources. Specifically, they will be specialized to the extent that those resources can only be obtained by, can only be used by, and/or are only valuable to, a single beneficial species. Examples of all of these are known. For example, figs, yuccas, and several other plant taxa are uniquely valuable to their pollinators because they offer oviposition sites (see ▶ Sect. 11.4.3); these sites are difficult to access, and require specific adaptations to use (Kato and Kawakita 2017). Mesoamerican ant-acacias offer chemically complex rewards that can only be digested effectively by high-quality ant mutualists (Heil et al. 2014). Legumes are able to selectively associate with more effective rhizobial symbionts in the soil; why there nevertheless is a wide range in mutualistic effectiveness within natural Rhizobia populations is poorly understood (Pahua et al. 2018).

These cases have been explored in depth, and have provided some of the best evidence we have of the coevolutionary process. However, most mutualisms lack fine-tuned filters. This is not to say that resources are universally accessible and acceptable. For example, plants solicit pollinators with nectar whose chemistry makes it more attractive or less attractive to certain taxa; the shape of the flower makes this nectar less accessible to some would-be consumers (Fenster et al. 2004). Floral odors too are relatively more or less detectable to certain taxa (Junker and Blüthgen 2010). These traits, however, restrict reward delivery not to specific mutualistic species, but to specific mutualistic guilds: long-tongued bees, long-billed hummingbirds, or night-flying moths are pollination guilds, for example. There may still be a great many species within that guild. For example, Johnson et al. (2016) looked at patterns of specialization within plant-hawkmoth (Sphingidae) pollinator communities in nine biogeographic regions worldwide. They found that within each community, moth tongue lengths and floral tube lengths are bimodal or multimodal; longer-tubed flowers tend to exclude shorter-tongued moths from accessing nectar. However, from the perspective of a short-tubed flower, there are many effective pollinators among short-tongued moths. Thus, while there are clearly adaptations that filter out many consumers, most resource delivery is still somewhat generalized.

As a consequence of their activities, many of these resource-consumers (although by no means all; see ▶ Sect. 11.2.4) will render mutualistic service to their partner. They are unlikely to deliver identical degrees of service, considered on a per-capita basis. For example, in a facultatively ant-defended Arizona cactus (*Ferocactus wislizeni*), ant occupants range from highly mutualistic to mildly detrimental. However, by simple force of numbers, some not-very-effective ants provide exceptionally good service (Ness et al. 2006). There are many similar cases in which the "quantity component" of mutualistic effectiveness outweighs the "quality component" (Schupp et al. 2017), that is, when the best partner species is simply the one that is locally most abundant.

These features of mutualism can provide some level of resilience to resource-providing species. Mutualist identities and abundances often vary greatly over time at a single location, and even more strikingly, vary across space. For example, using an 8-year data set comprised of over 20,000 records, Ponisio et al. (2017) showed that pollination network structure at agriculturally disturbed sites in California are highly dynamic in space and time, with the most persistent species acting a specialists at some points and generalists at others. Species able to attract a somewhat broader range of mutualists, even though they vary in the quality of service they render, will likely be at an advantage (Astegiano et al. 2015). The same is true on the other side of the interaction. Consumers able to locate and benefit from a relatively wider range of resource-providing partners should exhibit a degree of resiliency lacking in specialists (e.g., Wood et al. 2019). Although by no means a settled question (e.g., see Vázquez and Simberloff 2002; Rafferty et al. 2015), the advantage of generalization, even if slight, seems likely to be particularly important in an anthropogenically altered and rapidly changing world (Brodie et al. 2014; Aslan et al. 2016).

11.2.3 Mutualistic Resources Should Exist in Excess of Mutualists' Immediate Demands

In mutualisms in which one or both partners seek a resource, benefits will generally not be received if they cannot find it. In fact, if no resource is present, the interaction generally ends, either immediately or after a short period of searching in vain (Lichtenberg et al. 2020). There are certainly species that have evolved traits allowing them to attract partners and to solicit beneficial actions without providing any resource in return. Strikingly, over 7500 plant species are entirely nectarless (Renner 2006); many of these plants reproduce strictly due to errors by nectar-seeking foragers (Lichtenberg et al. 2020). These interactions can best be characterized as exploitative rather than mutualistic.

What about species that do provide resources? Some theoretical studies have proposed that selection should lead to production of an optimal amount or quality of resource (e.g., Pyke 1981; Klinkhamer and de Jong 1993; Cohen and Shmida 1993; Kim et al. 2011). This is the resource level that would maximize the difference between the benefit received by consumers and the cost of producing the resource. However, there are usually many resource-seekers consuming resources at different rates, conferring different levels of benefit when they do. Further, the resource-seeking community varies in both time and space. Under the circumstances, how much resource is the right amount of resource to offer? Existing models for the most part ignore environmental variability and divergent needs and behaviors of generalists. However, a set of thought experiments leads to the prediction that resources should be produced in excess.

In a variable world, resources risk being depleted before the benefits provided by mutualists are maximized. This could occur, for instance, if there is competition from species that offer shared partners more or better resources (Jones et al. 2012; Johnson and Bronstein 2019). Pollinators can be extremely effective at choosing among plants offering different amounts of nectar (e.g., Hodges 1995; Brandenburg et al. 2012), and abandon plants as they become drained (Hodges 1985; Cartar 2004). This competitive dynamic confers an advantage to species that can produce either more or higher-quality resources for mutualists, or else those that can replenish those resources quickly. An additional advantage of high resource production is that large visual displays of those resources can be a signal that attracts mutualists in the first place. This is particularly well-documented in pollination and seed dispersal mutualisms: although the relationship is not necessarily strong and is often nonlinear, plants with larger flower and fruit displays attract more partners (Ohashi and Yahara 2001; Palacio and Ordano 2018).

Adding to these advantages of overproducing resources is the complication of context-dependency, a ubiquitous feature of mutualism (Chamberlain et al. 2014; Hoeksema and Bruna 2015). The costs of producing mutualistic resources vary in both ecological space and time. Nectar production requires water, for example, and volumes are lower under drought conditions (Waser and Price 2016). Benefits received from the consumers of those resources are context-dependent as well. For example, Klinkhamer et al. (2001) show that the disadvantage of low nectar production in *Echium vulgare* can be countered by the presence of high-nectar plants growing nearby. Context-dependent costs and benefits of producing mutualistic resources are well-documented in other mutualisms as well (e.g., Linsenmair et al. 2001; Vannette and Hunter 2011; Palacio and Ordano 2018). The variation that results is raw material for the evolution and coevolution of mutualism (Thompson 1988).

Thus, in the face of ecological variation and of competition for mutualistic resources, the idea that selection could result in an optimal quantity or quality of that resource seems highly unrealistic. At the same time, *underproducing* mutualistic resources seems to be a fairly sure way to fail (although cognitive limitations of consumers may prevent them from accurately assessing resource quantity and quality; Lichtenberg et al. 2020). This logic leads to the hypothesis that resource-providers should be favored to offer their partners an excess of resources on the "biological market".

This is only possible if costs of actually providing resources are not prohibitive. The magnitudes of these costs are surprisingly poorly known for any form of mutualism (Bronstein 2001). A reasonable hypothesis, so far largely untested, is that the costs associated with mutualistic resource production will decrease over evolutionary time. To date, virtually all literature on the evolution of mutualism focuses on how selection acts on mutualistic *benefits*. However, mutualism can be reinforced not only by an increase in benefits, but by a reduction in mutualism's costs.

In sum, then, mutualistic resources are likely to be relatively abundant in nature. They are not, and indeed cannot realistically be, so fine-tuned to mutualists' needs that they are fully consumed at the end of the interaction. The excess, I will argue below, opens opportunities for the accumulation of biodiversity.

11.2.4 Many Species Other than Mutualists Subsist on Mutualistic Resources

Mutualistic resources are valuable commodities; they must be, in order to entice other species to visit and perform beneficial actions. As a consequence, they attract other consumers as well. Flowers are attacked by nectar robbers and nectar-thieves, flower consumers, and pollen consumers; few of these consumers provide any mutualistic service (McCall and Irwin 2006; Irwin et al. 2010). Fruits are eaten by pulp-feeders and seed predators, not only mutualistic seed dispersers (Fedriani and Delibes 2013; Penn and Crist 2018). Extrafloral nectar and honeydew are fed upon not only by aggressive ants that protect their food sources, but by other consumers as well, including timid species that run from danger (Roux et al. 2011; Vidal et al. 2016). Even species that sequester their resources for highly specialized mutualists do not escape non-mutualistic consumers. For example, it is very difficult to "break the code" and obtain the specialized oviposition-space and seed resources that figs provide to fig wasp pollinators. However, this has occurred repeatedly, and has opened the door for a massive radiation of exploitative wasps able to do this (see ▶ Sect. 11.4.3).

In a wide array of systems, non-mutualistic consumers compete with, and sometimes outcompete, mutualists for shared resources (e.g., Bennett and Bever 2009, Palmer et al. 2010, Hanna et al. 2014, Pillai et al. 2014). While this might be imagined to reduce diversity, models suggest that even when mutualist-exploiter competition is intense, there is a broad range of conditions under which coexistence is possible (Bronstein et al. 2003; Jones et al. 2009; Wang et al. 2012). Beyond this, however, some mutualistic resources are simply left over at the end of the interaction, at which point they are free to be used by scroungers and grazers (Barker and Bronstein 2016). Sharing of resources is not costly if there is a surplus, a phenomenon that has been well-studied in intraspecific social interactions (Barnard 1984, Blurton Jones 1984) but much less so in mutualisms (but see, e.g., Heiling et al. 2018).

It is important to clarify that it is overly simplistic to divide consumers into two categories, mutualists and non-mutualists. Some species are mutualistic only in a subset of ecological conditions. Protective associations in particular confer high benefits when natural enemies are abundant; when enemies are rare or absent, the interaction is commensal or even parasitic (Chamberlain and Holland 2009; Lee et al. 2009).

11.3 Mutualistic Resources and Biodiversity Accumulation

I have highlighted above four key features of mutualism. most of them involve one-way or two-way exchange of resources; these resources are commonly used by many different mutualists; as they are difficult or impossible to modulate to mutualists' needs, there would appear to be an advantage to overproduction; and many species other than mutualists subsist on these resources. The picture this suggests is one in which a great many species are linked in a rather large interaction network.

Of course, this is exactly what two decades of research on mutualistic networks have revealed: nested, cross-linked relationships in which there are few or no isolated, specialized modules. The patterns of species richness these features generate have been laid out by Chomicki et al. (2019). What I am proposing here is an underlying rationale for such biodiversity accumulation, based on the fundamental resource-provisioning nature of most mutualisms.

The resource-provisioning nature of mutualism goes beyond shaping mutualistic networks, however, in three key ways.

First, the networks that form around mutualistic resources are not strictly mutualistic. Rather, they include all species that subsist on those resources. To date, few studies have attempted to include both mutualists and non-mutualists in an ecological network (but see Genini et al. 2010, Rodríguez-Rodríguez et al. 2017, Montesinos-Navarro et al. 2018, Mesquita-Neto et al. 2018). By my argument, however, the set of species we should focus on are *all those that make use of resources provided to mutualists*, not just those that reliably provide benefits in return. These networks are going to be substantially larger, and very likely significantly more complex, than those comprised of mutualists alone. If we are interested in conserving local diversity, it is particularly critical that we not overlook non-mutualists, nor species that are mutualistic only under a relatively narrow range of conditions. It is the mutualistic resource base that allows them to thrive. Further, the context-dependent nature of mutualistic benefit favors the maintenance of species diversity at a geographic scale (Thompson 2005), as well as genetic variation at a local scale (Heath and Stinchcombe 2013).

Second, the non-mutualists foraging on mutualistic resources might be highly effective mutualists of *other* species in the community. I provide an example in ▶ Sect. 11.4.2, in which a community of hawkmoth pollinators in southern Arizona, USA is sustained by the copious nectar of a local bat-pollinated species (Alarcón et al. 2008).

Third, mutualistic networks are linked, such that the resource base of one form of mutualism facilitates the formation and diversity accumulation by others. Research on multilayer networks, including networks linked by shared species, is just beginning (Pilosof et al. 2017; Astegiano et al. 2017; Mello et al. 2019; Hale et al. 2020). Theoretical work is suggesting that multilayer networks that link mutualisms, or that link mutualisms to antagonisms, show high diversity, high biomass accumulation, and high resistance to perturbation (Kéfi et al. 2016; Hale et al. 2020). Here I note that mutualistic resources are what often lead multilevel networks to emerge. Pollination mutualisms, with a resource base (usually) of nectar, lead to seed production, with a resource base (usually) of the nutritious fruit that surrounds those seeds. Ant-plant protection mutualisms, with a resource base of ant rewards (extrafloral nectar, pearl bodies, Beltian bodies, as well as housing structures), lead to increased plant growth, facilitating pollination and seed dispersal mutualisms. Belowground mutualistic symbioses enhance plant nutrient status, allowing plant to invest more in resources that reward other mutualistic guilds (e.g., Gange and Smith 2005). All of these phenomena start with the existence of mutualistic resources. All end, potentially, with the accumulation of biodiversity.

The Gift That Keeps on Giving: Why Does Biological Diversity...

293

11

☐ **Fig. 11.1** Pollination Mutualisms that Accumulate Biological Diversity. Panels across the top illustrate three pollination mutualisms. The three lower panels are examples of additional biological diversity that accumulates around the mutualism illustrated just above. (**a**) Osmia ribifloris pollinating pointleaf manzanita, Arctostaphylos pungens, in chaparral habitat in Arizona, USA; note the nectar-robbing hole in the flower. (**b**) Manduca sexta pollinating Datura wrightii in Arizona desert grasslands; after pollinating, females lay their eggs on the same plant. (**c**) The fig wasp Pegoscapus jimenezi pollinating the fig Ficus aurea in southern Florida, USA. (**d**) Nectar-robbing holes in manzanita. (**e**) Agave palmeri, a secondary nectar source for Manduca sexta and other large hawk-moths; these moths are all pollinators, but not of this plant. (**f**) A mutualistic seed disperser, the Emerald Toucanet, feeding upon Ficus pertusa fruits in Monteverde, Costa Rica

As we will see below, an advantage of identifying a mechanistic underpinning for mutualistic networks is that it generates specific, testable hypotheses about when biodiversity accumulation will and will not occur, as well as which species are critical to its maintenance. First, however, I will present examples of three mutualistic systems to illustrate the arguments I have laid out here. They are illustrated in ☐ Fig. 11.1.

11.4 Biodiversity Accumulation around Three Pollination Mutualisms

The pollination mutualisms I highlight below contrast in many ways. They range in distribution from the tropics to the temperate zone, differ strikingly in obligacy and specificity, and involve distantly related plants and insects. Yet, all exhibit the features I have highlighted above: mutualistic resources that are shared by extensive mutualistic and non-mutualistic communities, linked to other interaction networks, forming as a whole a nexus of biodiversity within their communities. Further, they are all in peril in an anthropogenically altered world.

11.4.1   Manzanita: A Generalized Insect Pollination Mutualism

Point leaf manzanita (*Arctostaphylos pungens*, Ericaceae) is a perennial shrub found at elevations between 1200–2500 m in chaparral habitats in the southwestern US and California. Flowers are approximately 5 mm long and produced in dense inflorescences (Richardson and Bronstein 2012). Each flower lasts approximately 3 days. Nectar volumes per flower are very low (less than 0.5 µl per flower) but is replenished overnight. Flowers are largely self-incompatible, and are buzz-pollinated (Richardson and Bronstein 2012): that is, pollen is only released when the anthers are vibrated at a specific frequency by visiting insects. The flowering season for *A. pungens* is typically earlier than other flowering plants in its range, beginning in late winter (the end of January) and ending in March (but see below).

The insect community associated with *A. pungens* is very large. In 1998, we observed 46 taxa visiting flowers in the Santa Catalina Mountains north of Tucson, Arizona. Twelve of these species, including 10 bees and two flies, were responsible for over 90% of these visits (Richardson and Bronstein 2012). Importantly, though, only a fraction of the nectar was being consumed by mutualist visitors. We found that almost all of its floral visitors were pursuing mixed foraging strategies. None of them consistently visited *A. pungens* in a way that might lead to pollination; rather, individuals or species act as both pollinators and nectar-robbers, or else rob flowers using more than one tactic to acquire nectar (Richardson and Bronstein 2012). ◨ Fig. 11.1a illustrates a pollinator visiting a flower that has already been robbed; ◨ Fig. 11.1d shows how common robbing damage can be. More recently, we have looked in detail at the reproductive consequences to manzanita of visits from the most common flower foragers (Eliyahu et al. 2015a; K. Mathis, A. Nogueira, and J.L. Bronstein, *unpublished*). We have found that legitimate nectar collection does consistently result in pollen transfer, whereas primary robbing (chewing a hole through the side of the corolla) does not. The best pollinator of *A. pungens*, however, is one that mixes mutualistic and antagonistic foraging tactics. Interestingly, it is the honey bee (*Apis mellifera*), an invasive species in these deceptively pristine chaparral habitats.

Nectar consumers are not the only foragers at manzanita flowers at our study site. A large number of species collect, either in addition to nectar or exclusively, a second resource: pollen. Some of these species are effective pollinators, including *A. mellifera* and *Osmia ribifloris* (Megachilidae; ◨ Fig. 11.1a). Interestingly, three species of thrips (Thysanoptera) live and reproduce on *A. pungens*. Although voracious pollen consumers, they transfer pollen successfully as they move between flowers (Eliyahu et al. 2015b).

Arctostaphylos pungens clearly is a very popular resource. A key reason is its phenology: it is typically the first shrub to flower in its habitat, but its flowering period begins after many insects have already emerged and are seeking food (Richardson and Bronstein 2012; Rafferty et al. 2016). The earliest individuals to flower attract the greatest attention from floral visitors. Once other species come into flower, *A. pungens* is typically abandoned for the richer nectar sources available elsewhere (Rafferty et al. 2016). Its flowering position gives it a particularly striking role in the maintenance of biological diversity in its chaparral habitat.

Specifically, it is a critical early resource that maintains a generalist pollinator (and non-pollinator!) community for later flowering species.

This situation is changing, however. Recent studies have revealed two concerning patterns. First, over the past 25 years, the flowering time of *A. pungens* has shifted later by more than 4 weeks (Rafferty et al. 2016). (A climate-change induced shift towards later flowering has been apparent in many species within southern Arizona low montane habitats in recent years; Crimmins et al. 2010). As a consequence, *A. pungens* flowers are being visited less and less, as flowering comes to overlap temporally with superior resources more and more. Meanwhile, the earliest flower-foraging insects in the community likely have fewer and fewer resources to rely upon (Rafferty et al. 2016). Secondly, the diversity of manzanita's flower-foraging community has plummeted in recent years. In 1998, 46 species were recorded at the flowers; in 2016–2017, only 15 species were found. In 1998, 93% of floral visits were made by a group of 12 taxa; in 2016–2017, 90% of visits were by invasive honey bees alone. Some visitors that were common in 1998 are now completely absent.

Thus, generalist mutualistic systems can be exceptionally diverse. However, such rich biological diversity can be, and clearly is, easily lost as a consequences of anthropogenic activities.

11.4.2 Datura: A Specialized Pollination Mutualism that Is Also a Specialized Antagonism

Manzanita, discussed above, bears very small flowers that each produce very little nectar at a constrained period in late winter; yet, it provides a critical resource that likely boosts diversity in its community. In contrast, *Datura wrightii* (Solanaceae), found in arid habitats in the southwest US and Mexico at somewhat lower elevations than manzanita, presents nectar for many months each year – yet, it has only a single truly effective pollinator. As a whole, however, *D. wrightii* presents a bounty of resources for animals in a water-limited environment, and is likely critical for their persistence.

Datura wrightii is a perennial herb that flowers from mid-April to early November. Flowers are massive (maximum length is ~25 cm), white, and tubular. The flowers open at dusk and remain open for a single night, although flowers can remain open in the morning hours if the temperatures are cool (McCall et al. 2018). Flowers are self-compatible and frequently set fruit without pollinator activity; yet, nectar is sugar-rich (25% sucrose equivalents) and produced in massive quantity, averaging 65 µl (Riffell et al. 2008; Bronstein et al. 2009). Although very striking on the landscape, extensive studies have revealed that *D. wrightii* flowers in our southern Arizona study site are visited almost exclusively by two species: the hawkmoth *Manduca sexta* (Sphingidae), shown in ◘ Fig. 11.1b, and honeybees (Riffell et al. 2008; Alarcón et al. 2008, 2010; McCall et al. 2018). *Manduca sexta* individuals forage for nectar at night, and are highly effective pollinators at this site (Bronstein et al. 2009). Honeybees, in contrast, avoid *D. wrightii* nectar entirely. Rather, they collect large amounts of pollen at dusk, just before hawkmoths become active, as well as at dawn. Moving between flowers, they act as surprisingly effective pollinators (McCall et al. 2018).

Thus, *D. wrightii* is associated with a single, nectar-reliant pollinator; in a water-limited environment, it produces vast amounts of nectar upon which that pollinator subsists. Yet, even if the nectar is not removed, the plant produces large numbers of fully germinable seeds (Bronstein et al. 2009). These observations suggest that the cost of producing nectar might be surprisingly low.

From the biodiversity perspective, it is worth considering this interaction not only from the plant side but the floral visitor side as well. Honey bees swarm *Datura* flowers in large numbers, suggesting that its pollen is a critical resource sustaining larvae of this invasive species. *Manduca sexta*, in turn, relies heavily but not exclusively on *Datura* nectar in this region. In particular, it consumes large quantities of nectar from a co-occurring species, *Agave palmeri* (Riffell et al. 2008; Alarcón et al. 2008), shown in ▢ Fig. 11.1e. While the availability of *Agave* nectar sustain this specialist *Datura* pollinator, as well as other local hawkmoths that pollinate other local plants, these insects likely compete with *Agave*'s own bat pollinators for nectar. Feeding activities that help *Datura* and possibly other hawkmoth-pollinated plants to persist may thus come at a reproductive cost to *Agave*.

Agave is a rich and abundant nectar source, raising the question of why *M. sexta* is associated with *Datura* at all. The key appears not to be *M. sexta*'s nectar consumption at the adult state, but rather its leaf consumption at the juvenile stage. Throughout its large range (which extends across most of North and Central America), *M. sexta* is a Solanaceae specialist. In our study region, two *Datura* species, including *D. wrightii,* are nearly the only Solanaceous species that it will accept. *Manduca sexta* larvae are voracious, with the ability to entirely defoliate a plant during the final days of the larval stage (McCall et al. 2020). Thus, *D. wrightii* has a highly reliable, specialist pollinator, and feeds it copiously, yet its activities come at a potentially high cost. Apparently, *D. wrightii* has evolved to tolerate high levels of damage. If it repelled this consumer, it would likely lose its pollinator as well; by tolerating its damage, it retains its pollinator, and indeed sustains its population (McCall et al. 2020).

I have portrayed the *Datura-Manduca* system as a low-diversity, relatively closed system. This is not really accurate: diverse other feeding activities, and hence many other species, are sustained by this specialized association. *Datura wrightii* flowers contain a rich community of nectar microbes (von Arx et al. 2019), an element of biological diversity that has hardly been examined in any mutualistic system. This nectar is occasionally taken by other hawkmoths (Alarcón et al. 2008), although none appears to be an effective pollinator. The massive buds are chewed open by *Cyclocephala* beetles, which then spend the day in the cool interior (A.C. McCall and J.L. Bronstein, *unpublished*). Datura leaves are high in toxic alkaloids, but two other specialist herbivores, a Chrysomelid beetle and a weevil, are common. *Manduca sexta* itself must contend with a community of predators, parasitoids, and pathogens at the larval stage (Miranda et al. 2013; Wilson and Woods 2015, 2016). Finally, pollination results in production of spiky pods, which split open during the day and spill out seeds. Attached to each seed is an elaiosome,

11

a lipid-rich food body attractive to ants. Ants remove the seeds to their nests and, although we have not yet studied this, at least some members of the extensive desert ant community are likely to be effective seed dispersers. Not only ants collect the seeds, however: they are also attractive to nocturnal rodents more interested in consuming the seed than the elaiosome (Ness and Bressmer 2005). Thus, the vast nectar resource offered by *D. wrightii* ultimately ends up provisioning a second consumer community, composed of both mutualists and antagonists, centered upon the seeds.

What is the fate of these interaction networks in a changing world? The most obvious concern is that the loss of *Datura* would imperil the many consumers, both mutualists and non-mutualists, that depend upon it in various ways. Other effects are more subtle, but real. The southwestern US has been growing hotter and hotter, but also drier and drier. Hawkmoth abundance has plummeted in recent years. Although *D. wrightii*'s high self-compatibility and ability to employ honeybees as pollinators might sustain its population, this is less likely to be true for *Manduca*'s own consumers. A further risk to hawkmoths is the loss of *Agave*. *Agave palmeri* nectar is fed upon not only by *M. sexta*, but by essentially all the large hawkmoths in southern Arizona (Alarcón et al. 2008). The Santa Rita Mountains, where we work, is at increasing threat from open-pit copper mining. Loss of *Agave* can thus be predicted to negatively impact the hawkmoth community, as well as community of bats that pollinate these massive plants (Slauson 2000), which reproduce once in their life and then die.

11.4.3 Figs and Fig Wasps: The Iconic Specialized, Obligate Pollination Mutualism

As discussed above, *Datura* rewards its specialized herbivorous pollinator lavishly, yet can reproduce without its assistance. The *Datura-Manduca* interaction, therefore, is highly specialized but not obligate, an important distinction (Chomicki et al. 2020). In contrast, figs and fig wasps are involved in a far closer relationship, one that is both obligate and specialized. It is not the only plant/insect interaction in which a specialist seed consumer also pollinates its plant (Kato and Kawakita 2017), but it is the best studied among them.

The mutualism between fig trees (about 750 *Ficus* species) and their pollinator wasps (family Agaonidae) has long served as a model for the intricate adaptations and extreme specialization that coevolution can produce (Bronstein 1992; Herre et al. 2008). Most fig species are pollinated exclusively by a single species of fig wasp, which in turn is associated with a single fig species (but see below). Pollen-laden female wasps enter the closed inflorescence (◻ Fig. 11.1c), pollinate the flowers within, then deposit their eggs in a subset of those flowers. Their offspring feed on the developing seeds. When the wasps are mature, they mate; the females collect pollen and then depart in search of an oviposition site. Trees within a fig population generally flower in tight within-tree synchrony, but out of synchrony with each other, which forces the wasps to depart their natal tree. Hence, figs sacrifice some proportion of their seeds to guarantee that their pollen will be dispersed effectively among individuals. Figs cannot produce seeds in the absence of fig wasps, and fig wasps can reproduce nowhere but inside a fig.

This classic view of the fig-fig wasp mutualism might lead to the impression that it is that rarest of interactions within a mutualistic network, a true one-to-one interaction that leads figs to interact with no other species in the community. This is not the case.

First, figs have long been recognized as being "keystone mutualists" (Terborgh 1986) in tropical forests. This is due to the extreme generalization of their mutualisms with seed dispersers (◘ Fig. 11.1f). Figs are relatively nutritious, easy to consume, and available year-round, and may well be eaten by all tropical forest vertebrates at some point in their lives (Janzen 1979; Shanahan et al. 2001). Since these critical fruits cannot be produced without pollination having taken place, it is fair to say that figs' keystone role is fundamentally attributable to their specialized, obligate pollinators.

Second, figs house a large community of other species, many of them as species-specific as the pollinators themselves. They exhibit a wide range of feeding niches. Some of the most specialized compete with the pollinators for developing seeds; others are gallers, and others still are parasites of other fig residents (Bronstein 1991, 1999). Nematodes are ubiquitous associates that move between resource patches on the bodies of fig wasps (Krishnan et al. 2010). Looser associations involve predators of fig wasps both inside and outside the fig, as well as predators on both developing and dispersed seeds (Bronstein 1988). How so many species are able to coexist on these small, odd, patchy resources is an open question (Duthie et al. 2015).

Third, it has been recognized in recent years that the fig pollination mutualism is not quite as species-specific as it was once portrayed (e.g., by Bronstein 1992). As better molecular techniques have emerged to identify the minute wasps, more cases of relaxed specificity have been documented (e.g., Moe et al. 2011; Wang et al. 2016; Sutton et al. 2017). Certain fig species are associated with more than one pollinator wasp, either at a single site or across its range. Conversely, certain fig wasp species can reproduce in more than one species of fig.

Finally, figs are a wellspring of biological diversity at an evolutionary time scale. Figs appear to have radiated explosively, and each new species appears to quickly acquire, by host shifts or by cospeciation, a diverse community of animal (and likely microbial) associates. To some extent these communities overlap (Farache et al. 2018); this is certainly true for the disperser communities. To another extent, they represent multiple replicate "experiments", each resulting in a flowering of biological diversity.

The risks to these tropical outposts of diversity are extensive and well-identified (e.g., McKey 1989; Koh et al. 2004). Climate change is a particular concern, since fig wasps survive poorly in drier and hotter conditions (Harrison 2000; Sutton et al. 2018). Further, fig wasp populations can only persist in association with large fig populations (Bronstein et al. 1990). Habitat fragmentation that leads to tree loss can result in numbers too low, at least in theory, to sustain their populations (Anstett et al. 1997).

The Gift That Keeps on Giving: Why Does Biological Diversity...

299

11

This example captures the overall thesis of this paper well. Even highly specialized mutualisms involve, and produce, resources depended upon by large numbers of other species. In this sense, mutualisms are "the gift that keeps on giving". But once the gift is gone, we can expect the diversity that it fostered and sustained to disappear as well.

Conclusions

We now know a great deal about how the structure of mutualistic communities generates robustness and hence maintains diversity. However, the possible role of mutualisms themselves as a wellspring of diversity has not been well-appreciated (but see Chomicki et al. 2019). My goal in this chapter has been to build a general argument that mutualisms play key roles in the maintenance of biological diversity within natural communities. Consumer-resource interactions are the backbone of food webs. The history of the field of ecology has centered on questions such as how species can coexist on shared resources, how many trophic levels exist in natural communities, why consumers do not drive their prey extinct, and which species play key roles in structuring assemblages. I have stressed here that most mutualisms too are consumer-resource interactions – however, they are ones in which the consumers enhance rather than destroy the partner. Further, mutualistic resources are not, and, I argue, cannot be carefully modulated to match the needs of the best mutualist; who is best, and what "best" even means, varies greatly in space and time at every possible scale. As a consequence, large numbers of other species, both beneficial and detrimental, use these resources as well, leading to a progressive buildup of diversity that might be impossible in antagonistic consumer-resource interactions. I have supplemented this argument by discussing three contrasting pollination mutualisms that I have studied during my career. Beyond illustrating my major points, each of these interactions unfortunately demonstrates how easily the diversity that accumulates around mutualisms can be lost as a consequence of anthropogenic disruption.

My argument is conjectural at this point. It contains significant gaps, and raises many questions that need to be explored directly. I conclude by offering six predictions emerging from these ideas. None has yet been tested. This is an exciting challenge for the future.

1. Among all consumer-resource interactions, mutualisms will accumulate additional species around them to a greater extent than antagonisms will.

2. Among all mutualisms, those in which resources are offered to partners will accumulate more biodiversity than those that do not. Mutualisms in which partners exchange services but not resources, such as mixed-species feeding assemblages (see ▶ Sect. 11.2.1), should not be "gifts that keep on giving."

3. Biodiversity accumulation will be more notable in systems with functionally linked mutualisms, i.e., those in which mutualisms generate resources that are then the base for other mutualisms. Pollination and seed-dispersal mutualisms, for example, are functionally linked (fruits are the product of pollination and then the resource for dispersers).

4. Biodiversity accumulation will be more notable in seasonally ephemeral, generalized mutualisms. These interactions are those most likely to offer generally accessible resources at times when large numbers of species are seeking them.

5. Biodiversity accumulation will be more notable in systems in which resources can be produced inexpensively. Not all mutualistic resources are cheap (e.g., Heil et al. 2004; Ordano and Ornelas 2005). Those that are cheap are those most likely to be produced in excess and to be relatively poorly guarded chemically and physically, and hence to support other species.

6. Biodiversity accumulation will be more notable in systems in which resources are produced constitutively, rather than induced only in response to the presence of, or the need for, mutualists. Induced mutualistic resources (e.g., Holland et al. 2009) may foster more effective mutualisms; however, by effectively excluding non-mutualists and consumers at time when mutualists are not strictly required, they should much less likely to be hotspots of biological diversity.

Key Points
- Biological diversity tends to accumulate around mutualisms.
- The resources offered to mutualists in exchange for their services are available and used more widely, sustaining large numbers of species.
- Through these resources, mutualisms are linked to other mutualisms as well as to antagonistic interactions, building community structure, diversity, and resilience.
- The loss of mutualists and the consequent loss of the resources they offer to partners thus potentially results in much wider species losses.

❷ Questions
- What kinds of resources do mutualists provide to their partners?
- In nature, are mutualistic resources produced in the precise quantity necessary to attract and reward mutualists, and if not, why?
- What kinds of organisms other than mutualists take advantage of these resources?

Literature Cited

Alarcón R, Davidowitz G, Bronstein JL (2008) Nectar usage in a southern Arizona hawkmoth community. Ecological Entomology 33:503–509

Alarcón R, Riffell JA, Davidowitz G, Hildebrand JG, Bronstein JL (2010) Sex-dependent variation in the floral preferences of a hawkmoth (Manduca sexta). Anim Behav 80:289–296

Anstett M-C, Hossaert-McKey M, McKey D (1997) Modeling the persistence of small populations of strongly interdependent species: figs and fig wasps. Conserv Biol 11:204–213

Aslan CE, Bronstein JL, Rogers HS, Gedan KB, Brodie J, Palmer TM, Young TP (2016) Leveraging nature's backup plans to incorporate interspecific interactions and resilience into restoration. Restor Ecol 24:434–440

Astegiano J, Massol F, Vidal MM, Cheptou P-O, Guimarães Jr PR (2015) The robustness of plant-pollinator assemblages: linking plant interaction patterns and sensitivity to pollinator loss. PLoS One 10(2):e0117243

Astegiano J, Altermatt F, Massol F (2017) Disentangling the co-structure of multilayer interaction networks: degree distribution and module composition in two-layer bipartite networks. Sci Rep 7:15465

Aubier TG, Elias M (2020) Positive and negative interactions jointly determine the structure of Müllerian mimetic communities. Oikos 129:983–997

Barker JL, Bronstein JL (2016) Temporal structure in cooperative interactions: what does the timing of exploitation tell us about its cost? PLoS Biol 14:e1002371

Barnard CJ (1984) Producers and scroungers: strategies of exploitation and parasitism. Chapman and Hall, New York

Bascompte J, Jordano P (2014) Mutualistic networks. Princeton University Press, Princeton

Bascompte J, Jordano P, Melián CJ, Olesen JM (2003) The nested assembly of plant-animal mutualistic networks. Proceed Nat Acad Sci USA 100:9383–9387

Bennett AE, Bever JD (2009) Trade-offs between arbuscular mycorrhizal fungal competitive ability and host growth promotion in Plantagolanceolata. Oecologia 160:807–816

Blüthgen N, Menzel F, Hovestadt T, Fiala B (2007) Specialization, constraints, and conflicting interests in mutualistic networks. Curr Biol 17:341–346

Brandenburg A, Kuhlemeier C, Bshary R (2012) Hawkmoth pollinators decrease seed set of a low-nectar Petunia axillaris line through reduced probing time. Curr Biol 22:1635–1639

Brodie J, Aslan CE, Rogers HS, Redford KH, Maron JL, Bronstein JL, Groves CR (2014) Secondary extinctions of biodiversity. Trends Ecol Evol 29:664–672

Bronstein JL (1988) Predators of fig wasps. Biotropica 20:215–219

Bronstein JL (1991) The nonpollinating wasp fauna of Ficus pertusa: exploitation of a mutualism? Oikos 61:175–186

Bronstein JL (1992) Seed predators as mutualists: ecology and evolution of the fig/pollinator interaction. In: Bernays E (ed) Insect-plant interactions. Volume IV. CRC Press, Boca Raton, pp 1–44

Bronstein JL (1999) The biology of Anidarnes bicolor (Hymenoptera, Agaonidae, Sycophaginae), a galler of Ficus aurea. Florida Entomologist 82:454–461

Bronstein JL (2001) The costs of mutualism. Am Zool 41:127–141

Bronstein JL (2015) The study of mutualism. In: Bronstein JL (ed) Mutualism. Oxford University Press, New York, pp 3–19

Bronstein JL, Gouyon PH, Gliddon C, Kjellberg F, Michaloud G (1990) Ecological consequences of flowering asynchrony in monoecious figs: a simulation study. Ecology 71:2145–2156

Bronstein JL, Wilson WG, Morris WF (2003) The ecological dynamics of mutualist-exploiter communities. Am Nat 162:S24–S39

Bronstein JL, Huxman T, Horvath B, Farabee M, Davidowitz G (2009) Reproductive biology of Datura wrightii: the benefits of associating with an herbivorous pollinator. Ann Bot 103:1435–1443

Brooks WR, Gwaltney CL (1993) Protection of symbiotic cnidarians by their hermit crab hosts: evidence of mutualism. Symbiosis 15:1–13

Cartar RV (2004) Resource tracking by bumble bees: responses to plant-level differences in quality. Ecology 85:2764–2771

Chamberlain SA, Holland JN (2009) Quantitative synthesis of context dependency in ant-plant protection mutualisms. Ecology 90:2384–2392

Chamberlain SA, Bronstein JL, Rudgers JA (2014) How context dependent are species interactions? Ecol Lett 17:881–890

Chomicki G, Weber M, Antonelli A, Bascompte J, Kiers ET (2019) The impact of mutualisms on species richness. Trends Ecol Evol 34:698–711

Chomicki G, Kiers ET, Renner SS (2020) The evolution of mutualistic dependence. Annu Rev. Ecol Evol Syst 51:409–432

Cohen D, Shmida A (1993) The evolution of flower display and reward. Evol Biol 27:197–243

Crimmins TM, Crimmins MA, Bertelsen CD (2010) Complex responses to climate drivers in onset of spring flowering across a semi-arid elevation gradient. J Ecol 98:1042–1051

Duthie AB, Abbott KC, Nason JD (2015) Trade-offs and coexistence in fluctuating environments: evidence for a key dispersal-fecundity trade-off in five nonpollinating fig wasps. Am Nat 186:151–158

Eliyahu D, McCall AC, Lauck M, Trakhtenbrot A (2015a) Florivory and nectar-robbing perforations in flowers of pointleaf manzanita Arctostaphylos pungens (Ericaceae) and their effects on plant reproductive success. Arthropod Plant Interact 9:613–622

Eliyahu D, McCall AC, Lauck M, Trakhtenbrot A, Bronstein JL (2015b) Minute pollinators: the role of thrips (Thysanoptera) as pollinators of pointleaf manzanita, Arctostaphylos pungens (Ericaceae). J Pollination Ecol 16:64–71

Farache FHA, Cruaud A, Rasplus J-Y, Cerezini MT, Rattis L, Kjellberg F, Rereira RAS (2018) Insights into the structure of plant-insect communities: specialism and generalism in a regional set of non-pollinating fig wasp communities. Acta Oecologica 90:49–59

Fedriani JM, Delibes M (2013) Pulp feeders alter plant interactions with subsequent animal associates. J Ecol 101:1581–1588

Fenster CB, Armbruster WS, Wilson P, Dudash MR, Thomson JD (2004) Pollination syndromes and floral specialization. Annu Rev. Ecol Evol Syst 35:375–403

Fisher RM, Henry LM, Cornwallis CK, Kiers ET, West SA (2017) The evolution of host-symbiont dependence. Nat Commun 8:15973

Gange AC, Smith AK (2005) Arbuscularmycorrhizal fungi influence visitation rates of pollinating insects. Ecol Entomol 30:600–606

Genini J, Morellato LPC, Guimarães Jr PR, Olesen JM (2010) Cheaters in mutualism networks. Biol Lett 6(4):494–497

Glasier JRN, Poore AGB, Eldridge DJ (2018) Do mutualistic associations have broader host ranges than neutral or antagonistic associations? A test using myrmecophiles as model organisms. Insectes Sociaux 65:639–648

Grutter AS, Blomberg SP, Box S, Bshary R, Ho O, Madin EMP, McClure EC, Meekan MG, Murphy JM, Richardson MA, Sikkel PC, Sims CA, Sun D, Warner RR (2019) Changes in local free-living parasite populations in response to cleaner manipulation over 12 years. Oecologia 190:783–797

Guimarães Jr PR, Rico-Gray V, Furtado dos Reis S, Thompson JN (2006) Asymmetries in specialization in ant-plant networks. Proc R Soc Lond Ser B 273:2041–2047

Hale KRS, Valdovinos FS, Martinez ND (2020) Mutualism increases diversity, stability, and function of multiplex networks that integrate pollinators into food webs. Nat Commun 11:2182

Hanna C, Foote D, Kremen C (2014) Competitive impacts of an invasive nectar thief on plant-pollinator mutualisms. Ecology 95:1622–1632

Harrison RD (2000) Repercussions of El Nino: drought causes extinction and the breakdown of mutualism in Borneo. Proc R Soc Lond Ser B-Biol Sci 267:911–915

Heath KD, Stinchcombe JR (2013) Explaining mutualism variation: a new evolutionary paradox? Evolution 68:309–317

Heil M, Baumann B, Kruger R, Linsenmair K (2004) Main nutrient compounds in food bodies of Mexican Acacia ant-plants. Chemoecology 14:45–52

Heil M, Barajas-Barron A, Orona-Tamayo D, Wielsch N, Svatos A (2014) Partner manipulation stabilises a horizontally transmitted mutualism. Ecol Lett 17:185–192

Heiling JM, Ledbetter TA, Richman SK, Ellison HK, Bronstein JL, Irwin RE (2018) Why are some plant-nectar robber interactions commensalisms? Oikos 127:1679–1689

Herre EA, Jandér KC, Machado CA (2008) Evolutionary ecology of figs and their associates: recent progress and outstanding puzzles. Annu Rev. Ecol Evol Syst 39:439–458

Hodges CM (1985) Bumble bee foraging: the threshold departure rule. Ecology 66:179–187

Hodges SA (1995) The influence of nectar production on hawkmoth behavior, self pollination, and seed production in Mirabilis multiflora (Nyctaginaceae). Am J Bot 82:197–204

Hoeksema JD, Bruna EM (2015) Context-dependent outcomes of mutualistic interactions. In: Bronstein JL (ed) Mutualism. Oxford University Press, New York, pp 181–202

Hoeksema JD, Schwartz MW (2001) Modeling interspecific mutualisms as biological markets. In: Noë R, van Hooff R, Hammerstein P (eds) Economics in nature: the evolutionary biology of economic behaviour. Cambridge University Press, Cambridge, pp 173–183

Holland JN, Ness JH, Boyle A, Bronstein JL (2005) Mutualisms as consumer-resource interactions. In: Barbosa P (ed) Ecology of predator-prey interactions. Oxford University Press, Oxford, pp 17–33

Holland JN, Chamberlain SA, Horn KC (2009) Optimal defence theory predicts investment in extra-floral nectar resources in an ant-plant mutualism. J Ecol 97:89–96

Howe HF (1993) Specialized and generalized dispersal systems: where does "the paradigm" stand? Vegetatio 107(/108):3–13

11

Irwin R, Bronstein JL, Manson J, Richardson LE (2010) Nectar-robbing: ecological and evolutionary perspectives. Annu Rev Ecol Evol Syst 41:271–292

Janzen DH (1979) How to be a fig. Annu Rev Ecol Syst 10:13–51

Johnson NC (2015) Mutualisms and ecosystem-level processes. In: Bronstein JL (ed) Mutualism. Oxford University Press, New York, pp 221–238

Johnson CA, Bronstein JL (2019) Coexistence and competitive exclusion in mutualism. Ecology 100:e02708

Johnson SD, Moré M, Amorim FW, Haber WA, Frankie GW, Stanley DA, Cocucci AA, Raguso RA (2016) The long and the short of it: a global analysis of hawkmoth pollination niches and interaction networks. Funct Ecol 31:101–115

Jones EI, Ferrière R, Bronstein JL (2009) Eco-evolutionary dynamics of mutualists and exploiters. Am Nat 174:780–794

Jones EI, Ferrière R, Bronstein JL (2012) The fundamental role of competition in the ecology and evolution of mutualisms. Ann New York Acad Sci 1256:66–88

Junker RR, Blüthgen N (2010) Floral scents repel facultative flower visitors, but attract obligate ones. Ann Bot 105:777–782

Kato M, Kawakita A (eds) (2017) Obligate pollination mutualism. Ecological research monographs. Springer, Japan

Kéfi S, Miele V, Wieters EA, Navarrete SA, Berlow EL (2016) How structured is the entangled bank? The surprisingly simple organization of multiplex ecological networks leads to increased persistence and resilience. PLoS Biol 14(8):e1002527

Kim W, Gilet T, Bush JWM (2011) Optimal concentrations in nectar feeding. Proc Nat Acad Sci USA 108:16618–16621

Klinkhamer PGL, de Jong TJ (1993) Attractiveness to pollinators: a plant's dilemma. Oikos 66:180–184

Koh LP, Dunn RR, Sodhi NS, Colwell RK, Proctor HC, Smith VS (2004) Species coextinctions and the biodiversity crisis. Science 305:1632–1634

Krishnan A, Muralidharan S, Sharma L, Borges RM (2010) A hitchhiker's guide to a crowded syconium: how do fig nematodes find the right ride? Funct Ecol 24:741–749

Lee J, Kim T, Choe J (2009) Commensalism or mutualism: conditional outcomes in a branchiobdellid-crayfish symbiosis. Oecologia 159:217–224

Lever JJ, van de Leemput IA, Weinans E, Quax R, Dakos V, Bascompte J, Scheffer M (2020) Foreseeing the future of mutualistic communities beyond collapse. Ecol Lett 23:2–15

Lichtenberg EM, Heiling JM, Bronstein JL, Barker JL (2020) Noisy communities and signal detection: why do foragers visit rewardless flowers? Philos Trans R Soc Lond B 375:20190486

Linsenmair KE, Heil M, Kaiser WM, Fiala B, Koch T, Boland W (2001) Adaptations to biotic and abiotic stress: Macaranga-ant plants optimize investment in biotic defence. J Exp Bot 52:2057–2065

McCall AC, Irwin RE (2006) Florivory: the intersection of pollination and herbivory. Ecol Lett 9:1351–1365

McCall AC, Richman S, Thomson E, Edgerton M, Jordan S, Bronstein JL (2018) Do honeybees act as pollen thieves or pollinators of Daturawrightii? J Pollination Ecol 24:164–171

McCall AC, Davidowicz G, Bronstein JL (2020) How high are the costs inflicted by an herbivorous pollinator? Arthropod Plant Interact 14:387–397

McKey D (1989) Population biology of figs: applications for conservation. Experientia 45:661–673

Mello MAR, Felix GM, Pinheiro RBP, Muylaert RL, Geiselman C, Santana SE, Tschapka M, Lotfi N, Rodrigues FA, Stevens RD (2019) Insights into the assembly rules of a continent-wide multilayer network. Nat Ecol Evol 3:1525–1532

Mesquita-Neto JN, Blüthgen N, Schlindwein C (2018) Flowers with poricidal anthers and their complex interaction networks-disentangling legitimate pollinators and illegitimate visitors. Funct Ecol 32:2321–2332

Miranda VA, Navarro PD, Davidowitz G, Bronstein J, Stock SP (2013) Effect of insect host age and diet on the fitness of the entomopathogenic nematode-bacteria mutualism. Symbiosis 61:145–153

Moe AM, Rossi DR, Weiblen GD (2011) Pollinator sharing in dioecious figs (Ficus: Moraceae). Biol J Linn Soc 103:546–558

Montesinos-Navarro A, Hiraldo F, Tella JL, Blanco G (2018) Network structure embracing mutualism-antagonism continuums increases community robustness. Nat Ecol Evol 1:1661–1669

Ness JH, Bressmer K (2005) Abiotic influences on the behaviour of rodents, ants, and plants affect an ant-seed mutualism. Ecoscience 12:76–81

Noë R, Hammerstein P (1994) Biological markets: supply and demand determine the effect of partner choice in cooperation, mutualism and mating. Behav Ecol Sociobiol 35:1–11

Ohashi K, Yahara T (2001) Behavioral responses of pollinators to variation in floral display size and their influences on the evolution of floral traits. In: Chittka L, Thomson JD (eds) Cognitive ecology of pollination. Cambridge University Press, Cambridge, UK, pp 274–296

Ollerton J (2006) "Biological barter": patterns of specialization compared across different mutualisms. In: Waser NM, Ollerton J (eds) Plant-pollinator interactions: from specialization to generalization. University of Chicago Press, Chicago, pp 411–435

Ordano M, Ornelas JF (2005) The cost of nectar replenishment in two epiphytic bromeliads. J Trop Ecol 21:541–547

Pahua VJ, Stokes PJN, Hollowell AC, Regus JU, Gano-Cohen KA, Wendlandt CE, Quides KW, Lyu JY, Sachs JL (2018) Fitness variation among host species and the paradox of ineffective rhizobia. J Evol Biol 31:599–610

Palacio FX, Ordano M (2018) The strength and drivers of bird-mediated selection on fruit crop size: a meta-analysis. Front Ecol Evol 6:18

Palmer TM, Doak DF, Stanton ML, Bronstein JL, Kiers ET, Young TP, Goheen JR, Pringle RM (2010) Synergy of multiple partners, including freeloaders, increases host fitness in a multispecies mutualism. Proc Nat Acad USA 107:17234–17239

Parachnowitsch AL, Manson JS, Sletvold N (2019) Evolutionary ecology of nectar. Ann Bot 123:247–261

Pellmyr O, Thompson JN (1992) Multiple occurrences of mutualism in the yucca moth lineage. Proc Nat Acad Sci USA 89:2927–2929

Penn HJ, Crist TO (2018) From dispersal to predation: a global synthesis of ant-seed interactions. Ecol Evol 8:9122–9138

Pillai P, Gouhier TC, Vollmer SV (2014) The cryptic role of biodiversity in the emergence of host-microbial mutualisms. Ecol Lett 17:1437–1446

Pilosof S, Porter MA, Pascual M, Kéfi S (2017) The multilayer nature of ecological networks. Nat Ecol Evol 1:0101

Pires MM, O'Donnell JL, Burkle LA, Díaz-Castelazo C, Hembry DH, Yeakel JD, Newman EA, Medeiros LP, de Aguiar MAM, Guimarães PA (2020) The indirect paths to cascading effects of extinctions in mutualistic networks. Ecology 101:e03080

Ponisio LC, Gaiarsa MP, Kremen C (2017) Opportunistc attachment assembles plant-pollinator networks. Ecol Lett. 20:1261–1272

Pyke GH (1981) Optimal nectar production in a hummingbird pollinated plant. Theor Popul Biol 20:326–343

Rafferty NE, CaraDonna PJ, Bronstein JL (2015) Phenological shifts and the fate of mutualisms. Oikos 124:14–21

Rafferty NE, Bertelsen CD, Bronstein JL (2016) Later flowering is associated with a compressed flowering season and reduced reproductive output in an early season floral resource. Oikos 125:821–828

Renner SS (2006) Rewardless flowers in the angiosperms and the role of insect cognition in their evolution. In: Waser NM, Ollerton J (eds) Plant-pollinator interactions: from specialization to generalization. University of Chicago Press, Chicago, pp 123–144

Richardson LL, Bronstein JL (2012) Reproductive biology of pointleaf manzanita (Arctostaphylos pungens) and the pollinator-nectar robber spectrum. J Pollination Ecol 9:115–123

Riffell JA, Alarcón R, Abrell L, Davidowitz G, Bronstein JL, Hildebrand JG (2008) Behavioral consequences of innate preferences and olfactory learning in hawkmoth-flower interactions. Proc Nat Acad Sci USA 105:3404–3409

Rodríguez-Rodríguez MC, Jordano P, Valido A (2017) Functional consequences of plant-animal interactions along the mutualism-antagonism gradient. Ecology 98:1266–1276

11

Roux O, Céréghino R, Solano PJ, Dejean A (2011) Caterpillars and fungal pathogens: two co-occurring parasites of an ant-plant mutualism. PLoS ONE 6:e20538

Schupp EW, Jordano P, Gómez JM (2017) A general framework for effectiveness concepts in mutualisms. Ecol Lett 20:577–590

Schwartz MW, Hoeksema JD (1998) Specialization and resource trade: biological markets as a model of mutualisms. Ecology 79:1029–1038

Shanahan M, So S, Compton S, Corlett R (2001) Fig-eating by vertebrate frugivores: a global review. Biol Rev 76:529–572

Slauson LA (2000) Pollination biology of two chiropterophilous Agaves in Arizona. Am J Bot 87:825–836

Stanton ML (2003) Interacting guilds: moving beyond the pairwise perspective on mutualisms. Am Nat 162:S10–S23

Sutton TL, DeGabrieli JL, Riegler M, Cook JM (2017) Local coexistence and genetic isolation of three pollinator species on the same fig tree species. Heredity 118:486–490

Sutton TM, Degabriel JL, Riegler M, Cook JM (2018) A temperate pollinator with high thermal tolerance is still susceptible to heat events predicted under future climate change. Ecol Entomol 43:506–512

Taylor BN, Simms EL, Komatsu KJ (2020) More than a functional group: diversity within the legume-Rhizobia mutualism and its relationship with ecosystem function. Diversity 12:50

Terborgh J (1986) Keystone plant resources in the tropical forest. In: Soulé ME (ed) Conservation biology: the science of scarcity and diversity. Sinauer, Sunderland, pp 330–344

Thompson JN (1988) Variation in interspecific interactions. Annu Rev. Ecol Syst 19:65–87

Thompson JN (2005) The geographic mosaic of coevolution. University of Chicago Press, Chicago

Thomson DM (2019) Effects of long-term variation in pollinator abundance and diversity on reproduction of a generalist plant. J Ecol 107:491–502

Vannette RL, Hunter MD (2011) Plant defence theory re-examined: nonlinear expectations based on the costs and benefits of resource mutualisms. J Ecol 99:66–76

Vázquez DP, Simberloff D (2002) Ecological specialization and susceptibility to disturbance: conjectures and refutations. Am Nat 159:606–623

Vidal MC, Sendoya SF, Oliveira PS (2016) Mutualism exploitation: predatory drosophilid larvae sugar-trap ants and jeopardize facultative ant-plant mutualism. Ecology 97:1650–1657

von Arx M, Moore A, Davidowitz G, Arnold AE (2019) Diversity and distribution of microbial communities in floral nectar of two night-blooming plants of the Sonoran Desert. PLoS One 14(2):e0225309

Wang Y, DeAngelis DL, Holland JN (2012) Uni-directional interaction and plant-pollinator-robber coexistence. Bull Math Biol 74:2142–2164

Wang G, Cannon CH, Chen J (2016) Pollinator sharing and gene flow among closely related sympatric dioecious fig taxa. Proc R Soc Lond B 283:20152963

Waser NM, Price MV (2016) Drought, pollen and nectar availability, and pollination success. Ecology 97:1400–1409

Waser NM, Chittka L, Price MV, Williams NM, Ollerton J (1996) Generalization in pollination systems, and why it matters. Ecology 77:1043–1060

Wilson JK, Woods HA (2015) Protection via parasitism: Datura odors attract parasitoid flies, which inhibit Manduca larvae from feeding and growing but may not help plants. Oecologia 179:1159–1171

Wilson JK, Woods HA (2016) Innate and learned olfactory responses in a wild population of the egg parasitoid Trichogramma (Hymenoptera: Trichogrammatidae). J Insect Sci 16:110. 111–118

Wood TJ, Gibbs J, Graham KK, Isaacs R (2019) Narrow pollen diets are associated with declining Midwestern bumble bee species. Ecology 100:e02697

Yule KM, Johnson CA, Bronstein JL, Ferrière R (2020) Interactions among interactions: the dynamical consequences of antagonism between mutualists. J Theor Biol 501:110334

Further Reading/Additional Resources

Bronstein JL (ed) Mutualism. Oxford University Press, New York, pp 221–238

Blurton Jones NG (1984) A selfish origin for human food sharing: tolerated theft. Ethology and Sociobiology 5:1–3

Hoek TA, Axelrod K, Biancalani T, Yurtsev EA, Liu J, Gore J (2016) Resource availability modulates the cooperative and competitive nature of a microbial cross-feeding mutualism. PLoS Biology 14:e1002540

Goodale E, Sridhar H, Sieving KE, Bangal P, Z. GJC, Farine DR, Heymann EW, Jones HH, Krams I, Martínez AE, Montaño-Centellas F, Muñoz J, Srinavasan U, Theo A, Shanker K (2020) Mixed company: a framework for understanding the composition and organization of mixed-species animal groups. Biological Reviews 95:889–910

Klinkhamer PGL, Jong TJd, Linnebank LA (2001) Small-scale spatial patterns determine ecological relationships: an experimental example using nectar production rates. Ecology Letters 4:559–567

Ness JH, Morris WF, Bronstein JL (2006) Integrating quality and quantity of mutualistic service to contrast ant species visiting Ferocactus wislizeni, a plant with extrafloral nectaries. Ecology 87:912–921

11

Mutualism as a Source of Evolutionary Innovation: Insights from Insect-Plant Interactions

Rodrigo Augusto Santinelo Pereira
and Finn Kjellberg

Contents

© Springer Nature Switzerland AG 2021
K. Del-Claro, H. M. Torezan-Silingardi (eds.), *Plant-Animal
Interactions*, https://doi.org/10.1007/978-3-030-66877-8_12

⊜ **Learning Objectives**
After completing this chapter, you should understand the following:
- To understand the evolution of mutualisms we need to look at natural selection operating at different levels
- The general architecture of mutualistic associations
- Mutualisms are ecologically and evolutionarily stable
- Mutualistic associations may boost species diversification by creating new ecological opportunities (innovations) for the interacting species

12.1 Introduction

Mutualisms, defined as interspecific interactions that are beneficial to all the involved partners, are ubiquitous in nature. Probably all species in the world are involved in some form of mutualistic interaction (Bronstein et al. 2006). In this chapter we will address direct mutualisms in which the partners are in physical contact. Direct mutualisms may be further divided into symbiotic and non-symbiotic mutualisms (Boucher et al. 1982). In symbiotic mutualisms, individuals are physiologically integrated, whereas in non-symbiotic mutualism, despite some physiological co-adaptation, individuals of the interacting species are at least at some stage of their life cycle physiologically independent (Boucher et al. 1982). Many symbiotic mutualisms involve the exchange of nutritional and energetic services (*e.g.* photosynthesis). Mutualisms, and particularly non-symbiotic ones, can also involve protection and dispersal of gametes or propagules.

Often, individuals benefit directly from their mutualistic behaviour. Many definitions restrict mutualism to these cases. However, in a number of interactions that constitute undisputed examples of mutualisms, the individuals performing the act that is beneficial to the other species do not benefit directly from this act. This is typically the case in gynodioecious *Ficus* species (▶ Box 12.1). Pollinator individuals that ensure pollination by entering functionally female inflorescences (figs) die without reproducing: their reproductive value is equal to zero (Kjellberg et al. 2005; Pereira 2014). A definition of mutualism that would exclude half of *Ficus* species and many other undisputed cases of mutualisms would be useless. In the case of the fig pollinating wasp mutualism, the wasps maximise their fitness by trying to reach receptive figs as fast as possible, *i.e.* without engaging in the difficult task of avoiding female figs, and this is beneficial for the fig trees as it results in pollination. In this situation, individual selection on the fig wasp is beneficial for the host *Ficus* species, not to a particular individual of that species. Reciprocally we expect the interaction to be stable because individual selection on fig trees results in phenotypes that allow the survival of populations of its associated wasps. Hence, in many situations, mutualism benefits cannot be defined at the level of individual interactions. Defining mutualism at the population level could in a number of cases be operational, as the an interaction may vary from mutualistic to parasitic across populations (Addicott 1986; Thompson 1999; Friberg et al. 2019). However, the structuring of mutualists into populations and species may be so different that a population level approach is not operational. For instance, *Ficus hirta* present gradual genetic variation among populations while its pollinating wasp in South-East China constitutes a single population (Yu et al. 2019). In such cases, selective processes do not occur at the same geographic scales for the partners. Different geographic population structure between mutualists is probably the rule rather than the exception in non-symbiotic mutualisms (Alvarez et al. 2010). Therefore, in this chapter, we define mutualisms as interspecific interactions that are beneficial for the species involved and that result in adaptive innovation. Obligate interactions resulting from infection of a host and subsequent adaptation of the host to its pathogen do not enter within the limits of this definition if they do not result in biological innovation. Such interactions can become obligate when curing the host from its pathogen results in self-poisoning (Dedeine et al. 2001).

Generally, organisms that associate in a mutualism differ radically in biological traits and life habits. The combination of these traits confers them new biological capacities, allowing them to colonise new ecological niches. At a macro-evolutionary scale, mutualisms have been at the source of major biological innovation (Margulis and Nealson 1989; Wheat et al. 2007; Leigh Jr 2010). A classic example is the endosymbiontic origin of mitochondria and plastids through the inclusion of prokaryotic organisms into the pre-eukaryotic cell (Gray 2017). This association improved the energetic machinery of eukaryotes and allowed some of them to become primary producers (Margulis 1996). Several other symbiotic mutualisms have been the source of key innovations at the origin of major new modes of life. For instance, representatives of at least seven phyla (Annelida, Arthropoda, Ciliophora, Mollusca, Nematoda, Platyhelminthes and Porifera) are known to associate with chemosynthetic bacteria (Dubilier et al. 2008). This type of mutualism allows representatives from these diverse lineages of animals to colonize habitats presenting high concentrations of reduced energy sources, such as sulphide and methane. These habitats are generally transient but at different time scales ranging from whale carcasses and sunken wood on the deep-sea floor to somewhat more stable habitats, such as hydrothermal vents, cold seeps, shallow-water coastal sediments and continental margins (*e.g.* mangrove areas) (Dubilier et al. 2008). Herbivores, including insects of different orders and vertebrates derive their digestive capacities from bacteria, fungi, protozoa and other microorganisms. The fine mechanisms involved have been investigated for cockroaches, termites, attine ants, sap-feeding insects and ruminant vertebrates (Mueller et al. 2001; Nalepa et al. 2001; Aanen et al. 2002; Kamra 2005; Baumann 2005; Koike and Kobayashi 2009; Douglas 2009; Caldera et al. 2009). While taxonomically and biologically highly heterogeneous, these digestive mutualisms are all based on the capacity of the animal to collect carbon rich resources that they cannot digest by themselves. The associated microorganisms provide the metabolic capacities to degrade theses carbon rich resources, such as cellulose, hemicelluloses and lignin, to detoxify secondary plant compounds and to convert nitrogen into available amino acids. The association of plants with mycorrhizal fungi (*e.g.* fungi belonging to the Phylum Glomeromycota and forming arbuscular mycorrhiza) is most often mutualistic, improving water and nutrient uptake by the plant, especially the uptake of phosphate and nitrogen (Brundrett 2004). Plant fine roots are constrained in how thin they can become because of the size of their genome which results in large nuclei. Fungi have much smaller genomes allowing smaller nuclei and hence allowing fungal filaments to be much thinner than plant fine roots. Therefore, mycelia explore the soil at a much finer scale and at lower constitutive costs than plant roots. Mycorrhizal fungi are associated with roots of 70–90% of land plant species, representing one of the most extensive terrestrial symbioses (Parniske 2008). It may have played a central role in the colonisation of terrestrial habitats by enabling plants to exploit mineral soils. The association between plants and nitrogen-fixing bacteria is another example of mutualism allowing ecological innovation. The bacteria metabolise inert N_2 present in the atmosphere into the utilizable form ammonia (NH_3) through the action of nitrogenases. This mutualism is present in a diversity of plant lineages including ferns, gymnosperms, mono- and eudicots. The plants host the endophytic bacteria in their roots. Among them, the root nodule symbiosis is particularly sophisticated and complex. The root nodules house high bacteria densities in a structure that provides an anaerobic microenvironment favourable for nitrogenase activity (Markmann and Parniske 2009; Soto et al. 2009). The mutualisms exemplified above are just a few examples of known mutualisms that represent biological innovations central to the colonisation of earth and the functioning of extant ecosystems. There are many other examples, as for instance lichens associating fungi and algae/cyanobacteria (Nascimbene and Nimis 2006), or sessile/slow-moving marine animals and algae/cyanobacteria (Venn et al. 2008).

In the following, we first present some theoretical considerations on mutualisms. Indeed, developing a comprehensive theory of mutualisms is still an open endeavour. We propose a framework within which a theory may emerge. Then we develop some examples of insect-plant and insect-fungi mutualisms that are at the origin of major evolutionary innovation. Finally, we focus on the fig tree – associated animal interactions that allowed this plant lineage to diversify in subtropical and tropical ecosystems and to become keystone species for the functioning of some forest ecosystems.

12

Box 12.1   Pollination by Deceit in Fig Trees

An intriguing example is pollination by deceit in gynodioecious *Ficus* species (Kjellberg et al. 1987; ◘ **Fig. 12.1**). Functionally "male" trees produce pollinating wasps (pollen vectors) and pollen, and female trees produce seeds but neither pollen nor wasps. Female trees bear figs in which all the styles of pistillate flowers are too long to allow wasp oviposition. Wasps entering figs of female trees pollinate but die without reproducing. Functionally male trees bear figs in which all pistillate flowers have short styles. These flowers may receive an egg and produce a wasp, but they rarely produce seeds. As a result, although male figs present both male and female flowers, they are functionally male. This is a case of pollination by deceit and figs on female trees constitute a lethal trap for the wasps. Considering the difference in generation time between fig trees (counted in years) and fig wasps (counted in months) and the huge difference in population sizes (ranging from

10th of thousands to millions of times larger insect population sizes), a simple prediction is that the capacity to recognise female trees and avoid them will evolve rapidly. Nevertheless, approximately half of the 800 *Ficus* species are functionally dioecious and they correspond to a limited set of monophyletic lineages. Why is it so? The pollinating fig wasps are attracted to the fig tree by volatile compounds released by the receptive inflorescences. For the compounds that are perceived by the pollinators, the inflorescences of female and "male" trees emit exactly the same relative proportions (Hossaert-McKey et al. 2016; Proffit et al. 2020). Hence, a first answer is male-female mimicry. However, a better formulation is probably, male-female mimicry makes distinguishing between male and female trees complicated. Then the question becomes, could it be worthwhile to take the time and efforts to distinguish male and female trees, can being choosy be selected? In some *Ficus* species, there is hardly any selection for avoiding female trees because figs on female and "male" trees are not receptive at the same time of the year, so that the wasps never get a chance to choose between sexes. Most pollinators emerge from figs on "male" trees at a time when there are no receptive male figs. They will fail to reproduce irrespective of whether they avoid female trees or not (Kjellberg et al. 1987). In *Ficus* species in which "male" and female trees are receptive at the same time, the race to enter receptive figs (Conchou et al. 2014) selects against taking the time required to distinguish highly similar phenotypes.

▣ **Fig. 12.1** Reproductive systems in fig trees. **a** in monoecious species seeds and pollen-loaded wasps are produced in the same fig. **b** in gynodioecious (functionally dioecious) species, functionally male trees produce pollen-loaded wasps and female trees produce seeds

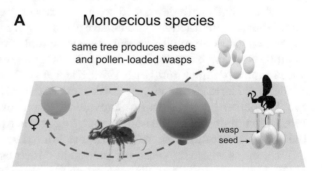

A   Monoecious species

same tree produces seeds
and pollen-loaded wasps

wasp →
seed →

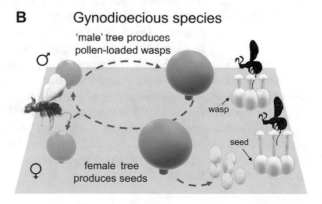

B   Gynodioecious species

'male' tree produces
pollen-loaded wasps

wasp

seed

female tree
produces seeds

12.2 Theoretical Remarks

The main pitfall of a vision in which adaptive innovation drives the evolution of mutualistic associations is formulations suggesting that a species acts for the sole benefit of another one and reciprocally. Such formulations cannot be defended. There is no way escaping that individuals are selected to maximize their fitness. In a mutualism this may, or may not, be at the expense of the partnerspecies. In this context, the evolution of mutualisms is often envisioned in a framework where individual selection leads to reciprocal exploitation that results in net benefits to all the interacting parties, *i.e.* in a framework in which individual selection intrinsically destabilises the mutualistic association (Axelrold and Hamilton 1981; Herre and West 1997). Therefore, it has been proposed that host sanction against non-cooperative partners may be important for mutualism stability. Host sanctions would be any traits evolved specifically to reduce the fitness of a partner that becomes non-cooperative.

Lists of mechanisms limiting non-cooperative behaviour or partner overexploitation and derived from intraspecific cooperation theory have been proposed (Herre et al. 1999; Yu 2001; Sachs et al. 2004; Doebeli and Hauert 2005; West et al. 2007; Bergmüller et al. 2007). While cooperation models have stimulated theoretical studies on mutualism, they do not find consistent empirical support to explain the stability of a range of mutualistic associations. For example, models based on retaliation (sanctions or refusing future interactions) are probably only of importance for animals with developed cognition capacity, as they require the ability to remember past interactions and behave accordingly. Sanction has been invoked to explain cooperation reinforcement in some mutualistic association, such as legume plants – N_2-fixing bacteria, fig trees – pollinating fig wasps and yucca – pollinating yucca moths (Pellmyr and Huth 1994; Kiers et al. 2006; Kiers and Denison 2008; Jandér and Herre 2010; Leigh Jr 2010; Jandér et al. 2012). In these examples, when the *Rhizobium* bacteria do not provide nitrogen, or when the fig wasps/yucca moths do not pollinate, the non-cooperating partner pays a cost, as the O_2 flux to the radicular modules where bacteria are housed stops or the flower ovaries or inflorescences where fig/yucca pollinators laid their eggs provide less nutrients or abort. However, there is no demonstration that these host traits have been selected to respond to non-cooperative partners, *i.e.* that they qualify as sanctions (Frederickson 2013). Further no theoretical framework has been proposed within which sanction could evolve. Indeed, there is an intrinsic problem with the strong differences in generation time and populations sizes between the interacting species (see ▶ Box 12.1). Individuals of the slow-reproduction small population size species interact with many individuals and many generations of their associated species. Such asymmetry is constitutive of most mutualistic associations. As a consequence, in visions of mutualism stability based on an evolutionary race between species, the partner with larger population sizes and the shorter generation time should win the race and reap the highest profits possible from the association. In the process, it would overcome any adaptation of the other species to limit overexploitation.

Contrary to the predictions based on ideas stemming from within species cooperation theory, mutualistic associations are robust to the presence of noncooperative partners. Indeed, phylogenetic evidences from a range of biological systems evidences that mutualisms rarely evolve into parasitic interactions in nature (Frederickson 2017). A striking example of this robustness is provided by the diffuse mutualism involving seed dispersion by vertebrates. Seed dispersion networks are structured into complex mosaics of frugivorous guilds associated with plant guilds and involve plant traits such as seed size, fruit colour, fruit flesh nutrient content, etc., with few specialist species (Shanahan et al. 2001; Albert et al. 2013; Sarmento et al. 2014). Poor dispersers and seed predators are frequent among members of the frugivorous guilds (Howe 1986; Compton et al. 1996; Shanahan et al. 2001). Nevertheless, even with this widespread occurrence of non-cooperating partners, about 70–90% of tropical trees present adaptions favouring animal seed dispersion (Howe and Smallwood 1982). Thinking about the stability of mutualisms and their role in biological innovation obliges to envision their evolution from two different perspectives. The individual selection perspective is a truism: only genes that increase in frequency in a population become fixed. Genes that decrease in frequency will disappear. Therefore, individual selection may lead to population size decrease and to species extinction by increasing the relative frequency of a gene that reduces population size. On the other hand, one of the defining characters of life is that living organisms have a history. Indeed, life on earth has survived for over 3 billion years. Species that have survived and given rise to new ones are those in which short-term individual selection was compatible with, or even favored, long-term species survival. Consequently, extant species derive from species in which some intrinsic biological traits resulted in lack of short-term selection favoring genes that would lead to species extinction. Only lineages in which there was no such destructive short-term selection have survived. If a lineage loses these biological traits, it will go extinct. This intrinsic property of life is illustrated by sexual reproduction. Most species on earth engage in some form of sex. Nevertheless, asexual lineages regularly appear. In the phylogenies, there are no asexual lineages but only isolated asexual species at the end of phylogenetic branches. This means that asexual species are evolutionary dead ends (Maynard-Smith 1978). Why is loss of sex so rare that it does not drive life on earth to extinction? Simply because extant lineages derive from species in which some intrinsic trait resulted in short term selection for sex and most often species inherit this trait from their ancestors (Gouyon 1999). As a result, loss of sex remains exceptional. We can apply the same line of thinking to mutualisms. Only mutualisms in which there is no short-term selection against the mutualism, or mutualisms in which no mutation that destabilizes the mutualism can arise, survive over evolutionary times. This explains why mutualisms are intrinsically stable (Frederickson 2017) and, hence, why we do not expect to find adaptations specifically evolved to punish cheating mutualists (i.e., that would take the benefits from the mutualism without reciprocating). Therefore, sanctions do not explain mutualism stability (Frederickson 2013).

The mutualism between *Yucca* and *Yucca*-moths provides an example of the pitfalls associated with thinking in terms of sanctions. Yuccas are pollinated by

Tegeticula moths that lay eggs in the flowers. The moth collects pollen with its modified labial palps and deposits some of it on the flower stigma after oviposition. The moth larva feeds on the developing young seeds. If the larva eats too many developing seeds in the ovary, the flower aborts leading to the death of the larva. From these observations, it is tempting to suggest that *Yucca* plants have developed a mechanism to limit oviposition by its mutualistic pollinator. If the pollinator turns into a parasite eating too many seeds, it is punished. However, *Tegeticula* belong to a group of moths that do limited damages in the ovaries. Lack of host overexploitation predates the mutualism (Yoder et al. 2010). Similarly, *Yucca* belong to a lineage of plants in which there is abundant early abortion of developing fruits. Abortion rates of fruits containing few developing seeds and of damaged flowers are high. Hence, the mechanisms limiting over-exploitation predated the evolution of the mutualism and do not qualify as sanctions. We propose that the evolution of the mutualism was made possible by pre-existing traits of the associates (Frederickson 2013). This is further demonstrated that one *Tegeticula* pollinated species, *Hesperoyucca whipplei*, is not a *Yucca*. The moth has jumped host, and has become the active pollinator of a new lineage (Pellmyr 2003). This lineage did not have a history of co-adaptation with the moth and therefore could not have developed adaptations against overexploitation by mutualist *Tegeticula* moths. Despite this, the association has thrived demonstrating that the adaptations limiting exploitation pre-existed to the mutualism. Successful host switching has occurred in several active pollination mutualisms. It has also been shown that *Epicephala* moths have colonised and pollinate actively four different lineages of plants within Phyllanthaceae converting pollinated lineages pollinated by generalist insects into lineages engaged in active pollination mutualisms (Kawakita 2010). Further, some chalcid fig-wasp lineages that do not belong to the Agaonidae have become fig-pollinators despite 70 Ma year co-diversification of the mutualistic association between *Ficus* and Agaonidae (Jousselin et al. 2001). Hence, we have numerous examples of new associations that became mutualisms because of pre-exiting traits of one or both partners involved in the new mutualism.

The complementarity of species traits in mutualisms are generally stricking. In both symbiotic and non-symbiotic mutualism, in general, there is a combination of a structural component, representing goods provided by a long-lived partner (*e.g.*, 3D structure, sheltering or food), and a service supplied by a short-lived partner (*e.g.*, gamete transport, resource transport from the surroundings, protection against natural enemies or biosynthetic capacities) (Leigh Jr 2010). In addition, mutualisms involve members of distant taxonomic groups, as exemplified by marine invertebrates – bacteria, mammals/insects – gut microorganisms, plant – root bacteria/fungi, fungi – algae/cyanobacteria, and plant – insect associations. Large phylogenetic distance between the partners is probably important for mutualism stability, as it enhances the combination of contrasting abilities, and decreases niche overlap. As a consequence, the potential for selective conflicts between partners is reduced, and the potential for overcoming them is enhanced. Indeed, the larger organism generally controls the arena in which the life of the smaller, shorter-lived organism is plaid out. Controlling the arena may canalise selective forces acting on the smaller partner. In this topic we argue that mechanisms to reinforce cooperation, acting exclusively at individual (or gene) level are not sufficient to stabilize mutualistic associations, although these mechanisms play a role in reducing conflicts between partners. One example of such mechanism is the vertical transmission of the symbiont that occurs in sap-sucking insects – gut bacteria (Baumann 2005) and ant/beetle – cultivated fungi (Aanen et al. 2002; Mueller et al. 2005), which increases the symbiont endogamy, limiting the selection of non-cooperative traits via kin selection. In other types of mutualism, specially the non-symbiotic ones, it is necessary to consider spatio-temporal dynamics at community level and the multilevel selection to accommodate an evolutionary framework of mutualisms (Gomulkiewicz et al. 2003; Wilson and Wilson 2008; O'Gorman et al. 2008; Nowak et al. 2010). Mutualism as source of evolutionary innovation that expands the partners capacities appears as a general aspect of + / + interactions. Such emergent attribute can sometimes transcend pure genetic determinism if one considers cultural evolution in animals with developed cognitive capacities, as for example corvids and primates, that have allowed them to better explore the ecosystem, and eventually use new niches (Wimsatt 1999; Castro and Toro 2004; Marzluff and Angell 2005; Vale et al. 2017).

12.3 Some Examples of Mutualisms Involving Insects and Plants

Most seed plants are sessile during their sporophytic phase, *i.e.* during most of their life cycle. Hence, many of their responses to ecological challenges, such as defence against herbivores, gamete transfer and seed dispersal, may rely on mutualistic associations with mobile animals (see previous topics). On the other hand, animals may expand their ability to use plant resources thanks to mutualistic interactions with microorganisms (Douglas 2009) or by tending phytophagous insects (Heil and McKey 2003). Those mutualisms can be classified according to the services provided by one of the partners, usually the one smaller in size and with shorter generation time (◘ Table 12.1). In the following, we present some examples of insect-plant mutualisms involving defence against herbivores, digestion of plant products, and pollination by insects (◘ Fig. 12.2).

12.3.1 Protection Mutualisms

Animals and plants can obtain protection against natural enemies from mutualistic associations. Particularly, two interrelated systems have been extensively studied: ant-plant-herbivore and ant-hemipteran interactions. Insect herbivory is a key factor in plant communities that effects plant productivity, survival and reproduction (Showalter 2000). Plants can derive protection against herbivores by associating with predatory ants. As ants are among the most important predators of arthropods, they can constitute an effective plant defence. Indeed, while many herbivores

Table 12.1 Examples of mutualisms involving insects and plants

Mutualism Classes	Provided services	Systems	Partner integrations	Sub-divisions	Some examples	References
Protection mutualism	Protection against herbivores	Predatory insects – plants	Non-symbiotic	Extra-floral nectar, domatia and food bodies	Myrmecophilic plants (several of flowering plants and ferns)	Heil and McKey (2003)
				Honey dew from phloem-feeding hemipterans (trophobionts)	Hemiptera – Formicidae mainly, but also Anthribidae, Coccinellidae, Apoidea, Tachinidae, Syrphidae and Neuroptera	Delabie (2001)
Digestive mutualism	Degrade carbon rich sources (*i.e.* cellulose, hemicelluloses and lignin), detoxify secondary plant compounds and convert nitrogen in available amino acids	Insects – microorganisms	Symbiotic	General feeders	Blattodea, Coleoptera and Psocoptera	Nalepa et al. (2001), Douglas (2009)
				Plant sap feeders	Hemiptera – *Baumannia, Buchnera, Carsonella, Portiera, Sulcia* and *Tremblayabactéria,* and clavicipitacean fungi	Douglas (2009)
				Fungus – growing insects	Attine ants, Macrotermitinae termites and Scolytinae (ambrosia beetles) – several fungi	Mueller et al. (2005)
Pollination mutualism	Gamete transport	Several insect orders – seed plants	Non-symbiotic	Nectar, pollen and sheltering as reward	bees, beetles, dipterans, lepidopterans, thrips and wasps	Rech et al. (2014)
			Symbiotic	Brood-site pollination	Obligate: fig trees – fig wasps, leafflowers – leafflower moths and yucca – yucca moths	Sakai (2002), Weiblen (2002)
					senita cactus – senita moths, Silene – Hadena/Perizoma, Lithophragma – Greya and globeflower – globeflower flies	Kephart et al. (2006), Hembry and Althoff (2016)

12

◘ Fig. 12.2 Examples of mutualisms involving insects and plants. Protection mutualism: **a** *Ectatomma tuberculatum* ant probing an extrafloral nectary in Leguminosae. Digestive mutualisms: **b** *Aconophora* sp. (Auchenorrhyncha:Membracidae), associated with *Baumannia* (γ-proteobacteria) and *Sulcia* (Bacteroidetes); **c** *Atta laevigatta* carring a floral bud. Pollination mutualisms: **d** *Editha magnifica* wasp visiting *Vernonia* flowers; non-specialized insects may eventually act as pollinators. **e** immature seeds and fig wasps in the monoecious *F. maxima*; note that seeds (s) are produced in flowers closer to the fig wall, while wasps (w) are produced closer to the fig cavity. **f** fig wasps collecting pollen (arrows) in the monoecious *F. albert-smithii*. Photocredits: **a** – **d** (Kleber Del-Claro), **e** – **f** (Finn Kjellberg)

have overcome plant chemical defences, few have evolved efficient adaptations against predatory ants (Heil and McKey 2003).

Plants can attract facultative mutualist ants by providing direct rewards (energy-rich extrafloral nectar and food bodies), as well as indirect rewards such as carbohydrate-rich excretions (honeydew) of phloem-feeding hemipterans (trophobionts). In this later tripartite mutualism, ants defend their trophobionts from predatory and parasitoid insects, resulting in another nested defensive interaction (Delabie 2001; Heil and McKey 2003; McKey et al. 2005). The direct and indirect rewards provided by the plants for the ants are poor in nitrogen and present highly unbalanced amino acid compositions. Ants may complement this unbalanced diet by hunting herbivorous insects or by harvesting some of their hemipteran trophobionts (Del-Claro et al. 2016; Calixto et al. 2018). Some ants, especially species adapted for living in the canopy, are associated with endosymbiotic microbes that help them to cope with such nutritional imbalance. The need of supplementation of nitrogen from external sources, seems to have driven the selection of prey-foraging strategies and physiological adaptations (Heil and McKey 2003; McKey et al. 2005).

In over 100 genera of tropical angiosperms, plants are involved in more permanent, in some cases obligatory, mutualisms with protective ants. These plants have specialised structures, called domatia, and used as nest by the ants. Domatia can occur in hollow stems (*e.g.*, *Cecropia*, *Leonardoxa* and *Macaranga*), thorns (*Acacia*), petioles (*Piper*), or leaf pouches (*Hirtella, Maieta, Scaphopetalum* and *Tococa*). In addition to domatia, plants may offer food rewards, such as extrafloral nectar, food bodies or both. Plants benefit from protection services and, additionally, from nutrients mobilised by the ants. It is estimated that 80% of the carbon in *Azteca* ant's bodies can be derived from their *Cecropia* host tree, whereas about 90% of the plant's nitrogen comes from ants debris (*i.e.*, exuviae, dead larvae, workers, and remains of arthropod prey) (Heil and McKey 2003).

Defensive mutualisms have evolved many times between insect species. Ants show trophobiotic relationships with lepidopteran larvae of the familes Lycaenidae, Riodinidae, and Tortricidae, and heteropteran species of the families Coreidae, Pentatomidae and Plataspidae. In addition to ants, hemipteran species have trophobiotic relationships with a range of other insect groups, such as Anthribidae, Coccinellidae, Apoidea (and other aculeate Hymenoptera), Tachinidae, Syrphidae and Neuroptera (Delabie 2001). However, protection mutualism is more frequent and better studied in ant-hemipteran interactions. Sap-feeding hemipterans are particularly vulnerable to predation. Some species are sessile at some life-stages. Even in more mobile species, to access the phloem, hemipteran insects introduce their stylets deep into plant tissues; a process that takes minutes to hours. While they are attached to the host plant, they are not able to remove the mouthparts quickly to escape predators. Moreover, because of their feeding mode, hemipteran insects excrete large quantities of honeydew. These insects have one additional problem; they have to get rid of such excretion that can accumulate on their bodies and serve as substrate for fungus growing. Thus, trophobiosis has this "cleaning" benefit for sap feeding insects (Delabie 2001).

12.3.2 Digestive Mutualism

Mutualism with microorganisms has allowed several groups of animals to feed on plants. Indeed, plant tissues constitute a source of energy and nutrients rich in cellulose and allelochemicals, which are primarily inaccessible to animal digestion. Plant feeding insects and vertebrates derive their digestive capabilities from bacteria, fungi, protozoa and other microorganisms. The most studied groups include cockroaches, termites, attine ants, plant sap-feeding insects and ruminants (Mueller et al. 2001; Nalepa et al. 2001; Aanen et al. 2002; Kamra 2005; Baumann 2005; Koike and Kobayashi 2009; Douglas 2009; Caldera et al. 2009). Despite their remarkable taxonomic and ecological diversity, digestive mutualisms all share the attribute of making use of the large metabolic capacities of microorganisms to degrade carbon rich sources (*i.e.* cellulose, hemicelluloses and lignin), detoxify secondary plant compounds and convert nitrogen in available amino acids.

Detritivorous termites and cockroaches are associated with bacteria, fungi and protozoa that degrade cellulosic material and recycle nitrogen from insect waste (Nalepa et al. 2001; Douglas 2009). Analogous interactions occur in ruminant vertebrates, allowing them to assimilate carbon present in cellulosic compounds and metabolise anti-nutritional and toxic substances of plants. Cattle support an impressively diverse and complex microorganism community. One millilitre of rumen liquor can enclose as many as 10^{10}–10^{11} cells of 50 bacterium genera, 10^4–10^6 ciliate protozoa from 25 genera, 10^3–10^5 zoospores of five anaerobic fungus genera and 10^8–10^9 bacteriophages (Kamra 2005). Nevertheless, the bacterial diversity is largely underestimated, as many rumen bacteria cannot be cultivated in the laboratory. Species survey based on DNA sequences suggest that 300–400 species of bacteria are present in the rumen (Koike and Kobayashi 2009).

Despite of the large metabolic capacity of microorganisms, the amount of vegetal material that insects can process is ultimately constrained by their gut volume. This limitation is overcome by insects that cultivate fungi in their nests (Douglas 2009). Fungus cultivation is well known in the Neotropical Attini ant tribe and in the Old-World termite subfamily Macrotermitinae (Mueller et al. 2001; Aanen et al. 2002; Caldera et al. 2009). However, fungus cultivation is much more widespread and is carried out by siricid woodwasps, cerambycid beetles and plant-ants (Douglas 2009; Defossez et al. 2009). All the 210 plus attine ant species rely on cultivation of fungi of the tribe Leucocoprineae as their main food. The majority of cultivated fungi belong to two genera, *Leucoagaricus* and *Leucocoprinus*. Basal attine species cultivate fungi on dead-plant matter and caterpillar frass, while derived lineages collect fresh leaves and flowers as substrate for fungi, suggesting that the development of a new mode of fungi cultivationis an innovation that has allowed ant species diversification. The fungi cultivated by derived attine ants are also highly specialized on the mutualism. They have evolved aspecialised structure rich in nutrients, the gongylidia, that serves as ant food. This structure is formed by densely packed clusters of hyphae that are easily harvested by the ants (Mueller et al. 2001). The mutualism involving fungus-growing termites is analogous with the attine-fungus association, as it allows termites to make use of a diversity of vegetal food sources such as wood, dry grass, and leaf litter. This mutualism is restricted to a single termite subfamily, Macrotermitinae, which is associated with fungi of the genus *Termitomyces*. In contrast with fungal symbionts of the attine ants that are usually propagated clonally and vertically by dispersing queens, *Termitomyces* reproduce sexually and are most often horizontally transmitted (Aanen et al. 2002).

A particular mutualism involves sap-feeding insects and intracellular bacteria. This system is one of the most specialized digestive mutualisms in terms of partner integration. Intracellular bacteria reside in specialized host cells (bacteriocytes) that constitute an organ called the bacteriome and are vertically transmitted - the symbionts migrate to the ovaries and enter the germ cells. This specialised mutualism is restricted to the monophyletic clade constituted by members of the two hemipteran suborders, Sternorrhyncha (psyllids, whiteflies, aphids, and mealybugs) and Auchenorrhyncha (sharpshooters). This clade comprises over 10,000 representatives. Within this clade different insect lineages are associated with different bacterium genera: *Buchnera* (aphids), *Carsonella* (psyllids), *Portiera* (whiteflies), *Tremblaya* (mealybugs) and *Baumannia* (sharpshooters). Plant sap is an unbalanced diet for insects as it is rich in carbohydrates relative to free amino acids and it is deficient in essential amino acids. The symbiotic bacteria convert non-essential amino acids into essential ones, allowing sap-feeding insects to circumvent their intrinsic nutritional limitation. In contrast to other sap-feeders that feed on sap circulating in the phloem, sharpshooters feed on xylem. The bacteriome in *Homalodisca* sharpshooters is bilobed, with one portion hosting *Baumannia* bacteria that can synthesize vitamins and cofactors, and the other portion hosting *Sulcia* bacteria that can synthesize essential amino acids (Baumann 2005; Moya et al. 2008; Douglas 2009).

12.3.3 Pollination Mutualism

Gamete transfer in seed plants is strongly constrained by the sessile nature of their predominant life stage, the sporophyte. Therefore, seed plants must rely on animal and abiotic pollen vectors (wind and sometimes water) to achieve cross fertilisation. Several groups of insects including bees, wasps, beetles, dipterans, lepidopterans and thrips, are involved in non-symbiotic and symbiotic pollination mutualisms. Non-symbiotic mutualisms include cases of diffuse reciprocal adaptation, where independent plant lineages share convergent floral attributes called pollination syndromes associated with pollinator guilds that present particular sensory biases and particular pollen transfer attributes (Rech et al. 2014; Dellinger 2020). In those mutualisms, insects benefit from food (nectar, pollen, oil), shelter (mainly for mate encounter) as well as other resources used for nest construction or their reproductive behaviour (oil, resin and floral scent) (Agostini et al. 2014).

Symbiotic pollination mutualisms are associations in which the plant provide breeding sites for their pollinators. Most of them qualify as nursery pollination mutualism. Larvae of pollinators feed on ovules/seeds or other floral parts (Sakai 2002; Dufaÿ and Anstett 2003; Hembry and Althoff 2016). Among these, broodsite pollination mutualism (*sensu* Hembry and Althoff 2016, *i.e.* association where pollinators feed on developing plant ovules) has evolved independently in several groups of plants and insects (Dufaÿ and Anstett 2003). It is often obligate (*e.g.*, fig trees – fig wasps, yucca – yucca moths and leafflowers – leafflower moths, senita cactus – senita moths), but can also be facultative when the plant is pollinated both by insects whose offspring develop feeding on seeds and by generalist insect pollinators that visit flowers to feed on pollen and/or nectar; e.g. *Silene – Hadena/ Perizoma*, *Lithophragma – Greya* and globeflower – globeflower flies (Kephart et al.

2006; Hembry and Althoff 2016). In these cases, the relationship with the seed eating taxa varies locally from mutualism to antagonism depending on the local pollination efficiency of generalist pollinators.

In brood-site pollination mutualisms the plants are selected to attract pollinators that will feed on developing seeds. We propose that the pollinating insects are selected to be host specialists so that they collect pollen from plants of and oviposit in plants of the same species, thus ensuring successful seed to feed their offspring. In most cases, the plants are selected to try to kill the developing pollinator larvae as the pollinator offspring will not carry pollen from their natal plant, and the system remains stable because this selection has not succeeded (Addicott et al. 1990). For instance, in the fig tree – fig wasp system, about 50% of the plant ovules are consumed by the pollinator larvae (Kjellberg et al. 2005). However, breeding pollinators does not come at a cost in *Ficus* as most consumed ovules produce female pollinating wasps, that will carry pollen from their natal fig (Jousselin and Kjellberg 2001; Kjellberg et al. 2001; Jousselin et al. 2003a).

12.4 Brood-Site Pollination Mutualism in Fig Trees: Did It Boost Species and Life Form Diversifications?

Fifty years ago, William Ramirez and Jacob Galil/Dan Eisikowitch independently discovered active pollination by some fig wasps (Hymenoptera: Agaonidae). Female actively pollinating wasps collect pollen into specialised pockets before leaving their natal figs, and later remove pollen from the pockets and deposit it on the stigmas in another fig containing receptive pistillate flowers (Galil and Eisikowitch 1969; Ramírez 1969). These observations marked the modern evolutionary studies of fig trees (*Ficus* spp) and their associated animals, making it a model system to investigate the evolution of mutualisms (Borges et al. 2018). The fig trees belong to Moraceae family, which includes ca. 39 genera. With a pantropical/subtropical distribution, *Ficus* encompasses approximately 70% of all 1100 described Moraceae species (Zerega et al. 2005; Gardner et al. 2017). Wind pollination is ancestral in Moraceae (Gardner et al. *Taxon, accepted*). Nevertheless, representatives of a series of genera are involved in nursery pollination mutualisms. They include representatives of genera *Artocarpus* (gall midges), *Antiaropsis/Castilla* (thrips), *Dorstenia* (flies), *Mesogyne* (bees), and *Ficus* (chalcid wasps) (Sakai et al. 2000; Sakai 2001; Zerega et al. 2004; Olotu et al. 2011; Araújo et al. 2017). Among these, only in *Ficus* do the wasp larvae feed on developing plant ovules. This is probably the sole Moraceae genus in which fertilisation female flowers generally directly benefits the wasp's offspring. Fig trees stand out for their obligate-specialized pollination mutualism and their taxonomic diversity. *Ficus* is about seven times more specious than the second moracean genus in number of species (*i.e., Dortenia* with 113 spp., which is followed by *Artocarpus* with ca. 70 spp) (◻ Fig. 12.3). *Ficus* is, in addition, functionally diversified. The genus presents diverse life-forms (freestanding and hemi-epiphytic trees, shrubs and climbers), breeding systems (monoecious and gynodioecious), pollination mode (active and passive), seed dispersion syndromes (birds and volant and non-volant

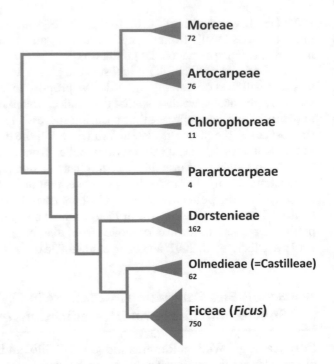

Fig. 12.3 Phylogeny of Moraceae tribes according to Gardner et al. (*in press*). Terminal widths are proportional to the squared root (number of species)

mammals, as well as reptiles and fishes) that allow its representatives to occupy a range of (micro)habitats (Shanahan et al. 2001; Jousselin et al. 2003b; Harrison 2005; Coelho et al. 2014).

Each *Ficus* species is pollinated by one or some species of host species-specific agaonid wasp (Ramírez 1970; Kjellberg et al. 2005). Approximately half of *Ficus* species are monoecious, bearing urn shaped inflorescences (also called syconium or fig) that enclose both pistillate and staminate flowers. The other species are structurally gynodioecious (see next paragraph), but functionally dioecious (Berg 1989). Fertilised pollen-loaded pollinating female wasps are attracted by a blend of volatile substances released by the receptive figs (Grison-Pigé et al. 2002; Souza et al. 2015). The pollinating wasp enters the fig through the ostiole, a pore closed by floral involucral bracts and pollinates (actively or passively) the pistillate flowers. In some of them, the wasp inserts its ovipositor through the style to lay one egg precisely between the nucellus and the inner integument (Jansen-Gonzalez et al. 2012). Ovaries that receive a wasp egg turn into galls where the pollinator larvae will develop feeding on endosperm, and those that have been pollinated and have escaped wasp oviposition will produce seeds. Approximately 4–8 weeks later (up to 9 months later for the pollinator of the common fig!), depending on the temperature (Pereira et al. 2007), the offspring complete their development. The males emerge first, locate and copulate with the females still enclosed in their natal galls. Then, the fertilised females emerge from their galls, actively collect pollen from staminate flowers at anthesis (in passively pollinated species the pollen from dehiscing anthers adheres spontaneously to the wasp's body) and leave the natal fig to search for another tree bearing receptive figs (Frank 1984). Subsequently the figs ripen and become attractive to a diversity of vertebrate frugivores that can act as seed dispersers (Shanahan et al. 2001). The figs that are not consumed by frugivores fall to the ground and turn into a resource for a diverse range of more or less specialised animals, mainly insects (Palmieri and Pereira 2018).

Gynodioecius *Ficus* species present two types of plants – one that produces figs containing only pistillate flower (*i.e.*, female tree). The pollinator attracted to a fig in a female tree pollinates but cannot lay eggs, as the styles are too long for the wasp ovipositor to reach the ovary of these flowers (◘ Fig. 12.1b). Thus, the pistillate flowers of female trees are pollinated by deceit, as the pollinator is attracted by a deceptive resource signalling. Pollen production, on the other hand, occurs in the functionally male trees, which have figs that enclose both pistillate and staminate flowers (◘ Fig. 12.1b). Wasp can oviposit in 'male' trees as the pistilate flowers have short styles. 'Male' trees do not produce seeds because either the wasps deposit pollen precisely on the stigmas of the flowers into which they oviposit (active pollination) or because of poor germination of pollen in 'male' trees (Jousselin and Kjellberg 2001). However, pollen is dispersed by the pollinator offspring in the same way of in monoecious *Ficus* species (◘ Fig. 12.1a; Weiblen 2002).

The ancestral mode of pollination of Moraceae and their closest relatives Cannabaceae and Urticaceae is wind pollination involving an explosive mechanism of pollen dispersal with stamens inflexed in the flower bud that, when they distend cause a rapid anther movement, releasing large amounts of pollen (Pedersoli et al. 2019). This pollen release mechanismcan achieve an exceptional initial velocity of Mach 0.7 (232 m.s^{-1}) in mulberry plants (Taylor et al. 2006). This mechanism may be an adaptation to circumvent the wind limitation in the tropical forest understory (Bawa and Crisp 1980). In this context, the fig – fig wasp mutualism brings out a singular innovation in pollen dispersion, by combininginitial wind dispersal of the wasps followed by wasp chemotaxis allowing cross-pollination at amazingly low population densities (Ware and Compton 1994). This is achieved despite the dispersing wasps only surviving 24–48 hours outside figs (Kjellberg et al. 1988; Jevanandam et al. 2013). Despite this temporal constraint, agaonid wasps can regularly disperse pollen over remarkably long distances comparatively to usual insect pollinated plants. It is demonstrated that average pollination distances by agaonid wasps within a fig population may reach several tens of kilometres (Compton et al. 1988; Nason et al. 1996; Nazareno and Carvalho 2009; Ahmed et al. 2009). Long-distance pollen dispersal, at least in monoecious *Ficus* species, is mediated by wind. Collection of aerial plankton shows that wasps pollinating monoecious *Ficus* species are dispersed by the wind above the forest canopy over long distances (Compton et al. 2000; Harrison and Rasplus 2006). When a wasp detects the plume of receptive fig scentreleased from a receptive tree, it moves down, out of the main wind current and then flies upwind to reach the tree (Ware and Compton 1994).

The pollination mutualism seems to have opened other evolutionary opportunities in *Ficus*, such as the active pollination that increases the efficiency of pollen transfer from the plant to the insect. As a consequence, the plants can invest in other aspects of male function such as producing more male inflorescences or in the case of *Ficus*, breeding more pollinator offspring (Sakai 2002; Pellmyr et al. 2020). Active pollination is present in two thirds of the *Ficus* species. Agaonid wasps that actively pollinate fig trees transport the pollen clumped into body containers (*i.e.*, pollen pockets), which apparently shares functional analogies with the cohesive pollen dispersion in pollinia observed in Asclepiadaceae and Orchidaceae (Ramírez 2007). It is postulated that pollen clumping can improvethe pollination success as it decreases the pollen waste during transport, and increases its probability of being deposited on a conspecific stigma (Johnson and Edwards 2000). However, this is not ever true in *Ficus* as actively pollinating wasps may carry limiting quantities of pollen (Kjellberg et al. 2014). Indeed, in the fig system as in other actively pollinated systems, the wasps, not the tree decide how much pollen they load into their pockets. In general, canopy fig tree species produce synchronous crops within trees but asynchronous among trees, making fig crops available at population level year-round (Milton et al. 1982; Windsor et al. 1989; Figueiredo and Sazima 1997; Pereira et al. 2007). Trees are selected to produce synchronous crops for wasp attraction as in some *Ficus* species producing small figs consumed by local animals (*e.g. F. guianensis, F. caulocarpa, F. subpisocarpa*), fig receptivity is synchronised within crop, wasp emergence is synchronised within crop, but fig ripening is scattered over a longer period of time (Chiang et al. 2018). Synchronous fig crops lead to year-round fruiting because of strong within fig protogyny associated with the production of synchronised crops. If receptive figs are particularly abundant at one period of the year, then there is selection on trees to produce crops that will pollinate these figs, *i.e.* that will release wasps at that time. These pollinating figs were receptive several weeks earlier, leading to the selection of figs producing crops that would pollinate them. Hence, there is frequency dependent selection favouring year-round production of crops due to the strong protogyny of figs and long wasp development time. Year-round fig production makes figs keystone resources for the year round survival of highly diversified frugivorous vertebrates, allowing different seed-dispersion mutualism to arise (Shanahan et al. 2001). Moreover, monoecious fig trees, which are adapted to long-distance pollen dispersion (Compton et al. 2000; Harrison 2003), usually produce huge fig crops that consequently result in a massive production of small seeds. Thus, capacity to be pollinated at very low densities coupled with the massive seed dispersion by canopy frugivores allow fig trees to colonise highly transient habitats and unlikely (in terms of frequency) habitats.

We postulate that brood-site pollination mutualism has boosted species diversification in *Ficus*, by opening new adaptive opportunities. As a matter of comparison the sister group of *Ficus* (*i.e.*, tribe Castilleae, represented by 10 genera and approximately 60 species) is nearly 13 times less speciose than *Ficus* (Gardner et al. 2017). Pollination biology in Castilleae is not well known, but pollination by thrips is reported for *Antiaropsis decipiens* and *Castilla elastica*, and potential bee/vespid pollination for *Mesogyne insignis* (Sakai 2001; Zerega et al. 2004; Olotu et al. 2011). These three genera have 1–3 species each, and there is no evidence that those insect pollinations parallel the fig – fig wasp mutualism in terms of pollination efficiency. Patterns of diversification/extinctions in *Ficus* lineages support a diversification hypothesis based on new ecological opportunities. Bruun-Lund et al. (2018), based

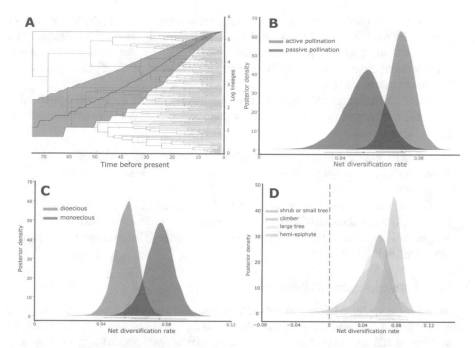

◘ Fig. 12.4 Diversification rate in *Ficus* (Bruun-Lund et al. 2018). **a** lineage-though-time plot depicted on the dated phylogenetic tree from Cruaud et al. (2012). **b–d**: net diversification rate in *Ficus* lineages, according to pollination modes, reproductive systems and lifeforms. Bruun-Lund et al. (2018) was published under the terms of the Creative Commons Attribution-Non Commercial-No Derivatives License (CC BY NC ND)

on a dated and comprehensive phylogenetic hypothesis, demonstrated that fig trees follow the evolutionary model of 'museum of diversity', with gradual accumulation of species over time coupled with very low extinction rates (Fig. 12.4a). They showed that key innovations directly or indirectly associated with the reproductive biology of fig trees correlated with higher diversification rates in clades where those features were present. For instance, actively pollinated *Ficus* species diversify faster and present lower extinction rates than passively pollinated ones (Fig. 12.4b). Similarly, monoecious and hemi-epiphytic species diversified faster than gynodioecious and other life forms (◘ Fig. 12.4c, d).

Bruun-Lund et al. (2018) hypothesized the following scenario to describe the success of fig trees. As the fig – fig wasp mutualism emerged by the late Cretaceous (75–90 Ma) (Cruaud et al. 2012), fig trees expanded into vacant niches left by the putative massive plant extinctions at the Cretaceous–Paleogene boundary (~65 Ma) (Wilf and Johnson 2004), thanks to their capacity to colonize unlikely suitable habitats, and to cross-pollinate at low population density. Then, *Ficus* would have diversified at a constant rate, as fig trees have several attributes of pioneer plants, such as fast growth, small seeds, high fecundity and flexible rooting habits. Those features make the fig – fig wasp association very robust and successful. Indeed, fossil evidences support that the fig – fig wasp mutualism is stable along its evolutionary history. Fossil specimens of a pollinating fig wasp from a limestone bed in England (~34 Ma), and pollinating and non-pollinating fig wasps from Dominican amber (10–20 Ma), as well as pollen morphology, display the same set of associated anatomical characters as modern species (Peñalver et al. 2006; Compton et al. 2010; Farache et al. 2016).

A major challenge in the study of mutualisms is how to go beyond particular examples and biological models and draw generalisations. We may list some challenges:

» "Most fig papers start with the statement that the fig-fig wasp mutualism is a model system. Is it a model system of anything else than figs?" Richard T. Corlett.

» "What will matter a lot to whether the paper is accepted (…) is how you frame the results. (…) They absolutely have to convincingly inform broader concepts in ecology, evolution and/or behavior that aren't system specific. That's always a little hard with figs." Judith L. Bronstein.

» "Fig wasps are wonderful" E. Allen Herre.

What we hope to have shown in this chapter is that figs and fig wasps and more generally brood-site pollination mutualisms, and case studies of mutualisms have something to tell us, beyond fascinating stories. We have tried to approach generalisation. This is a first sketch, and as such, we have made a number of provocative statements. The aim was to suggest alternative perspectives, to stir reflexion. A very important point is that as long as we do not know enough about a biological model it is very easy to make false inferences. To understand mutualisms we need strong biological data on the different systems.

12

Conclusion

The view that mutualism represents reciprocal exploitation leads to (1) a false assumption of an inherent conflict of interest between interacting parties, and (2) that the evolution of strategies that limit non-cooperative behaviours, such as sanctions, is condition for mutualism stability. However, phylogenetic evidences from a range of mutualistic systems point out that traits claimed as sanctioning acts were preexisting adaptations. Indeed, mutualisms are stable at macroevolutionary timescales, and rarely evolve to parasitic associations. In general, mutualisms involve members of distant taxonomic groups that combines a structural (goods provided by a long-lived partner) and a service component supplied by a short-lived partner. Thus, mutualism is a source evolutionary innovations that expand the partners ecological capacities. The brood-site pollination in the fig tree – fig wasp mutualism resulted in a set of innovations, such as efficient long-distance pollination and year-round massive seed rain, that opened new ecological niches and allowed a remarkable net diversification at a macroevolutionary timescale.

Key Points
- Cooperation theory
- Mechanisms to limit non-cooperative behaviour
- Mutualism as source of evolutionary innovations

❓ Questions
- Do the architecture "goods *vs.* services" occur in other examples of mutualisms?
- In addition to *Ficus*, are lineages involved in mutualistic associations in general more diverse than close-related lineages not engaged in mutualisms?
- What should be the insights resulted from expanding mutualism theory to non-pure biological interactions (*i.e.* culture *vs.* animals)?

References

Aanen DK, Eggleton P, Rouland-Lefevre C et al (2002) The evolution of fungus-growing termites and their mutualistic fungal symbionts. Proc Natl Acad Sci 99:14887–14892. https://doi.org/10.1073/pnas.222313099

Addicott JF (1986) Variation in the costs and benefits of mutualism: the interaction between yuccas and yuccas months (Tegeticula yuccasella). Oecologia 70:486–494

Addicott JF, Bronstein J, Kjellberg F (1990) Ev·oolution of mutualistic life-cycles: yucca moths and fig wasps. In: Gilbert F (ed) Insect life cycles. Springer, London, pp 143–161

Agostini K, Lopes AVF, Machado IC (2014) Recursos florais. In: Rech AR, Agostini K, Oliveira PE, Machado IC (eds) Biologia da Polinização. Projeto Cultural, Rio de Janeiro, pp 129–150

Ahmed S, Compton SG, Butlin RK, Gilmartin PM (2009) Wind-borne insects mediate directional pollen transfer between desert fig trees 160 kilometers apart. Proc Natl Acad Sci 106:20342–20347

Albert A, Savini T, Huynen M-C (2013) The role of macaca spp. (Primates: Cercopithecidae) in seed dispersal networks. Raffles Bull Zool 61:423–434

Alvarez N, McKey D, Kjellberg F, Hossaert-McKey M (2010) Phylogeography and historical biogeography of obligate specific mutualisms. In: Morand S, Krasnov BR (eds) The biogeography of host-parasite interactions. Oxford University Press, Oxford, pp 31–39

Araújo LM, Valentin-Silva A, Fernandes GW, Vieira MF (2017) From anthesis to diaspore dispersal: reproductive mechanisms of rare herbaceous Moraceae species endemic to Brazil. Darwin 5:83–92. https://doi.org/10.14522/darwiniana.2017.51.723

Axelrold R, Hamilton WD (1981) The evolution of cooperation. Science 211:1390–1396

Baumann P (2005) Biology of bacteriocyte-associated endosymbionts of plant sap-sucking insects. Annu Rev Microbiol 59:155–189. https://doi.org/10.1146/annurev.micro.59.030804.121041

Bawa KS, Crisp JE (1980) Wind-pollination in the understorey of a rain forest in Costa Rica. The Journal of Ecology 68:871. https://doi.org/10.2307/2259462

Berg CC (1989) Classification and distribution of Ficus. Experientia (Basel) 45:605–611

Bergmüller R, Russell AF, Johnstone RA, Bshary R (2007) On the further integration of cooperative breeding and cooperation theory. Behav Process 76:170–181. https://doi.org/10.1016/j.beproc.2007.06.013

Borges RM, Compton SG, Kjellberg F (2018) Fifty years later, figs and their associated communities. Acta Oecol 90:1–3. https://doi.org/10.1016/j.actao.2018.05.006

Boucher DH, James S, Keeler KH (1982) The ecology of mutualism. Annu Rev Ecol Syst 13:315–347

Bronstein JL, Alarcón R, Geber M (2006) The evolution of plant-insect mutualisms. New Phytol 172:412–428. https://doi.org/10.1111/j.1469-8137.2006.01864.x

Brundrett M (2004) Diversity and classification of mycorrhizal associations. Biol Rev 79:473–495. https://doi.org/10.1017/S1464793103006316

Bruun-Lund S, Verstraete B, Kjellberg F, Rønsted N (2018) Rush hour at the museum – diversification patterns provide new clues for the success of figs (Ficus L., Moraceae). Acta Oecol 90:4–11. https://doi.org/10.1016/j.actao.2017.11.001

Caldera EJ, Poulsen M, Suen G, Currie CR (2009) Insect symbioses: a case study of past, present, and future fungus-growing ant research. Environ Entomol 38:78–92. https://doi.org/10.1603/022.038.0110

Calixto ES, Lange D, Del-Claro K (2018) Protection mutualism: an overview of ant-plant interactions mediated by extrafloral nectaries. Oecologia Australis 22:410–425. https://doi.org/10.4257/oeco.2018.2204.05

Castro L, Toro MA (2004) The evolution of culture: from primate social learning to human culture. Proc Natl Acad Sci 101:10235–10240. https://doi.org/10.1073/pnas.0400156101

Chiang Y-P, Bain A, Wu W-J, Chou L-S (2018) Adaptive phenology of Ficus subpisocarpa and Ficus caulocarpa in Taipei, Taiwan. Acta Oecol 90:35–45. https://doi.org/10.1016/j.actao.2017.11.013

Coelho LFM, Ribeiro MC, Pereira RAS (2014) Water availability determines the richness and density of fig trees within Brazilian semideciduous forest landscapes. Acta Oecol 57:109–116

Compton SG, Thornton IWB, New TR, Underhill L (1988) The colonization of the Krakatau Islands (Indonesia) by fig wasps and other chalcids (Hymenoptera, Chalcidoidea). Philos Transact Royal Soc London B Biol Sci 322:459–470

Compton SG, Craig AJFK, Waters IWR (1996) Seed dispersal in an African fig tree: birds as high quantity, low quality dispersers? J Biogeogr 23:553–563

Compton SG, Ellwood MDF, Davis AJ, Welch K (2000) The flight heights of chalcid wasps (Hymenoptera, Chalcidoidea) in a lowland bornean rain forest: fig wasps are the high fliers. Biotropica 32:515–522

Compton SG, Ball AD, Collinson ME et al (2010) Ancient fig wasps indicate at least 34 Myr of stasis in their mutualism with fig trees. Biol Lett 6:838–842

Conchou L, Cabioch L, Rodriguez LJV, Kjellberg F (2014) Daily rhythm of mutualistic pollinator activity and scent emission in Ficus septica: ecological differentiation between co-occurring pollinators and potential consequences for chemical communication and facilitation of host speciation. PLoS One 9. https://doi.org/10.1371/journal.pone.0103581

Cruaud A, Ronsted N, Chantarasuwan B et al (2012) An extreme case of plant-insect codiversification: figs and fig-pollinating wasps. Syst Biol 61:1029–1047

Dedeine F, Vavre F, Fleury F et al (2001) Removing symbiotic Wolbachia bacteria specifically inhibits oogenesis in a parasitic wasp. Proc Natl Acad Sci 98:6247–6252. https://doi.org/10.1073/pnas.101304298

Defossez E, Selosse M-A, Dubois M-P et al (2009) Ant-plants and fungi: a new threeway symbiosis. New Phytol 182:942–949. https://doi.org/10.1111/j.1469-8137.2009.02793.x

Delabie JHC (2001) Trophobiosis between formicidae and hemiptera (Sternorrhyncha and Auchenorrhyncha): an overview. Neotrop Entomol 30:501–516. https://doi.org/10.1590/S1519-566X2001000400001

Del-Claro K, Rico-Gray V, Torezan-Silingardi HM et al (2016) Loss and gains in ant–plant interactions mediated by extrafloral nectar: fidelity, cheats, and lies. Insect Soc 63:207–221. https://doi.org/10.1007/s00040-016-0466-2

Dellinger AS (2020) Pollination syndromes in the 21st century: where do we stand and where may we go? New Phytol nph:16793. https://doi.org/10.1111/nph.16793

Doebeli M, Hauert C (2005) Models of cooperation based on the Prisoner's Dilemma and the Snowdrift game: Prisoner's Dilemma and the Snowdrift game. Ecol Lett 8:748–766. https://doi.org/10.1111/j.1461-0248.2005.00773.x

Douglas AE (2009) The microbial dimension in insect nutritional ecology. Funct Ecol 23:38–47. https://doi.org/10.1111/j.1365-2435.2008.01442.x

Dubilier N, Bergin C, Lott C (2008) Symbiotic diversity in marine animals: the art of harnessing chemosynthesis. Nat Rev Microbiol 6:725–740. https://doi.org/10.1038/nrmicro1992

Dufaÿ M, Anstett MC (2003) Conflicts between plants and pollinators that reproduce within inflorescences: evolutionary variations on a theme. Oikos 100:3–14

Farache FHA, Rasplus JY, Azar D et al (2016) First record of a non-pollinating fig wasp (Hymenoptera: Sycophaginae) from Dominican amber, with estimation of the size of its host figs. J Nat Hist 50:2237–2247. https://doi.org/10.1080/00222933.2016.1193646

Figueiredo RA, Sazima M (1997) Phenology and pollination ecology of three Brazilian fig species (Moraceae). Botanica Acta 110:73–78

Frank SA (1984) The behavior and morphology of the fig wasps Pegoscapus assuetus and P. jimenezi: descriptions and suggested behavioral characters for phylogenetic studies. Psyche 91:289–308

Frederickson ME (2013) Rethinking mutualism stability: cheaters and the evolution of sanctions. Q Rev Biol 88:269–295. https://doi.org/10.1086/673757

Frederickson ME (2017) Mutualisms are not on the verge of breakdown. Trends Ecol Evol 32:727–734. https://doi.org/10.1016/j.tree.2017.07.001

Friberg M, Schwind C, Guimarães PR et al (2019) Extreme diversification of floral volatiles within and among species of *Lithophragma* (Saxifragaceae). Proc Natl Acad Sci U S A 116:4406–4415. https://doi.org/10.1073/pnas.1809007116

Galil J, Eisikowitch D (1969) Note on pollen transport, pollination and protection of ovaries in Ficus sycomorus. New Phytol 68:1243–1244

Gardner EM, Sarraf P, Williams EW, Zerega NJC (2017) Phylogeny and biogeography of Maclura (Moraceae) and the origin of an anachronistic fruit. Mol Phylogenet Evol 117:49–59. https://doi.org/10.1016/j.ympev.2017.06.021

Gomulkiewicz R, Nuismer SL, Thompson JN (2003) Coevolution in variable mutualisms. Am Nat 162:S80–S93. https://doi.org/10.1086/378705

Gardner EM, Garner M, Cowan R, Dodsworth S, Epitawalage N, Maurin O, Arifiani A, Sahromi S, Baker WJ, Forest F, Zerega NJC, Monro AK, Hipp AL (In press) Repeated parallel losses of inflexed stamens in Moraceae: phylogenomics and generic revision of the tribe Moreae and the reinstatement of the tribe Olmedieae (Moraceae). Taxon

Gouyon (1999) Sex: a pluralist approach includes species selection. (one step beyond and it's good.). J Evolution Biol 12:1029–1030. https://doi.org/10.1046/j.1420-9101.1999.00130.x

Gray MW (2017) Lynn Margulis and the endosymbiont hypothesis: 50 years later. MBoC 28:1285–1287. https://doi.org/10.1091/mbc.e16-07-0509

Grison-Pigé L, Bessiere JM, Hossaert-Mckey M (2002) Specific attraction of fig-pollinating wasps: role of volatile compounds released by tropical figs. J Chem Ecol 28:283–295

Harrison RD (2003) Fig wasp dispersal and the stability of a keystone plant resource in Borneo. Proc Royal Soc London Series B-Biol Sci 270:S76–S79

Harrison RD (2005) Figs and the diversity of tropical rainforests. Bioscience 55:1053–1064

Harrison RD, Rasplus JY (2006) Dispersal of fig pollinators in Asian tropical rain forests. J Trop Ecol 22:631–639

Heil M, McKey D (2003) Protective ant-plant interactions as model systems in ecological and evolutionary research. Annu Rev Ecol Evol Syst 34:425–553. https://doi.org/10.1146/annurev.ecolsys.34.011802.132410

Hembry DH, Althoff DM (2016) Diversification and coevolution in brood pollination mutualisms: windows into the role of biotic interactions in generating biological diversity. Am J Bot 103:1783–1792. https://doi.org/10.3732/ajb.1600056

Herre EA, West SA (1997) Conflict of interest in a mutualism: documenting the elusive fig wasp-seed trade-off. Proc Royal Soc London Series B-Biol Sci 264:1501–1507

Herre EA, Knowlton N, Mueller UG, Rehner SA (1999) The evolution of mutualisms: exploring the paths between conflict and cooperation. Trends Ecol Evol 14:49–53

Hossaert-McKey M, Proffit M, Soler CCL et al (2016) How to be a dioecious fig: chemical mimicry between sexes matters only when both sexes flower synchronously. Sci Rep 6:21236. https://doi.org/10.1038/srep21236

Howe HF (1986) Seed dispersal by fruit-eating birds and mammals. In: Murray DP (ed) Seed dispersal. Academic, New York, pp 123–190

Howe HF, Smallwood J (1982) Ecology of seed dispersal. Annu Rev Ecol Syst 13:201–228. https://doi.org/10.1146/annurev.es.13.110182.001221

Jandér KC, Herre EA (2010) Host sanctions and pollinator cheating in the fig tree–fig wasp mutualism. Proc R Soc B 277:1481–1488. https://doi.org/10.1098/rspb.2009.2157

Jandér KC, Herre EA, Simms EL (2012) Precision of host sanctions in the fig tree-fig wasp mutualism: consequences for uncooperative symbionts. Ecol Lett 15:1362–1369. https://doi.org/10.1111/j.1461-0248.2012.01857.x

Jansen-Gonzalez S, Teixeira SP, Pereira RAS (2012) Mutualism from the inside: coordinated development of plant and insect in an active pollinating fig wasp. Arthropod Plant Interact 6:601–609

Jevanandam N, Goh AGR, Corlett RT (2013) Climate warming and the potential extinction of fig wasps, the obligate pollinators of figs. Biol Lett 9:20130041. https://doi.org/10.1098/rsbl.2013.0041

Johnson SD, Edwards TJ (2000) The structure and function of orchid pollinaria. Plant Syst Evol 222:243–269

Jousselin E, Kjellberg F (2001) The functional implications of active and passive pollination in dioecious figs. Ecol Lett 4:151–158

Jousselin E, Rasplus JY, Kjellberg F (2001) Shift to mutualism in parasitic lineages of the fig/fig wasp interaction. Oikos 94:287–294

Jousselin E, Hossaert-Mckey M, Herre EA, Kjellberg F (2003a) Why do fig wasps actively pollinate monoecious figs? Oecologia 134:381–387

Jousselin E, Rasplus JY, Kjellberg F (2003b) Convergence and coevolution in a mutualism: evidence from a molecular phylogeny of Ficus. Evolution 57:1255–1269

Kamra DN (2005) Rumen microbial ecosystem. Curr Sci 89:124–135

Kawakita A (2010) Evolution of obligate pollination mutualism in the tribe Phyllantheae (Phyllanthaceae). Plant Species Biol 25:3–19. https://doi.org/10.1111/j.1442-1984.2009.00266.x

Kephart S, Reynolds RJ, Rutter MT et al (2006) Pollination and seed predation by moths on Silene and allied Caryophyllaceae: evaluating a model system to study the evolution of mutualisms. New Phytol 169:667–680. https://doi.org/10.1111/j.1469-8137.2005.01619.x

Kiers ET, Denison RF (2008) Sanctions, cooperation, and the stability of plant-rhizosphere mutualisms. Annu Rev Ecol Evol Syst 39:215–236. https://doi.org/10.1146/annurev.ecolsys.39.110707.173423

Kiers ET, Rousseau RA, Denison RF (2006) Measured sanctions: legume hosts detect quantitative variation in rhizobium cooperation and punish accordingly. Evol Ecol Res 8:1077–1086

Kjellberg F, Gouyon PH, Ibrahim M et al (1987) The stability of the symbiosis between dioecious figs and their pollinators: a study of Ficus carica L. and Blastophaga psenes L. Evolution 41:693–704

Kjellberg F, Doumesche B, Bronstein JL (1988) Longevity of a fig wasp (Blastophaga psenes). Proceedings of the Koninklijke Nederlandse Akademie Van Wetenschappen series C biological and medical. Sciences 91:117–122

Kjellberg F, Jousselin E, Bronstein JL et al (2001) Pollination mode in fig wasps: the predictive power of correlated traits. Proc Royal Soc London Series B-Biol Sci 268:1113–1121

Kjellberg F, Jousselin E, Hossaert-Mckey M, Rasplus JY (2005) Biology, ecology and evolution of fig-pollinating wasps (Chalcidoidea, Agaonidae). In: Raman A, Schaefer CW, Withers TM (eds) Biology, ecology and evolution of gall-inducing arthropods. Science publishers, Inc., New Hampshire, pp 539–571

Kjellberg F, Suleman N, Raja S et al (2014) Some pollinators are more equal than others: factors influencing pollen loads and seed set capacity of two actively and passively pollinating fig wasps. Acta Oecol 57:73–79. https://doi.org/10.1016/j.actao.2013.11.002

Koike S, Kobayashi Y (2009) Fibrolytic rumen bacteria: their ecology and functions. Asian-Australas J Anim Sci 22:131–138. https://doi.org/10.5713/ajas.2009.r.01

Leigh EG Jr (2010) The evolution of mutualism. J Evol Biol 23:2507–2528. https://doi.org/10.1111/j.1420-9101.2010.02114.x

Margulis L (1996) Archaeal-eubacterial mergers in the origin of Eukarya: phylogenetic classification of life. Proc Natl Acad Sci U S A 93:1071–1076

Margulis L, Nealson KH (1989) Symbiosis as the source of evolutionary innovation. Endocytobiosis Cell Res:235–239

Markmann K, Parniske M (2009) Evolution of root endosymbiosis with bacteria: how novel are nodules? Trends Plant Sci 14:77–86. https://doi.org/10.1016/j.tplants.2008.11.009

Marzluff J, Angell T (2005) Cultural coevolution: how the human bond with crows and ravens extends theory and raises new questions. JEA 9:69–75. https://doi.org/10.5038/2162-4593.9.1.5

Maynard-Smith J (1978) The evolution of sex. Cambridge University Press, Cambridge

McKey D, Gaume L, Brouat C et al (2005) The trophic structure of tropical ant–plant–herbivore interactions: community consequences and coevolutionary dynamics. In: Burslem D, Pinard M, Hartley S (eds) Biotic interactions in the tropics, 1st edn. Cambridge University Press, pp 386–413

Milton K, Windsor DM, Morrison DW, Estribi MA (1982) Fruiting phenologies of two netropical Ficus species. Ecology 63:752–762

12

Moya A, Peretó J, Gil R, Latorre A (2008) Learning how to live together: genomic insights into pro-karyote–animal symbioses. Nat Rev Genet 9:218–229. https://doi.org/10.1038/nrg2319

Mueller UG, Schultz TR, Currie CR, Malloch D (2001) The origin of the Attine ant-fungus mutual-ism. Q Rev Biol 76:169–197. https://doi.org/10.1086/393867

Mueller UG, Gerardo NM, Aanen DK et al (2005) The evolution of agriculture in insects. Annu Rev Ecol Evol Syst 36:563–595. https://doi.org/10.1146/annurev.ecolsys.36.102003.152626

Nalepa CA, Bignell DE, Bandi C (2001) Detritivory, coprophagy, and the evolution of digestive mutualisms in Dictyoptera. Insectes Soc 48:194–201. https://doi.org/10.1007/PL00001767

Nascimbene J, Nimis PL (2006) Freshwater lichens of the Italian Alps : a review. Ann Limnol - Int J Lim 42:27–32. https://doi.org/10.1051/limn/2006003

Nason JD, Herre EA, Hamrick JL (1996) Paternity analysis of the breeding structure of strangler fig populations: evidence for substantial long-distance wasp dispersal. J Biogeogr 23:501–512

Nazareno AG, Carvalho D (2009) What the reasons for no inbreeding and high genetic diversity of the neotropical fig tree Ficus arpazusa? Conserv Genet 10:1789–1793

Nowak MA, Tarnita CE, Wilson EO (2010) The evolution of eusociality. Nature 466:1057–1062. https://doi.org/10.1038/nature09205

O'Gorman R, Sheldon KM, Wilson DS (2008) For the good of the group? Exploring group-level evolutionary adaptations using multilevel selection theory. Group Dyn Theory Res Pract 12:17–26. https://doi.org/10.1037/1089-2699.12.1.17

Olotu MI, Ndangalasi HJ, Nyundo BA (2011) Effects of forest fragmentation on pollination of Mesogyne insignis (Moraceae) in Amani nature reserve forests, Tanzania: effects of forest frag-mentation on pollination of Mesogyne insignis. Afr J Ecol 50:109–116. https://doi.org/10.1111/j.1365-2028.2011.01302.x

Palmieri L, Pereira RAS (2018) The role of non-fig-wasp insects on fig tree biology, with a proposal of the F phase (fallen figs). Acta Oecol 90:132–139. https://doi.org/10.1016/j.actao.2017.10.006

Parniske M (2008) Arbuscular mycorrhiza: the mother of plant root endosymbioses. Nat Rev Microbiol 6:763–775. https://doi.org/10.1038/nrmicro1987

Pedersoli GD, Leme FM, Leite VG, Teixeira SP (2019) Anatomy solves the puzzle of explosive pollen release in wind-pollinated urticalean rosids. Am J Bot 106:489–506. https://doi.org/10.1002/ajb2.1254

Pellmyr O (2003) Yuccas, yucca moths, and coevolution: a review. Ann Mo Bot Gard 90:35–55. https://doi.org/10.2307/3298524

Pellmyr O, Huth CJ (1994) Evolutionary stability of mutualism between yuccas and yucca moths. Nature 372:257–260. https://doi.org/10.1038/372257a0

Pellmyr O, Kjellberg F, Herre EA et al (2020) Active pollination drives selection for reduced pollen-ovule ratios. Am J Bot 107:164–170. https://doi.org/10.1002/ajb2.1412

Peñalver E, Engel MS, Grimaldi DA (2006) Fig wasps in dominican amber (Hymenoptera: Agaonidae). Am Museum Novitates 3541:1. https://doi.org/10.1206/0003-0082(2006)3541[1:FWI DAH]2.0.CO;2

Pereira RAS (2014) Polinização por vespas. In: Biologia da Polinização. Projeto Cultural, Rio de Janeiro, pp 291–309

Pereira RAS, Rodrigues E, Menezes AO Jr (2007) Phenological patterns of Ficus citrifolia (Moraceae) in a seasonal humid-subtropical region in Southern Brazil. Plant Ecol 188:265–275

Proffit M, Lapeyre B, Buatois B et al (2020) Chemical signal is in the blend: bases of plant-pollinator encounter in a highly specialized interaction. Sci Rep 10:10071. https://doi.org/10.1038/s41598-020-66655-w

Ramírez W (1969) Fig wasps: mechanism of pollen transfer. Science (Washington D C) 163:580–581

Ramírez W (1970) Host specificity of fig wasps (Agaonidae). Evolution 24:680–691

Ramírez W (2007) Pollination analogies between Orchidaceae, Ficus (Moraceae) and Asclepiadaceae. Lankesteriana 7. https://doi.org/10.15517/lank.v7i1-2.19668

Rech AR, Avila RSJr, Schlindwein C (2014) Síndromes de polinização, especialização e generaliza-ção. In: Rech AR, Agostini K, Oliveira PE, Machado IC (eds) Biologia da PolinizaçãoProjeto Cultural, Rio de Janeiro, pp. 171–182

Sachs JL, Mueller UG, Wilcox TP, Bull JJ (2004) The evolution of cooperation. Q Rev Biol 79:135–160

Sakai S (2001) Thrips pollination of androdioecious Castilla elastica (Moraceae) in a seasonal tropi-cal forest. Am J Bot 88:1527–1534

Sakai S (2002) A review of brood-site pollination mutualism: plants providing breeding sites for their pollinators. J Plant Res 115:161–168. https://doi.org/10.1007/s102650200021

Sakai S, Kato M, Nagamasu H (2000) *Artocarpus* (Moraceae)-gall midge pollination mutualism mediated by a male-flower parasitic fungus. Am J Bot 87:440–445. https://doi.org/10.2307/2656640

Sarmento R, Alves-Costa CP, Ayub A, Mello MAR (2014) Partitioning of seed dispersal services between birds and bats in a fragment of the Brazilian Atlantic Forest. Zoologia (Curitiba) 31:245–255. https://doi.org/10.1590/S1984-46702014000300006

Shanahan M, So S, Compton SG, Corlett RT (2001) Fig-eating by vertebrate frugivores: a global review. Biol Rev 76:529–572

Showalter TD (2000) Insect ecology: an ecosystem approach. Academic, New York

Soto MJ, Domínguez-Ferreras A, Pérez-Mendoza D et al (2009) Mutualism versus pathogenesis: the give-and-take in plant-bacteria interactions. Cell Microbiol 11:381–388. https://doi.org/10.1111/j.1462-5822.2009.01282.x

Souza CD, Pereira RAS, Marinho CR et al (2015) Diversity of fig glands is associated with nursery mutualism in fig trees. Am J Bot 102:1564–1577. https://doi.org/10.3732/ajb.1500279

Taylor PE, Card G, House J, et al (2006) High-speed pollen release in the white mulberry tree, Morus alba L 6

Thompson JN (1999) The evolution of species interactions. Science 284:2116–2118. https://doi.org/10.1126/science.284.5423.2116

Vale GL, Carr K, Dean LG, Kendal RL (2017) The cultural capacity of human and nonhuman primates: social learning, innovation, and cumulative cultural evolution. In: Kaas JH (ed) Evolution of nervous systems. Elsevier, Amsterdam, pp 475–508

Venn AA, Loram JE, Douglas AE (2008) Photosynthetic symbioses in animals. J Exp Bot 59:1069–1080. https://doi.org/10.1093/jxb/erm328

Ware AB, Compton SG (1994) Dispersal of adult female fig wasps: 2. Movements between trees. Entomol Exp Appl 73:231–238

Weiblen GD (2002) How to be a fig wasp. Annu Rev Entomol 47:299–330

West SA, Griffin AS, Gardner A (2007) Social semantics: altruism, cooperation, mutualism, strong reciprocity and group selection. J Evolution Biol 20:415–432. https://doi.org/10.1111/j.1420-9101.2006.01258.x

Wheat CW, Vogel H, Wittstock U et al (2007) The genetic basis of a plant insect coevolutionary key innovation. Proc Natl Acad Sci 104:20427–20431. https://doi.org/10.1073/pnas.0706229104

Wilf P, Johnson KR (2004) Land plant extinction at the end of the Cretaceous: a quantitative analysis of the North Dakota megafloral record. Paleobiology 30:347–368

Wilson DS, Wilson EO (2008) Evolution "for the good of the group". Am Sci 96:380. https://doi.org/10.1511/2008.74.1

Wimsatt WC (1999) Genes, memes, and cultural heredity. Biol Philos 14:279–310. https://doi.org/10.1023/A:1006646703557

Windsor DM, Morrison DW, Estribi MA, De Leon B (1989) Phenology of fruit and leaf production by "strangler" figs on Barro Colorado Island, Panama. Experientia (Basel) 45:647–653

Yoder JB, Smith CI, Pellmyr O (2010) How to become a yucca moth: minimal trait evolution needed to establish the obligate pollination mutualism: HOW TO BECOME A YUCCA MOTH. Biol J Linn Soc 100:847–855. https://doi.org/10.1111/j.1095-8312.2010.01478.x

Yu DW (2001) Parasites of mutualisms. Biol J Linn Soc 72:529–546

Yu H, Tian E, Zheng L et al (2019) Multiple parapatric pollinators have radiated across a continental fig tree displaying clinal genetic variation. Mol Ecol 28:2391–2405. https://doi.org/10.1111/mec.15046

Zerega NJC, Mound LA, Weiblen GD (2004) Pollination in the New Guinea endemic *Antiaropsis decipiens* (Moraceae) is mediated by a new species of Thrips, *Thrips antiaropsidis* sp. nov. (Thysanoptera: Thripidae). Int J Plant Sci 165:1017–1026. https://doi.org/10.1086/423869

Zerega NJC, Clement WL, Datwyler SL, Weiblen GD (2005) Biogeography and divergence times in the mulberry family (Moraceae). Mol Phylogenet Evol 37:402–416. https://doi.org/10.1016/j.ympev.2005.07.004

12

Impacts of Anthropocene Defaunation on Plant-Animal Interactions

Kleber Del-Claro and Rodolfo Dirzo

Contents

© Springer Nature Switzerland AG 2021
K. Del-Claro, H. M. Torezan-Silingardi (eds.), *Plant-Animal
Interactions*, https://doi.org/10.1007/978-3-030-66877-8_13

🔒 Learning Objectives

This chapter will help readers to:

- Become familiar with the term Anthropocene and some of its impacts on plant-animal interactions
- Understand what contemporary defaunation means and why it has considerable consequences on the diversity of interactions between plants and animals
- Foment awareness of the anthropogenic threats to biodiversity and species interactions and the significance thereof for human wellbeing

13.1 The Anthropocene, an Epoch of Distinguishable Human Footprint on the Planet

The 1977 Nobel Laureate in Chemistry, Paul Crutzen, has championed the use of the term "Anthropocene". In a short paper (Crutzen 2002), this prominent scientist argued that the omnipresent and intense impact of the human enterprise on the planet demanded global recognition that the characteristics of the Holocene bore such a formidable anthropogenic footprint that it should be replaced with the term Anthropocene. Although not free of debate (see Zalasiewicz et al. 2011), the term has percolated considerably not only among the scientific community, but also in other branches of academy and, indeed, society at large. Presently, for example, at least two periodicals carry the term: Anthropocene (a print and digital magazine produced under the auspices of *Futurearth* – ▶ https://www.anthropocenemagazine. org) and The Anthropocene Review (from *Sage Journals* – ▶ https://journals. sagepub.com/home/anr). Among the 171 neologisms appearing in the 2014 edition of the Oxford English Dictionary, Anthropocene is featured therein, and the entry reads "*the era of geological time during which human activity is considered to be the dominant influence on the environment, climate and ecology of Earth.*" Dictionaries for other major languages have followed suit, further confirming the penetration of the term into the lexicon of society at large. We propose, therefore, that an examination of our understanding of the status of biodiversity in contemporary time needs to consider plant-animal interactions in the Anthropocene—as the chapter's title indicates.

The manifestations of the Anthropocene are multiple and, naturally, encompass a host of both socio-economic and environmental variables. A comprehensive review of the manifestations and trajectories of the Anthropocene (Steffan 2015) depicts an evident pattern of what therein is referred to as the "Anthropocene's great acceleration," whereby such variables exhibit a steep increase, particularly manifested roughly over the last five decades of the last century and until now. Leaving the socio-economic variables aside, the environmental variables of this prominent paper, include several trends of biogeochemical nature (particularly relevant to climate change) and three that are most critically related to the subject matter of this chapter:

1. tropical forest loss (measured as percent forest coverage loss);

2. "domestication of land" (total terrestrial area converted to human-dominated landscapes);
3. "terrestrial bio-degradation" (the estimated mean percent decline in species abundance).

As we will elaborate below, these have important bearings on the disruption of ecosystem processes, but none of the trends of this important review considers the trajectory of species interactions (such as plant-animal interactions) in the Anthropocene.

Let us consider, first, the trends of tropical forest loss and land domestication of Steffen's (2015) great Anthropocene acceleration. When biologists, ecologists and conservationists call attention to anthropogenic destruction or degradation of natural ecosystems and the loss of biodiversity, they are not exaggerating. Deforestation, soil degradation, water and air pollution, and uncontrolled fires caused (directly and indirectly) by human activities are nowadays at dramatic levels. Sadly, examples are vivid and abundant. For example, the Annual Report on Deforestation in Brazil (MapBiomes 2020) revealed that the Amazon lost an average of 2110 hectares of forest per day in 2019 – an area equivalent to about 2000 football fields. This biome was the most devastated in the country, representing 63% of the 3339 hectares felled per day in the entire country. Considering deforestation of all six major Brazilian biomes, Earth lost 1,218,708 ha of natural vegetation in 2019. In this assessment, the two best-monitored biomes in Brazil are the Amazon rainforest and Cerrado, and these accounted for 96.7% of the total deforested area detected for the country in 2019 (MapBiomes 2020; ◘ Figs. 13.1 and 13.2). This figure is alarming, given that the Amazon harbors the greatest density of biodiversity on Earth, and the Cerrado holds the greatest biodiversity among savanna-type ecosystems in the world (Oliveira and Marquis 2002).

Let us now consider the case of fires, an increasingly impactful driver of forest area loss. In the Brazilian state that concentrates the largest portion of preserved Amazonian forest, Amazonas State, Brazil's INPE (InstitutoNacional de PesquisasEspaciais) tracks fires and deforestation using multiple satellites equipped with optical thermal sensors (INPE 2020). A time-course of incidence of fires readily shows that they are increasing and consistently surpassing the 22-year average number of 10,000 events since 2015 (◘ Fig. 13.1).Furthermore, in the first 8 months of 2020, fires in the Brazilian Amazonian forest and in the forests of neighboring countries, as well as in the Cerrado savanna and the Pantanal wetlands are surpassing historical records (INPE 2020; MapBiomas 2020). As broadly publicized in international media, year 2019 (e.g., in Australia), and year 2020 (e.g., in California, Colorado, Brazil) was characterized by a dramatic increase in number and intensity of fires around the world, moving us into what appears to be a strong companion (and, perhaps, name-competitor) of the Anthropocene—a sort of "Pyrocene" epoch. Sadly, Brazil's case is not unique, and deforestation and fire also are ravaging extensive areas of Africa, North America, Europe, Australia and Asia.

In terms of the proximate drivers of land domestication and tropical forest loss, conversion to agriculture, cattle ranching, massive logging and mining are predom-

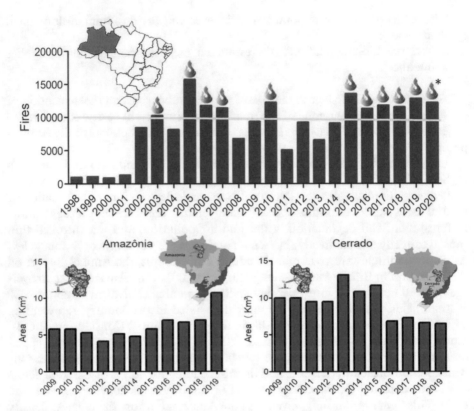

◘ Fig. 13.1 Annual incidence of fires (upper panel) monitored by INPE in the Amazonas state (color-highlighted on the map), Brazil. The original data were derived from ► http://queimadas.dgi. inpe.br/queimadas/portal-static/estatisticas_estados/. *Year 2020 includes only the period January– August. Deforestation trends (lower panel; in thousands of km^2) in the two best monitored Brazilian biomes: Amazonia (left) and Cerrado (right).The original data are available for public use and consultation at INPE/PRODES – Terra Brasilis (► http://terrabrasilis.dpi.inpe.br/app/dashboard/deforestation/biomes/amazon/increments)

13

inant and they have devastating effects on the diversity and structure of the targeted ecosystems.

The quantification and monitoring of the subsequent spatio-temporal dynamics of the areas impacted by these anthropogenic drivers should be technologically feasible given the increasing sophistication and availability of remote sensing tools. This, and the fate of areas currently set aside for biodiversity conservation should inform us of the status and trends of the vegetation in the Anthropocene. Such technology, however, can only partly or indirectly provide information of biodiversity as a whole, since the current situation and trajectories of animal communities –that is, the current and future trends of defaunation– are at least partly overlooked. Therefore, assessment of the conservation status of plant-animal interactions, needs to carefully examine the defaunation trends.

Regarding the immediate biodiversity consequences of the recent pulses of deforestation and fires, preliminary reports provide a gloomy picture. For example,

◘ Fig. 13.2 The Brazilian Tropical Savanna, Cerrado. The images show the same area of Vereda (above) and cerrado sensu stricto (below), before and after fire occurrence. Municipality of Uberlândia, MG, Brazil

Australian scientists who have considered the possible faunistic impacts of the 2019-early 2020 fires in that country reached an estimated number of over 800 million animals killed in New South Wales, and over 1 billion nationally (Ward et al. 2020). Long-term studies are needed for this and other fire- and deforestation-impacted parts of the world, and very likely will be conducted over the next few years, but it will be critical to also examine, via careful monitoring, the possible patterns of animal recolonization into the wild areas impacted by land use change (including fire, ◘ Fig. 13.2), as well as into the areas that are not permanently assigned to agricultural activities for food production and livestock maintenance. In the meantime, we can use the currently available information to gather a picture of our current understanding of defaunation in the Anthropocene, as we discuss next.

13.2 Defaunation in the Anthropocene

The word "defaunation" in a sense of animal removal from a community due to human activities ("anthropocene defaunation" sensu Dirzo et al. 2014), has its origins in the elegant work of Simberloff and Wilson (1969). In their classic experimental manipulative study, the authors applied insecticide onto small islands in the Florida, USA to remove the entire assemblage of arthropods, and examine the recolonization patterns and effects on the plant community. In the decade of 1980, Janzen and Martin (1982) suggested that the extinction of the Pleistocene megafauna generated "plant anachronisms" to explain the occurrence of traits (e.g., extremely hard fruits, or very spiny fruits and stems that protect relatively small-sized seeds) among many Neotropical trees that cannot be understood if one does not consider their interaction (i.e, seed dispersal) with such extinct megafauna. Such animals likely were important selective forces leading to the evolution of these anachronistic traits. These initial studies highlighted the impact that animal extinction could have on plant communities and in the biodiversity of entire communities. Several years later, in the 1990's, a study considering two tropical forests with contrasting degree of conservation, performed a detailed comparative analysis of the impacts of reduced abundance and diversity of understory mammalian herbivores on the plant community (Dirzo and Miranda 1990). This study presented for the first time the notion and the term "contemporary defaunation", show-casing the current negative impact of the human enterprise on the animal community, and its impacts in the structure and diversity of tropical plant communities. Presently, we see the use the term defaunation and its different facets to describe the status of animal communities under anthropogenic disturbance and the patterns and consequences thereof on plant communities and ecological processes, including species interactions at local and global scales (Dirzo et al. 2014; Young et al. 2016).

13.3 Drivers and Magnitude of Defaunation

Defaunation is a useful entry way to examine the "terrestrial bio-degradation" variable of Steffen's (2015) great acceleration of the Anthropocene. Recent reviews (e.g., Young et al. 2016) indicate that the main direct driver of defaunation is overexploitation (hunting, and illegal trading). Land use change, as we have discussed above, is the most critical indirect defaunation driver, followed by invasive taxa and toxication and, already detectable, but undoubtedly more prominently in the future, climate change. Although the number of globally extinct species of vertebrates may seem small, i.e., ~700 since year 1500, ca. 60% of the extinctions occurred since 1900, clearly revealing an accelerated trend, and representing values that are 2–3 orders of magnitude higher (depending on the group of vertebrates) than background extinction rates (Ceballos et al. 2015). Nevertheless, for the purposes of this chapter, our main concern here is the decline in the local abundance of animals, as this facet of defaunation can reveal the impacts on plant-animal interactions (Dirzo et al. 2014). Monitoring of a robust number of animal populations across the globe (although with important geographic biases, particularly under-representation of tropical regions), suggests that over the last four to five decades, abundance of vertebrates has declined by about 50% (Dirzo et al. 2014), with more recent reports elevating the figure to as much as 68% (Live Planet Report, WWF 2020). Although these analyses do not consider the populations of species that are increasing (which would be informative of species turnover and net changes in animal communities, see Dornelas et al. 2019), they provide a picture of considerable decline. When declines at given locations become extreme, they lead to an alarming problem of population extinctions. Assessments of the possible magnitude of global population loss are still in their infancy, yet available studies indicate an alarming situation. For example, using a sample of 177 species of mammals, Ceballos et al. (2017) estimated that over the period 1990–2015, close to 50% of the examined species had reduced their distribution range by 80% or more, therefore implying a dramatic loss of populations, including cases of species not deemed as threatened according to IUCN's assessments. Detailed studies at some particular regions are similarly alarming. For example, the annual magnitude of hunting in the Brazilian Amazon has been estimated in ~16 million mammals, or 23 million vertebrates in general.

We know much less about the loss of invertebrates, mainly because we do not know the major part of this fauna, especially the cryptic below-graund species. However, according to Dirzo et al. (2014, and sources therein) 67% of monitored populations of invertebrates show ~45% mean abundance decline. More recently, a string of studies documenting declines in abundance of invertebrates have been published. While some of the papers report considerable declines (e.g. Wagner 2020). The most updated and thorough examination of data (166 long-term studies, encompassing 1676 sites), focusing on insects (van Klink et al. 2020) reports an average decline of 9%, much lower than that of the previous "Armageddon-type" studies (e.g., Sanchez-Mayo and Wyckhuys 2019), while also showing substantial heterogeneity across space (both within and among regions), biogeographic region, metric of abundance used, and time of measurement. Another salient aspect of this monumental work is the tremendous paucity of tropical studies – a critical knowledge lacuna for the understanding of plant-animal interactions in the Anthropocene.

Beyond the quantitative aspect of defaunation, an important emerging pattern is that, for vertebrates, susceptibility to anthropogenic impact is not random but instead varies with life-history traits, with an animal body size signal being particularly strong: medium and large animals are considerably impacted ("losers") while small-bodied animals are frequently benefited ("winners"). A recent example in frugivorous birds and the associated plant species in the Atlantic forest of Brazil underscores this point (Emer et al. 2020).

What will be the impact of these human interferences in the structure of ecological networks that are based on plant-animal interactions? In this context, a broad, alarming issue is whether the Anthropocene may be a global forcing of a magnitude leading to mass, global disruptions of ecological interactions and potentially leading to the end of the biodiversity of interactions (sensu Thompson 2005, 2014)?

13.4 Contemporary Defauntion and the Consequences for Species Interactions

Biotic interactions are dynamic, and they vary in space and time. As was discussed in the introductory chapter, all types of such biotic interactions have existed for at least 300 million years. In the evolutionary process, species can be

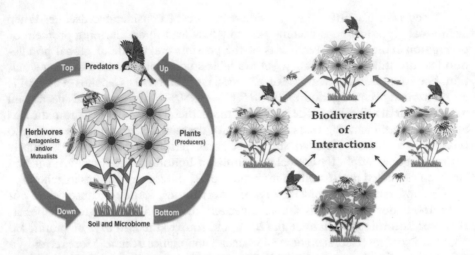

◘ Fig. 13.3 A simplified multitrophic system (left) composed of plants, herbivores and herbivores' natural enemies. The elements of such a system do not live in isolation, however; rather they are immersed in a network of multiple species from each level. Such constellations of interacting species represent nodes in an ecological network with linkages between plants, animal consumers (invertebrates and vertebrates) and consumers' consumers within communities characterized by a rich diversity of interactions (right), maintaining viable natural communities. See also ► Chap. 10 of this book, Figs. 4 and 5, which illustrates a multitrophic system according to interspecific ecological networks involving plants and animals

replaced (i.e., extinction and speciation), yet independently of the existing species at any particular time, the interactions typically remain (Thompson 2005; Del-Claro et al. 2016). Through millions of years of evolutionary and coevolutionary processes, the interactions have shaped and maintained biodiversity (Thompson 2013). However, in the recent history of the Earth, critically over the last two centuries, one species has impacted nature in such a strong way that the evolutionary theater and its evolutionary plays, as we know them, are endangered (Dirzo et al. 2014).

In terrestrial environments, a simple trophic chain is composed of the plants, whose survival, growth and reproduction are directly dependent of abiotic forces such as climate, soil, and substrate, and intrinsic biotic factors such as their microbiome (see Reverchon and Méndez-Bravo, ► Chap. 8). The immediately next trophic level above plants is composed by its herbivores (invertebrates or vertebrates) and pathogens, which may also exert a strong negative pressure affecting plant fitness, in terms of survival, growth or reproduction, or several combinations thereof (see Marquis and Moura, ► Chap. 3). In contrast, other types of plant consumers play a positive role for plant performance, particularly via pollination or seed dispersal (see ◘ Fig. 13.3; and see Figs. 4 and 5, in Luna and Dáttilo, ► Chap. 10 of this book). These animals are typically regarded as mutualistic interactors, since their positive role is generally rewarded by plant resources in the form of nectar, surplus pollen, or oils (for pollinators) and nutritious fruit (for dispersers). In addition, a number of plant species, encompassing a variety of lineages, interact with other consumers that serve as mutualists by affording anti-herbivore defenses to the plant (Del-Claro et al. 2016, see also Moura et al., ▸ Chap. 5), which in turn provide rewards in the form of extra floral nectar, or lipids produced in specialized structures and, additionally, in a few cases, "housing" for the defenders (domatia). The interactions of plants with the antagonistic consumers (the herbivores) is regulated by (i) bottom-up factors consisting on intrinsic plant traits, such as secondary metabolites or nutritional characteristics, or the combination thereof, and (ii) the natural enemies of the herbivores, thus representing the top-down regulating factors. Such top-down forces, predators or parasitoids, can control the abundance and diversity of herbivores (◙ Fig. 13.3). However, these interactions are much more nuanced and non-linear than this description suggests, as plants and their consumers, and the consumers' consumers do not live in a vacuum. For example, a bee species rarely is a specific pollinator of just one plant species; frequently it is the pollinator of several plant species (see Torezan-Silingardi et al., ▸ Chap. 6). Likewise, although some herbivores can be specialized on a host plant, many herbivores are generalists to some degree, polyphagous animals that feed on different plant species; and plants in turn can be the food source of one or multiple herbivores. Feeding on distinct plants and moving among different areas and populations, frugivores can disperse seeds and improve plant variability through genetic recombination (see Holeski, ▸ Chap. 7). Animal predators specialized in one or few prey items are not the general rule. So, each one of these plant-animal multitrophic systems are not isolated in single communities. Herbivores, microorganisms, predators, parasites, and facilitators, ultimately represent a large amount of nodes connected by links across different multitrophic systems (◙ Fig. 13.3). Considering a given community, one can envisage the complexity of the networks, even in the case, of, for example, two trophic levels, such as the plants in a community, and its links to their pollinators and their herbivores—hybrid networks. For example, Morrison et al. (2020) has constructed a network of plants and the associated communities of herbivores and pollinators, in a relatively simple temperate community. These connections between nodes of ecological networks structure the biodiversity of interactions, the multitude of different connections among species that maintain the viability and functionality of natural communities (Thompson 2005, 2013, 2014).

Studies of plant-animal interactions in light of contemporary defaunation have focused largely on mutualistic interactions (pollination, seed dispersal), while those examining antagonistic interactions of herbivory and seed predation remain appallingly under-represented in the literature (c.g. Gardner et al. 2019). However, the few available suggest that the richness of this sort of plant-animal interactions can also be dramatically affected. In one particular example in Mexican tropical forests, researches have shown that contemporary defaunation of the understory herbivores can actually lead to the local extinction of mammalian herbivory: a large sample of carefully monitored plants over multiple years, showed that in contrast to intact sites, plants in a heavily defaunated forest showed consistent absence of herbivore damage (Dirzo and Miranda 1990, Dirzo et al. 2020). The consistent absence of foliage herbivory led authors of this work to suggest that this represents a case of the local extinction of a plant-animal interaction.

The seminal coevolution paper by Ehrlich and Raven (1964) advocated that plant-animal interactions have played an essential role in the generation of Earth's biodiversity. Bascompte and Jordano (2007) extended this view pointing out that mutually beneficial interactions between plants and pollinators and seed dispersers have been paramount in the generation of Earth's biodiversity (see Bronstein, ▶ Chap. 11). These mutualistic interactions often involve dozens or even hundreds of species that form complex networks of interdependences, whose structure has important implications for the coexistence and stability of species as well as for the coevolutionary process (Bascompte and Jordano 2007). We know that more than 90% of tropical plant species rely on animals for the dispersal of their seeds (Jordano 2000), and similar numbers are reported for pollination interactions (Bawa 1990). In addition, in the food chain of higher plants, herbivorous insects and their natural enemies, comprise a significant fraction of Earth's known species richness (just shy of 80%) engaged in herbivory (Dirzo et al. 2020). If these animals disappear, their plant partners may follow suit and, likely, the associated microbiome too (see Reverchon and Méndez-Bravo, ▶ Chap. 8).

Obviously, the pulse of "Anthropocene defaunation" described above has cascading effects that will affect all organisms and ecosystems across the planet's ecosystems, impacting directly plants via herbivory, pollination and dispersal. But it also affects human health and economy, considering the enormous loss of ecological services (mainly pollination; see Torezan-Silingardi et al. ▶ Chap. 6). Defaunation, including local, regional or the complete extinction of vertebrates and invertebrates species, will disrupt uncountable plant-animal interactions, with direct effects in energy flow through communities and also gene flow, possibly producing new evolutionary pathways. The local or regional biodiversity of interactions will have to be rebuilt, if possible. Gardner et al. (2019) conducted a meta-analysis pointing out that real-world defaunation caused by hunting and habitat fragmentation leads to reduced forest regeneration. The elimination of seed dispersers and frugivores like primates and birds, may cause the greatest declines in forest regeneration, with impacts on carbon stores and climate change.

Defaunation precipitates the extinction of evolutionarily distinct interactions, as clearly demonstrated by Emer et al. (2020) in an elegant study with frugivorous birds and its associated plant species. These authors suggested that defaunation is driving evolution to the reduction in the size of seeds and dispersers. Additionally, they demonstrated that defaunation is provoking the loss of interactions involving unique lineages of bird and plant species. A direct result of this contemporary human-driven disturbance is the loss of an irreplaceable set of genetic variability.

Biodiversity is not a product of single interactions that occur in a determined place and time. On the contrary, it is a dynamic result of uncountable relationships, with extremely conditional outcomes and variable according to the mosaic of species present in the environment. In the network of ecological interactions that maintain viable natural communities, the beta diversity or the diversity of connected species among local and regional populations is the best indicator of the health of a natural system (e.g. Dáttilo and Rico-Gray 2018). In this scenario animals are the connectors per excellence, the mobile force of beta diversity in plant-animal ecological networks. Thus the human-driven negative impacts on nature,

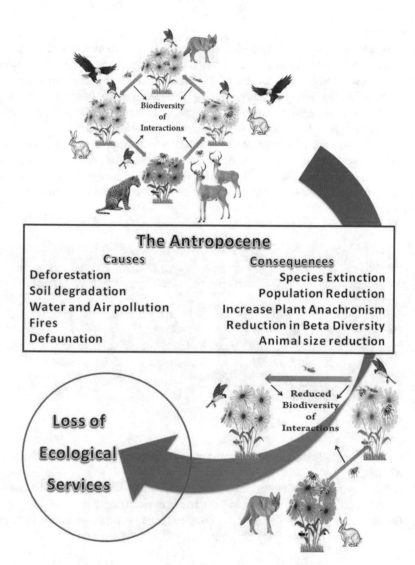

Fig. 13.4 The Anthropocene is producing strong negative impacts on nature, including deforestation, soil degradation (with loss of soil microbiomes), water and air pollution, and uncontrolled hunting and fires leading many ecosystems to a status of defaunation. The loss of animal links in the ecological networks (upper panel) due to extinction and population reduction cause the loss of connections (beta diversity) within and between communities (lower panel). Entire trophic chains and connections between distinct networks are being lost causing a severe reduction of biodiversity of interactions and of ecological services

resulting in defaunation, are indeed breaking the links between nodes of ecological networks so rapidly that it is unlikely that nature will have time to replace the actors of the evolutionary plays (e.g. Thompson 2005, 2013, 2014, ▣ Fig. 13.4).

Conclusion

Plant-animal interactions are dynamic, and they vary in space and time. The biodiversity of interactions they originated through evolutionary time is one of the main drivers of diversification and success of life on Earth. Anthropocene defaunation, accelerated since the last century, is producing disruptions in the diversity of interactions and ecological networks. Animals are declining in numbers, losing populations, and changing in size-distributions, which causes degradation of ecological services. The natural dynamics of speciation and extinction maintains, over time, constellations of interacting species. However, in the Anthropocene, defaunation is occurring at such a rapid pace and globally, that it begs the question: will we maintain and or have time to restore the constellations of species interaction networks that sustain humanity?

Key Points

1. Defaunation and its impacts in plant-animal interactions is omnipresent in the Ahthropcene;
2. Deforestation, defaunation, fires, and pollution, have negative impacts on the biodiversity of interactions;
3. Defaunation disrupts species interactions and plant-animal networks;
4. Defaunation-related degradation of species interactions has cascading consequences of significance for human wellbeing

❓ Questions

1. How does defaunation impact plant-animal interactions and biodiversity as a whole?
2. Are defaunation and deforestation accelerated in the Anthropocene as a result of humanity's negative impacts in the natural environment?
3. Discuss with your classmates or colleagues, what actions we could take to reverse or stop 'defaunation. Elaborate three viable suggestions.

Acknowledgements   We thank the financial support for our research from Sanford University (unrestricted funds to RD) and CNPq (KDC).

References

Bascompte J, Jordano P (2007) Plant-animal mutualistic networks: the architecture of biodiversity. Annu Rev Ecol Evol Syst 38:567–593

Bawa K (1990) Plant-pollinator interactions in tropical rain forests. Annu Rev Ecol Syst 21:399–422

Ceballos G, Ehrlich PR, Barnosky AD, Garcia A, Pringle RM, Palmer TM (2015) Accelerated modern human–induced species losses: entering the sixth mass extinction. Sci Adv e1400253:1

Ceballos G, Ehrlich P, Dirzo R (2017) Biological annihilation via the ongoing sixth mass extinction signaled by vertebrate population losses and declines. Proc Nat Acad Sci USA 114(30):e6089–e6096

13

Crutzen, P. J. 2002 Geology of mankind. Nature 415, 23. https://doi.org/10.1038/415023a

Dáttilo W, Rico-Gray V (2018) Ecological networks in the tropics: an integrative overview of species interactions from some of the most species-rich habitats on earth. In: Springer Nature

Del-Claro K, Rico-Gray V, Torezan-Silingardi HM, Alves-Silva E, Fagundes R, Lnge D, Dáttilo W, Vilela AA, Aguirre A, Rodriguez-Morales D (2016) Loss and gains in ant–plant interactions mediated by extrafloral nectar: fidelity, cheats, and lies. Insect Soc 63:207–221

Dirzo R, Guevara RE, Mendoza E (2020) Disruption of plant-herbivore interactions in light of the current defaunation crisis. In: Nunez-Farfan J, Valverde PL (eds) Evolutionary ecology of plant-herbivore interactions. Springer, pp 227–248

Dirzo R, Miranda A (1990) Contemporary neotropical defaunation and forest structure, function, and diversity—a sequel to John Terborgh. Conserv Biol 4:444–447

Dirzo R, Young HS, Galetti M, Ceballos G, Isaac NJ, Collen B (2014) Defaunation in the Anthropocene. Science 345:401–406

Dornelas M, Gotelli N, Shimadzu H, Moyes F, Magurran AE, McGill B (2019) A balance of winners and losers in the Anthropocene. Ecol Lett 22:847–854

Ehrlich P, Raven P (1964) Butterflies and plants: a study in coevolution. Evolution 18:586–608

Emer C, Jordano P, Pizo MA, Ribeiro MC, da Silva FR, Galetti M (2020) Seed dispersal networks in tropical forest fragments: area effects, remnant species, and interaction diversity. Biotropica 52:81–89

Gardner CJ, Bicknell JE, Baldwin-Cantello W, Struebig MJ, Davies ZG (2019) Quantifying the impacts of defaunation on natural forest regeneration in a global meta-analysis. Nat Commun 10:4590

INPE (2020) Instituto Nacional de Pesquisas Espaciais – Brasil – data http://queimadas.dgi.inpe.br/queimadas/portal-static/estatisticas_estados/

Jordano P (2000) Fruits and frugivory. In: Fenner M (ed) Seeds: the ecology of regeneration in natural plant communities. Commonw Agri Bur Int, Wallingford, pp 125–166

MapBiomas (2020) Annual report on deforestation, https://s3.amazonaws.com/alerta.mapbiomas.org/relatrios/MBI-deforestation-report-2019-en-final5.pdf

MapBiomes (2020) https://mapbiomas.org/en?cama_set_language=enDirzo

Morrison BML, Brosi BJ, Dirzo R (2020) Agricultural intensification drives changes in hybrid network robustness by modifying network structure. Ecol Lett 23:359–369

Oliveira P S, Marquis R J (2002) The cerrados of Brazil: ecology and natural history of a Neotropical savanna. Columbia University Press, New York, 398p

Sanchez-Mayo F, Wyckhuys KAG (2019) Worldwide decline of the entomofauna: a review of its drivers. Biol Conserv 232:8–27

Simberloff DS, Wilson EO (1969) Experimental zoogeography of islands: the colonization of empty islands. Ecology 50:278–296

Steffen W, Broadgate W, Deutsch L, Gaffney O, Ludwig C. The trajectory of the Anthropocene: The Great Acceleration. The Anthropocene Review. 2015;2(1):81–98. https://doi.org/10.1177/2053019614564785

Thompson JN (2005) The geographic mosaic of coevolution. University of Chicago Press, Chicago

Thompson JN (2013) Relentless evolution. University of Chicago Press, Chicago

Thompson JN (2014) Interaction and coevolution. University of Chicago Press, Chicago

Van Klink R, Bowler DE, Gongalsky K, Swengel A, Gentile A, Chase J (2020) Meta-analysis reveals declines in terrestrial but increases n freshwatrer insect abundances. Science 368:417–420

Wagner DL (2020) Insect declines in the anthropocene. Ann Rev Entomol 65(1):457–480

Ward M, Tulloch AIT, Radford JQ et al (2020) Impact of 2019–2020 mega-fires on Australian fauna habitat. Nat Ecol Evol 4:1321–1326

WWF (2020) Living planet report 2020 – bending the curve of biodiversity loss. In: Almond REA, Grooten M, Petersen T (eds) . WWF, Gland

Young HS, McCauley JD, Galetti M, Dirzo R (2016) Patterns, causes and consequences of contemporary defaunation. Ann Rev Ecol Evol Syst 47(1):333–358

Zalasiewicz J, Williams M, Haywood A, Ellis M (2011) The Anthropocene: a new epoch of geological time? Phil Trans R Soc A 369:835–841

Supplementary Information

© Springer Nature Switzerland AG 2021
K. Del-Claro, H. M. Torezan-Silingardi (eds.), *Plant-Animal Interactions*, DOI https://doi.org/10.1007/978-3-030-66877-8

Index

Printed in the United States
by Baker & Taylor Publisher Services